STUDENT WORKBOOK

HUMAN PHYSIOLOGY

AN INTEGRATED APPROACH

FIFTH EDITION

Dee Unglaub Silverthorn

Richard Damian Hill

Benjamin Cummings

San Francisco   Boston   New York
Cape Town   Hong Kong   London   Madrid   Mexico City
Montreal   Munich   Paris   Singapore   Sydney   Tokyo   Toronto

Executive Editor: Deirdre McGill Espinoza
Assistant Editor: Shannon Cutt
Managing Editor: Wendy Earl
Production Editor: Leslie Austin
Copyeditor: John Hammett
Compositor: Cecelia G. Morales
Interior Designer: Cecelia G. Morales
Cover Design: Riezebos Holzbaur Design Group
Image Rights and Permissions Manager: Zina Arabia
Senior Manufacturing Buyer: Stacey Weinberger
Marketing Manager: Derek Perrigo

Cover Photo Credit: Susumu Nishinaga/Science Photo Library

Benjamin Cummings
is an imprint of

www.pearsonhighered.com

ISBN 10: **0-321-59643-9**; ISBN 13: **978-0-321-59643-7**

2 3 4 5 6 7 8 9 10—MAL—13 12 11 10
Manufactured in the United States of America.

Contents

Owner's Manual

Welcome! Please take a moment to look through this introduction to familiarize yourself with the features contained in your workbook. The workbook is divided into the following sections to help you effectively study the chapters in the text.

LEARNING OBJECTIVES AND SUMMARY

For each chapter in the textbook, you will find a list of key learning objectives and a short narrative summary of the chapter's key themes.

TEACH YOURSELF THE BASICS

This section is organized using the section headers from the textbook chapter. Teach Yourself the Basics has a series of questions about each section, with figure numbers and cross-references so that you can refer back to the book for pertinent material.

There are two ways to use Teach Yourself the Basics:

- **Answer questions as you read**, using the workbook to actively direct your reading and notetaking. This is an excellent method for making sure you are getting the important information out of each section of the chapter.
- **Use the questions to test yourself.** Wait until you have studied the chapter, then see if you can answer the questions without referring to the textbook.

END OF CHAPTER MATERIAL

At the end of each chapter in this workbook, you will find the following sections:

- **Talk the Talk** is a vocabulary list of the important terms from the chapter. As you're studying, see if you can use each of these words without referring to your notes or the book.
- **Quantitative Thinking** highlights key quantitative problems and walks you through their solutions.
- **Practice Makes Perfect** questions allow you to practice material you learned in the chapter. They range from simple memorization questions to difficult application questions. The answers to these questions are contained in the answers section at the back of the workbook.

Some chapters contain a **Beyond the Pages** section, which gives additional material related to the chapter:

- **Running Problem** sections provide additional information about the Running Problem in the text chapter.
- **Try It** sections include activities that you may want to try, such as mini-experiments and demonstrations or interesting web sites.

A NOTE ON CONVENTIONS USED IN THE WORKBOOK

Ions in the workbook are often written in the following format: Na^+, Ca^{2+}, Cl^-, K^+, P_i, and PO_4^{3-}. If you are not familiar with the concept of ions, take a look at Ch. 2 in the textbook.

Body fluid compartments are abbreviated as follows:

ICF = intracellular fluid
ECF = extracellular fluid
IF   = interstitial fluid

STUDY HINTS FOR PHYSIOLOGY STUDENTS

There are several differences you will notice when studying and learning science:

- There is a large volume of unfamiliar and frequently intimidating vocabulary.
- Science textbooks have a different writing style and need to be read differently than textbooks in the humanities.
- The thought process for science requires linear thinking and the ability to trace a process in some detail from beginning to end. Humanities students are used to analyzing interrelationships and tend to think too broadly.

Here are some suggestions to use in class and when studying physiology.

NOTETAKING

- Do not try to write down every word the instructor says. Listen particularly for vocabulary, concepts, and points the instructor emphasizes. Develop your own shorthand so that you can get down more information. For example: Use up and down arrows for increase and decrease.

- Stop the instructor when a major point is unclear, or if the instructor has gone too fast.

- If you can't ask a question about something or if you get behind in your notetaking, put a "?" in the margin so that you know that your notes in that section are lacking. Check with a friend or the instructor to clear up what you missed.

- Develop one or two study buddies with whom you can compare your notes. If you have to miss class, try to get notes from more than one person. Notes are a memory aid rather than a verbatim copy of the lecture, so two different sets of notes are more likely to give you a complete overview of the lecture.

VOCABULARY

Most scientific words sound terribly complicated and difficult to remember. However, you can learn some common prefixes, suffixes, and roots that will help you to remember the words. In the textbook after some vocabulary words, there will be a note in ( ) that shows the roots and their meanings. Later in this introduction, there is a list of some of the most common roots you will encounter in physiology. You can also use a dictionary to find out the origin of a word. Start a list in your own notebook of other suffixes, prefixes, and roots.

READING THE TEXTBOOK

- Find out from your instructor if you are responsible for material that is in the text but has not been covered in lecture. If you are, you will need to add the extra information from the text to the information in your class notes. If you are not, then you need to be familiar with your notes so that you can pay less attention to material in the text that is not relevant.

- Science texts are written with important facts in every sentence. You cannot speed-read a science text. Go slowly and analyze each sentence as you read it. Ask yourself if you understand the concepts in that sentence, and how that sentence relates to the facts presented in previous sentences.

- Use the charts and diagrams in the book. Read the captions to the figures, as they will frequently explain what is in the diagram. Sometimes the figures in a physiology text provide a summary of material in a section. Make use of the textbook's figure questions to see if you really do understand what the figure is conveying.

ORGANIZING YOUR STUDYING

As soon as possible after class, you should glance over your notes and mark the points that are unclear or where your notes may be lacking.

Some students sit down and spend a lot of time rewriting their notes. Usually there are other, more productive ways of spending your study time. You will still be writing down the information in your notes, but you will also be reorganizing it into study notes of a form that you can remember. The following are some possible ways to do this. You probably won't have time to do all of them, so experiment until you find the method that works best for the way you learn.

1. Mark up your original notes with colored pens. Use colored pens to see if you can divide the notes into levels and sublevels for an outline. Assign a different color to each level of organization. Underline or draw a box around each word that fits into that heading. For example:

   SUBJECT HEADING: Purple
   I. Major topic = red
       A. Secondary topic = green
           1. Facts under that topic = turquoise

   Give vocabulary words and concepts their own color, such as yellow.

   Once you have marked your notes this way, go back and make a skeleton outline using the words you marked. It probably won't be a perfect outline (Topic I may have A but not B), but don't let that worry you. Use this outline to give yourself an overview of the material covered and the progression of the ideas or concepts covered.

2. Make a working vocabulary list.

   - Take a sheet of lined notebook paper and fold it in half lengthwise.
   - Down the left-hand side, list all the words and concepts you have marked in yellow in your notes.
   - Down the right-hand side (on the other side of the fold), list an abbreviated definition.
   - Study with the paper open so that you can see both sides. When you think you have learned the material, fold the paper in half so that you see only the list of words. Test yourself by going down the list and saying the definitions to yourself. If you don't know a definition, keep going. When you reach the end of the list, go back and look at the definitions of the words you missed.
   - Now turn the paper over so that you are looking at the definitions. Read the definitions and see if you can say and spell the word that fits each definition.

3. Get the big picture as well as the details. In physiology, for each system studied, you should make an outline, study sheet, or chart that answers the following questions:

   - What is the anatomical structure of the system? Can you trace a molecule involved in the system through all the parts? For example: trace a drop of blood from the aorta to various parts and back through the heart. What kinds of tissues or cells make up this system? What kind of muscle? Is there some structural entity that we can call the functional unit?
   - What are the functions of the system? Which parts carry out which function? How are the functions carried out?
   - How is the system regulated? Consider control by the nervous system and the endocrine system. Are there any reflexes? Know where any pertinent hormones are secreted and what controls their release.
   - How is the circulatory system involved with this system?

- Certain themes will keep popping up throughout the chapters. Make note of them. They include:

    movement of molecules across membranes
    pressure and flow
    biomolecules—the roles, transport, and metabolism of carbohydrates, fats, and proteins
    ions: Na^+, K^+, H^+, Ca^{2+}, HCO_3^-
    gases: oxygen and carbon dioxide
    energy use and storage

4. Make charts, diagrams, flow charts, and concept maps. One advantage of a chart is that it also allows you to compare and contrast different concepts that at first glance may not seem to have much relationship.

    To make a chart: Divide your paper into columns and rows. Across the top, write the topics you want to compare (for example: male and female reproduction). Down the left side, label the rows or blocks with the points you want to compare (for example: name of gamete, name of gonad, hormones). Go back and fill in the chart.

    One technique some students have used is to try to condense everything they have learned about a system onto a piece of poster paper. One effective way to do this is to make a giant drawing of the structure (anatomy) of the system and then add in all the physiological processes at or near the appropriate structure.

    For example, make a poster of the respiratory system: On the board you might draw a large upper body with the upper and lower respiratory systems drawn in and labeled. In the head you would also include the neurological control of ventilation. Add an enlarged cluster of alveoli just below the lung, and draw in the circulatory system going to a single cell.

5. Practice higher-level thinking. One objective of many physiology courses is to teach students how to use what are called higher-level thinking skills. How many levels do you usually use when you study?

    Experts* recognize four types of knowledge:

1.	Factual knowledge	Facts, terms, vocabulary, details
2.	Conceptual knowledge	Principles, theories, models, generalizations, categories: the big picture
3.	Procedural knowledge	How to do things: skills, research methods, algorithms
4.	Metacognitive knowledge	Understanding how you (and people in general) learn

    Each type of knowledge has a range of cognitive levels associated with it. These levels of knowing and understanding are given below, ranked from lowest to highest:

1.	Remember	Ability to recognize or recall information. ("I knew this material...")
2.	Understand	A deeper understanding of the information than simple memorization. Examples include the ability to classify items into similar groups, summarize information, explain a process, and give examples.
3.	Apply	Ability to solve or explain a problem by applying what the person has learned to the problem.
4.	Analyze	Ability to solve a problem by systematically examining facts and looking for patterns and relationships.
5.	Evaluate	Ability to make a judgment based on some standard or criteria.
6.	Create	Ability to use original, creative thinking to create something.

    If you want to use the higher levels of thinking skills, you must master the first two levels. In other words, you must have a memorized database of information upon which to act, and you must really *understand* it, not just *know* it.

---

*Anderson, L. W., *et al.* (eds.) *A taxonomy for learning, teaching, and assessing*. New York: Longman, 2001.

TEN TASKS FOR STUDENTS IN CLASSES THAT USE ACTIVE LEARNING

Written by Marilla Svinicki, Ph.D.
Director, University of Texas Center for Teaching Effectiveness

1. Make the switch from an authority-based conception of learning to a self-regulated conception of learning. Recognize and accept your own responsibility for learning.

2. Be willing to take risks and go beyond what is presented in class or the text.

3. Be able to tolerate ambiguity and frustration in the interest of understanding.

4. See errors as opportunities to learn rather than failures. Be willing to make mistakes in class or in study groups so that you can learn from them.

5. Engage in *active* listening to what's happening in class.

6. Trust the instructor's experience in designing class activities and participate willingly, if not enthusiastically.

7. Be willing to express an opinion or hazard a guess.

8. Accept feedback in the spirit of learning rather than as a reflection of you as a person.

9. Prepare for class physically, mentally, and materially (do the reading, work the problems, etc.).

10. Provide support for your classmates' attempts to learn. The best way to learn something well is to teach it to someone who doesn't understand.

DR. DEE'S ELEVENTH RULE

Don't Panic! Pushing yourself beyond the comfort zone is scary, but you have to do it in order to improve.

WORD ROOTS FOR PHYSIOLOGY

Root	Meaning	Root	Meaning	Root	Meaning
a- or *an-*	without; absence	*-cyte* or *cyto-*	cell	*gluco-, glyco-*	sugar or sweet
anti-	against	*de-*	without, lacking	*hemo-*	blood
-ase	signifies an enzyme	*di-*	two	*hemi-*	half
auto-	self	*dys-*	difficult, faulty	*hepato-*	liver
bi-	two	*-elle*	small	*homo-*	same
brady-	slow	*endo-*	inside or within	*hydro-*	water
cardio-	heart	*exo-*	outside	*hyper-*	above or excess
cephalo-	head	*extra-*	outside	*hypo-*	beneath or deficient
cerebro-	brain	*-emia*	blood	*inter-*	between
contra-	against	*epi-*	over	*intra-*	within
-crine	a secretion	*erythro-*	red	*-itis*	inflammation of
crypt-	hidden	*gastro-*	stomach	*kali-*	potassium
cutan-	skin	*-gen, -genic*	produce	*leuko-*	white

continues

Root	Meaning	Root	Meaning	Root	Meaning
lipo-	fat	patho-, -pathy	related to disease	re-	again
lumen	inside of a hollow tube	para-	near, close	retro-	backward or behind
-lysis	split apart or rupture	peri-	around	semi-	half
macro-	large	poly-	many	sub-	below
micro-	small	post-	after	super-	above, beyond
mono-	one	pre-	before	supra-	above, on top of
multi-	many	pro-	before	tachy-	rapid
myo-	muscle	pseudo-	false	trans-	across, through
oligo-	little, few				

Add your own here:

LITERATURE RESEARCH

Everyone uses the Internet now to look up information. But how valid is it? For scientific work such as physiology research, you will want to be sure that you are using authoritative sources. This section tells you a little about valid ways that research results are disseminated to other scientists and to the public at large.

JOURNALS

Scientific journals are usually sponsored by a scientific organization and consist of contributed papers that describe the **original scientific research** of an individual or group. When a scientist speaks of writing a paper, he/she is usually referring to the scientific paper published in a journal. Many journals will publish **review articles,** a synopsis of recent research on a particular topic. Reviews are an excellent place to begin a search for information.

Most articles published in scientific journals have gone through a screening process known as **peer review,** in which the article is read and critiqued by other specialists in a particular field. The purpose of peer review is to ensure good quality. In some cases, submitted articles are rejected by the journal editor, and in many cases the authors must make revisions to the article before it can be published. This process acts as a safeguard against the publication of poorly done research.

▶ Most articles on the Web are *not* peer-reviewed unless they are in an online journal!

Anyone can create a web page and publish information on the Web. There is no screening process, and the reader must decide how valid the information is. Web sites that are published by recognized universities (URLs that end in *.edu*) and not-for-profit organizations (URLs that end in *.org*) are likely to have good information. But an article about vitamins on the web page of a health food store (*name.com*) should be viewed with a skeptical eye unless the article cites published research.

CITATION FORMAT

Citation formats for published papers will vary but will usually include the following elements somewhere:

- **Title**. [Brackets around the title indicate an English translation of a foreign language paper.]
- **Year** the paper was published.
- **Name of author(s)**. Within a body of work, a multiauthor paper is usually cited as **first author,** *et al*. *Et al.* is the abbreviation for the Latin *et alii* meaning "and others," and indicates that there are additional authors.
- **Journal** abbreviation, **volume (issue)**: inclusive **pages**. A **volume** number is usually given to all issues published in one calendar year (six months for weekly journals). **Issue 1** would be the first issue published in a volume, Issue 2 would be the second, etc.

For example: Horiuchi, M., Nishiyama, H., and Katori, R. Aldosterone-specific membrane receptors and related rapid non-genomic effects. [Review] *Trends in Pharmacological Sciences 14*(1): 1–4, 1993.

In many citations, the name of a journal is abbreviated. Here is a list of commonly used abbreviations.

Abbreviation	Meaning	Abreviation	Meaning
Adv	advances	Am	American
Ann	annals	Annu	annual
Appl	applied	Arch	archives
Assoc	association	Behav	behavior
Biochem	biochemistry	Biol*	biology or biological
Biophys	biophysics	Br	British
Can	Canadian	Chem	chemistry or chemical
Clin	clinical	Commun	communications
Curr	current	Dev	developmental
Dis	disease	Eur	European
Exp	experimental	Gen	general
Hum	human	Int	internal
Intl	international	J	journal
Med	medicine or medical	Monogr	monograph
Nat	natural	Natl	national
Pharm	pharmacy	Physiol*	physiology or physiological
Proc	proceedings	Q	quarterly
Res	research	Rev	review
Sci	science	Soc	society or social
Surg	surgery or surgical	Symp	symposium
Ther	therapy		

* Most words ending with -ology or -ological will be abbreviated by stopping after the "l". Titles of one word such as "Nature" are never abbreviated.

Citing Sources Published on the World Wide Web

Citing sources from the Web requires a different format. Here is one suggested format:

Author. Title. Web page sponsor. URL. Date accessed.

For example:

> English, P. Birds of the Ecuadorian Rainforest. University of Texas.
> www.utexas.edu/depts/grg/gstudent/grg394k/spring97/english/english.html (1997 Nov 10).

MAPPING STRATEGIES FOR PHYSIOLOGY
INTRODUCTION

Mapping is a technique to improve a student's understanding and retention of subject material. It is based on the theory that each person has a memory bank of knowledge organized in a unique way based on prior experience. Learning occurs when you attach new ideas to your preexisting framework. By *actively interacting* with the information and by organizing it in your own way *before* you load it into memory, you will find that you remember the information longer and can recall it more easily.

Mapping is a nonlinear way of organizing material, closely related to the flowcharts used to explain many physiological processes. A map can take a variety of forms but usually consists of terms or concepts linked by explanatory arrows. The map may include diagrams or figures. The connecting arrows can be labeled to explain the type of linkage between the terms (structure/function, cause/effect) or may be labeled with explanatory phrases ("is composed of").

You will find a number of maps in the text that you can simply memorize, but the real benefit from using maps occurs when you create the maps yourself. By organizing the material yourself, you question the relationships between terms, organize concepts into a hierarchical structure, and look for similarities and differences between items. Such interaction with the material ensures that you process it into long-term memory instead of simply memorizing it for a test. Teaching you how to map is an important part of the process, as you may not know where to begin.

KEY ELEMENTS OF MAPS

A map has only two parts: the concepts and the linkages between them. A concept is an idea, event, or object. Concepts do not exist in isolation; they have associations to other relevant concepts. An example is the sentence "The heart pumps blood." "Heart" and "blood" are two concepts related by the verb "pumps." A map consists of a group of related terms that are hierarchically ranked and linked by explanatory arrows. In this Student Workbook, we have provided some groups of words to be mapped. You will probably want to develop your own groups of words to fit the material you are studying.

How to Make a Map

1. Choose the concepts to map. Begin at either the top or the center with the most general, important, or overriding concept from which all the others naturally stem. If you're mapping a reflex pathway, you would start with the stimulus. Next, use the other concepts to break down this one idea into progressively more specific parts or to follow the reflex pathway. Use horizontal cross-links to tie branches together. The downward development of the map may reflect the passage of time if the map represents a process or increasing levels of complexity if the map represents something like a cell. If you are trying to map a large number of terms, you might try writing each term on a small piece of paper. You can then lay out the papers on a table and rearrange them until you are satisfied with your map. Even an experienced physiologist may draw a map several times before being satisfied that it is the best representation of the information.

2. Think about the type of association between two concepts. Arrows will point the direction of the linkage, but you should also label the kind of linkage. You may label the line with linking words or by the type of link, such as cause or effect (CE). Color is very effective on maps. You can use colors for different types of links or for different sections.

3. Once you have your map, sit back and think about it. Are all the items in the right place? You may want to move them around once you see the big picture. Revise your map to expand the picture with new concepts or to correct wrong linkages. Review the information in the map by beginning with recall of the main concept and then moving to the more specific details. Ask yourself questions like, "What is the cause? Effect? Parts involved? Main characteristics?" to jog your memory.

4. The best way to study with a map is to trade maps with your study partner and see if you can understand each other's maps. You may want to find an empty classroom, put your maps on the blackboards, then step back and compare them. Did one of you put in something the other forgot? Did one of you have an incorrect relationship between two items?

Practice making maps. The study questions in each chapter of your textbook will give you some ideas of what you should be mapping. Your instructor can help you if you do not know how to get started.

1 Introduction to Physiology

LEARNING OBJECTIVES

When you complete this chapter, you should be able to:

- Create a diagram of the different levels of organization for living organisms.
- Name the physiological systems of the human body, their major organs, and their major function(s).
- Contrast the differences between teleological and mechanistic approaches in physiology.
- List and explain the key themes in physiology.
- Describe the key elements of a well-designed experiment and give scenarios for which each design is most appropriate.
- Describe how you would find reliable information on a physiological topic using search terms and web-based resources.
- Apply mapping strategies to simple concepts in order to prepare for more complex physiological maps you will encounter later.
- Practice interpreting graphs and evaluate when it is appropriate to use a particular graph type with a particular data set.

SUMMARY

Physiology is the study of how organisms function and adapt to a constantly changing environment. The human body comprises 10 organ systems. However, these organ systems do not act as isolated units. Instead, they communicate and cooperate to maintain homeostasis, a relatively stable internal environment.

Physiologists use a wide variety of techniques to study how the human body operates. Some of these techniques examine the activities of molecules and cells, while others focus on the response of entire organ systems. Physiology is usually approached from a functional, or mechanistic, viewpoint. Physiological events can also be explained in terms of their significance, which is considered a teleological approach to physiology.

Scientific experimentation includes formulation of a hypothesis, observation and experimentation, and data collection and analysis. Experiments using human subjects are difficult to perform and analyze because of tremendous variability within human populations and because of ethical problems.

TEACH YOURSELF THE BASICS

PHYSIOLOGICAL SYSTEMS

1. List the ten levels of organization, starting with atoms and ending with the biosphere. (Fig. 1-1)

2. List the ten human organ systems. (Table 1-1; Fig. 1-2)

FUNCTION AND PROCESS

3. Use possible answers to the question "Why do we breathe?" to explain the difference between a teleological approach and a mechanistic approach.

HOMEOSTASIS

4. Describe the importance of "adaptive significance." Include an example.

5. Humans are animals adapted to a terrestrial environment. What is the primary challenge of life on land?

6. Describe the teleological importance of extracellular fluid. How does it differ from intracellular fluid? (Fig. 1-3)

7. Define homeostasis. Describe it from both a teleological and a mechanistic perspective. Who coined the term homeostasis?

8. What happens when the body is unable to maintain homeostasis? What are some factors that might contribute to a failure of homeostasis? (Fig. 1-4)

PHYSIOLOGY: MOVING BEYOND THE GENOME

9. Define genome. How is it different from a proteome? What are some other "-omes" you might encounter?

10. What is the difference between the study of genomics and the study of proteomics?

THEMES IN PHYSIOLOGY

11. List and briefly describe the four major physiological themes discussed in this book. (Table 1-2)

Homeostasis and Control Systems

12. What are regulated variables, and how do they interact with control systems?

13. Describe and draw a simple control system. (Fig. 1-5)

Biological Energy Use

14. What types of physiological processes require energy?

Structure-Function Relationships

15. Describe the three main structural influences on function we'll be covering in this book.

Communication

16. Information flow in the body takes what forms?

17. Differentiate between local communication and long-distance communication in the body.

18. How/where is genetic information stored?

19. Communication between intracellular compartments and the extracellular fluid requires:

20. What do we mean when we say that a membrane is selectively permeable?

21. What role does signal transduction play in communication across a cell membrane?

22. What is mass flow? Cite some examples.

23. What role do gradients play in mass flow? What role does resistance play in mass flow?

PHYSIOLOGY IS AN INTEGRATIVE SCIENCE

24. What do we mean when we say that physiology is an integrative science?

25. What are emergent properties?

26. Describe translational research.

27. Practice mapping in Fig. 1-6 and in "Focus on Mapping."

THE SCIENCE OF PHYSIOLOGY

28. Describe the key steps of a scientific inquiry.

29. In an experiment, which are independent variables and which are dependent variables?

30. Why should every experiment have an experimental control?

31. How does a scientific theory differ from a hypothesis? How does a scientific theory differ from a model?

32. Why is a crossover study better than a study in which the experimental and control groups are composed of different organisms from a population?

33. Define and contrast placebo and nocebo effects.

34. What advantage is gained by having a blind study? A double-blind study? A double-blind crossover study? When might you use each type of study?

35. Briefly define and contrast the following types of studies:

    longitudinal study

    prospective study

    cross-sectional study

    retrospective study

36. What is accomplished by a meta-analysis?

SEARCHING AND READING THE SCIENTIFIC LITERATURE

37. What is meant by "peer-reviewed article"? Why do we have journals with a peer-review system?

38. What is a review article?

39. Refer to the textbook sections "Defining Search Terms" and "Citation Formats." Practice searching the literature for articles about diabetes-related cardiovascular disease. Also, practice writing a few citations from your search results.

TALK THE TALK

adaptive significance
anatomy
average values
bar graphs
biome
biosphere
blind study
cell
cell (or plasma) membrane
circulatory system
compartmentation
compliance
control system
crossover study
cross-sectional studies
data (plural; singular datum)
dependent variable
diabetes mellitus
digestive system
double-blind crossover study

double-blind studies
ecosystems
elastance
emergent properties
endocrine system
et al.
evidence-based medicine
experimental control
extracellular fluid
function
genome
gradient resistance
histogram
homeodynamics
homeostasis
hypothesis
identical
immune system
independent variable
information flow

integration
integumentary system
interpolate
intracellular fluid
journals
keyword search
levels of organization
line graphs
longitudinal studies
mass flow
mechanistic approach
membranes
meta-analysis
models
molecular interactions
musculoskeletal system
nervous system
nocebo effect
organ systems
organelles

organs
pathological
pathophysiology
peer-reviewed
pharmaco-genomics
physiological processes or
    mechanisms
physiology
placebo
placebo effect
populations

prospective study
proteomics
regulated variables
replication
reproductive system
respiratory system
retrospective studies
review articles
scatter plots
scientific inquiry
scientific theory

selectively permeable
signal transduction
similar
teleological approach
themes
tissues
translational research
urinary system
variability
variable(s)

PRACTICE MAKES PERFECT

1. How does physiology differ from anatomy?

2. A scientist wants to study the effects of the cholesterol-lowering drug pravastatin in rats.

    a. What might be the hypothesis in this experiment?

    b. The scientist decides to use four different doses of pravastatin. What would be an appropriate control for this experiment?

3. Write one or two sentences that summarize the graph below.

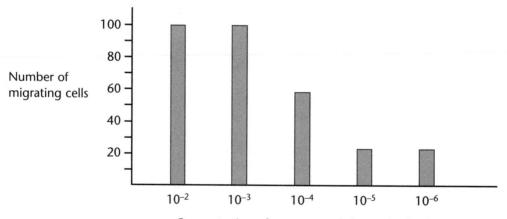

4. Glucose is transported into cells from the extracellular fluid. Construct a graph using the following data and label the axes. Summarize these results in the legend. Identify and distinguish between the independent and dependent variables.

Extracellular concentration of glucose (mM)	Intracellular concentration of glucose (mM)
150	100
140	101
130	100
120	91
110	83
100	75
90	61
80	52

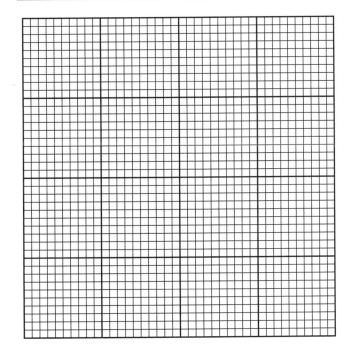

5. What kind of graph would be most appropriate for the following data sets? What labels go on the *x*-axis and *y*-axis?

   a.   oxygen consumption in summer-collected fish and winter-collected fish acclimated to 10°, 25°, and 30°

   b.   growth in boys and girls from ages 1 to 15 as indicated by measurements of height and weight

   c.   the effect of food intake on body weight, measured in a population of people selected at random

BEYOND THE PAGES

Find examples of good and bad graphs in your textbooks or in newspapers and magazines. *USA Today* is an excellent source of misleading graphs.

2 Molecular Interactions

LEARNING OBJECTIVES

When you complete this chapter, you should be able to:

- Diagram the general structure of atoms.
- Describe how subatomic composition affects the nature of atoms.
- Name and distinguish between the four common types of bonds.
- Contrast polar and nonpolar molecules and predict the behavior of each in hydrophobic and hydrophilic environments.
- Classify the groups of biomolecules according to chemical structure and representative biological functions.
- Characterize the behavior of solutes in solution and practice calculating expressions of solution concentration.
- Define pH (verbally and mathematically) and demonstrate the nature of acids, bases, and buffers in biological environments.
- Categorize protein interactions and identify the factors that can affect these interactions. Start thinking ahead: How might the regulation of protein interactions be important in the control of biological functions?

SUMMARY

Atoms are composed of protons, neutrons, and electrons, but it is the number of protons that distinguishes one element from another element. Atoms of a particular element that have differing numbers of neutrons are called isotopes. Electrons are located in shells that circle the nucleus. Covalent, ionic, and hydrogen bonds are formed by specific types of interactions between electrons of different atoms. Van der Waals forces are very weak bonds formed by nonspecific attractions and repulsions between atoms.

Water is essential for life. Molecules that easily dissolve in water are called hydrophilic, and molecules that do not dissolve are called hydrophobic. The concentration of a solution is important to function and can be expressed in molar, equivalent, and percent solutions. Understanding concentration allows you to predict water movement and cell volume changes. The body has many homeostatic mechanisms, to be discussed later in the book, that are specifically designed to maintain concentration and water balance.

The concentration of hydrogen ions (H^+) determines body pH. Acids are molecules that contribute H^+; bases bind H^+. The body contains specific buffers that act to maintain a normal body pH of 7.4.

There are four basic groups of biomolecules: carbohydrates, lipids, proteins, and nucleotides. Carbohydrates are structurally divided into monosaccharides, disaccharides, and polysaccharides. The most common lipid, the triglyceride, is composed of three fatty acids linked to a glycerol. Other lipid-related molecules include phospholipids, steroids, and eicosanoids. Proteins are composed of more than 100 amino acids that are arranged into chains, spirals, sheets, and complex globular configurations. These molecules have a wide variety of functions, ranging from structural components to carriers and enzymes.

Lipoproteins, glycoproteins, and glycolipids are frequently associated with cell membranes. Nucleotides consist of one or more phosphate groups, a five-carbon sugar, and a nitrogenous base. Examples of nucleotides include ATP, ADP, cAMP, NAD, FAD, DNA, and RNA.

Protein interactions are highly specific. Because proteins largely represent the commands of the genetic blueprint contained in our DNA, most of our biological reactions depend on this high degree of specificity in protein interaction. For these same reasons, protein interactions are also highly regulated. These statements apply to all proteins, including enzymes, membrane transporters, receptors, binding proteins, and immunoglobulins.

The induced-fit model describes the way proteins bind with ligands to carry out a specific function. Only specific ligands can bind to a particular protein. As an example, this accounts for the reason you can take a pill for a specific ill and have it "find its way" through your bloodstream to the part of your body where it can take its effect.

Multiple factors can alter protein binding and therefore alter protein behavior. Some proteins must be activated before they can carry out their functions. Isoforms of ligands can interact to affect protein activity. Various modulators can alter protein activity: chemical modulators, covalent modulators, allosteric modulators, competitive inhibitors, agonists, antagonists. Physical factors can also influence protein behavior: pH, temperature, salinity.

Reactions involving protein-ligand interactions are subject to a maximum rate at which all ligand binding sites are occupied. This maximum rate of reaction represents the point of saturation.

In summary, protein interactions exhibit specificity, competition, and saturation. It's very important that you understand the factors affecting protein interactions and reaction rates to successfully understand future topics in this book. Spend the time to become familiar with the generalizations presented here in Ch. 2.

TEACH YOURSELF THE BASICS

CHEMISTRY REVIEW

Atoms Are Composed of Protons, Neutrons, and Electrons

1. List three subatomic particles and describe their characteristics. (Fig. 2-1)

2. What is found in the nucleus of an atom?

The Number of Protons in the Nucleus Determines the Element

3. Spend some time familiarizing yourself with the periodic table of the elements, found in the inside back cover of your textbook.

4. What are essential elements, and why do we call them "essential"?

5. What determines the atomic number of an element?

6. What determines the atomic mass of an element?

Isotopes of an Element Contain Different Numbers of Neutrons

7. Describe isotope characteristics. Use hydrogen as an example.

8. What are some of the uses of isotopes?

9. What are three types of radiation emitted by radioisotopes?

Electrons Form Bonds Between Atoms and Capture Energy

10. Describe how electrons are arranged in an atom.

11. What are four important roles electrons play in physiology?

12. What is a high-energy electron?

13. What is a free radical?

14. How do antioxidants prevent damage to our cells? List some common antioxidants.

MOLECULAR BONDS AND SHAPES

15. What is a molecule?

16. Contrast molecule with compound. Give an example of each.

17. What are the four common bond types?

Covalent Bonds Are Formed When Adjacent Atoms Share Electrons

18. Explain how covalent bonds are formed. Use water as a simple example. (Fig. 2-2b)

19. How can you determine the number of covalent bonds that an atom can form? (Fig. 2-2a)

Polar and Nonpolar Molecules

20. What are the properties of a polar molecule? Give at least one example. (Fig. 2-3)

21. What are the properties of a nonpolar molecule? Give at least one example.

22. Contrast the terms hydrophobic and hydrophilic.

Ionic Bonds Form When Atoms Gain or Lose Electrons

23. What is an ion? Give examples.

24. How is an ionic bond different from a covalent bond? (Fig. 2-4)

Hydrogen Bonds and Van der Waals Forces Are Weak Interactions Between Atoms

25. How are hydrogen bonds formed and what elements form hydrogen bonds? (Fig. 2-5a)

26. What property of water is associated with hydrogen bonds between water molecules? (Fig. 2-5b)

27. Define van der Waals forces.

28. Name a function of van der Waals forces.

Molecular Shape Is Related to Molecular Function

29. How do molecular bonds affect a molecule's shape? What role does shape play in a molecule's function? (Fig. 2-6)

BIOMOLECULES

30. What is an organic molecule and what elements are most commonly found in these molecules?

31. Name the four major groups of biomolecules. How are they used by your body?

32. What are polymers?

33. Describe functional groups. Draw a few of the most common functional groups. (Use a separate sheet of paper, if necessary.) (Table 2-2)

Carbohydrates Are the Most Abundant Biomolecules (Fig. 2-7)

34. What is the basic formula of a carbohydrate?

35. Contrast monosaccharides, disaccharides, and polysaccharides. Include examples and discuss basic functionality in your answer.

36. In what form do animals store complex sugars? Plants? Yeasts and bacteria?

37. Name two structural polysaccharides.

Lipids Are Structurally the Most Diverse Biomolecules (Fig. 2-8)

38. What elements make up lipids? How do lipids differ from carbohydrates?

39.   Name three types of lipid-related molecules. Give at least one function for each type.

40.   Describe the chemical nature of glycerol. Contrast this with the chemical nature of fatty acids. How do these two chemical "players" contribute to the diversity of lipids?

41.   Among lipids, what is the distinguishing feature of a phospholipid?

42.   What is the difference between a saturated, a monounsaturated, and a polyunsaturated fatty acid?

43.   What are trans fats? What health implication is associated with trans fats?

44.   Describe a triglyceride. How do triglycerides differ from mono- and diglycerides? What are some significant facts about triglycerides?

45.   Describe steroids. Draw the basic molecular configuration of a steroid. What is the source of steroids in the body?

46.   Describe eicosanoids. Give some examples.

Proteins Are the Most Versatile Biomolecules (Fig. 2-9)

47.   Which biomolecules are the building blocks of proteins? Draw the molecular structure that represents this class of molecules. Describe the different functional groups represented in your drawing. (Use a separate sheet of paper if necessary.)

48.   What is an essential amino acid?

49. Name some amino acids that do not occur in proteins but have important physiological functions.

50. How do peptide bonds form?

51. Contrast the terms peptide, oligopeptide, polypeptide, and protein.

52. Describe the structural organization patterns of proteins, starting with primary structure and ending with quaternary structure. Define each level of protein organization. What shapes can arise at each organizational level? What roles do noncovalent bonds play in protein structure? (Use a separate sheet of paper if necessary.)

Some Molecules Combine Carbohydrates, Proteins, and Lipids

53. What is a conjugated protein?

54. Can a glycosolated molecule also be a conjugated protein? Illustrate the relationship between these two terms.

Nucleotides Transmit and Store Energy and Information (Fig. 2-11)

55. List the components of a nucleotide. Draw an example.

56. Contrast the options for specific nucleotide components.

57. List as many examples of nucleotides as you can.

58. Compare and contrast the structures and functions of DNA and RNA. Draw the possible base pair interactions for each. (Fig. 2-12; also see Appendix C)

AQUEOUS SOLUTIONS, ACIDS, BASES, AND BUFFERS

IP *Fluids & Electrolytes: Acid-Base Homeostasis*

Not All Molecules Dissolve in Aqueous Solution

59. What is the difference between a hydrophilic and a hydrophobic molecule? (Fig. 2-14)

60. What factors affect solubility?

There Are Several Ways to Express the Concentration of a Solution

61. Concentration is defined as:

62. When describing the concentration of a solution,

   a.   weight is usually expressed in what units?

   b.   number of solute molecules is usually expressed in what units?

   c.   solute ions can also be expressed in what units?

   d.   volume is usually expressed in what units?

Moles

63. 1 mole = _____ atoms

Molecular Mass

64. How do you calculate the molecular mass of a molecule?

65. How do you determine the gram molecular mass?

Molarity

66. Define molar solution.

67. Sodium chloride (NaCl) has a molecular mass of 58.5 Da. How would you make a 1 molar solution of NaCl? (See answer in workbook appendix.)

Equivalents

68. Define an equivalent.

69. One mole of magnesium ions (Mg^{2+}) contains _____ equivalents. (See answer in workbook appendix.)

Weight/Volume, Volume/Volume, and Percent Solutions

70. How would you make 250 mL of a 10% solution of NaCl? (See answer in workbook appendix.)

71. If a solution contains 200 mg NaCl/dL, what is its concentration in g/L? (See answer in workbook appendix.)

The Concentration of Hydrogen Ions in the Body Is Expressed in pH Units

72. Molecules that ionize and contribute a H^+ to a solution are called _____.

73. Molecules that combine with H^+ or produce OH^- are called _____.

74. Define pH in words. (Fig. 2-15)

75. Define pH mathematically. (See Appendix B.)

76. How do buffers in the body help maintain normal pH?

PROTEIN INTERACTIONS

77. Define proteome. Why has it become the focus of intense research?

78. List and briefly describe the seven broad categories of soluble proteins.

79. When referring to protein interactions, what is a binding site?

Proteins Are Selective About the Molecules They Bind

80. Define and compare ligands and substrates.

81. Do ligand and protein need to fit exactly? Describe the induced-fit model of protein-ligand interaction. (Fig. 2-16)

Protein Specificity

82. Discuss protein specificity. To what does it refer?

Affinity

83. What is meant when we say that a protein has an affinity for a particular ligand?

84. Using both words and a quantitative representation, describe equilibrium.

85. Now describe how a dissociation constant relates to protein-ligand interactions.

86. What is meant when we say there is competition between ligands?

87. What are agonists? Cite some examples.

Multiple Factors Can Alter Protein Binding

Isoforms

88. What are isoforms? Describe a scenario in which isoforms could have an effect on protein binding.

Activation

89. What is protein activation? How does this process affect protein binding? (Fig. 2-17)

90. What role do cofactors play in protein binding?

Modulation Alters Protein Binding and Activity

91. Define the term modulator. What do modulators do?

92. Briefly describe the actions of the following modulators. (Table 2-3)

chemical modulators

antagonists

competitive inhibitors (Fig. 2-19)

irreversible antagonists

allosteric modulators (Fig. 2-20)

covalent modulators

Physical Factors Modulate or Inactivate Proteins

93.   Describe how temperature and pH affect proteins. (Fig. 2-21)

The Body Regulates the Amount of Protein Present in Cells

94.   How does the body use up-regulation and down-regulation to maintain appropriate protein levels? (Fig. 2-22)

Reaction Rate Can Reach a Maximum

95.   Describe and draw the interaction between ligand concentration and reaction rate. (Fig. 2-23)

96.   What is saturation as it applies to protein-ligand interactions? How is it related to the reaction rate?

TALK THE TALK

acetylcholine
acid
acidic
acidity
adenine
ADP (adenosine diphosphate)
affinity
agonists
albumin
alkaline
allosteric activators
allosteric inhibitors
allosteric modulators
alpha-helix
amino acids
aminopeptidases
anion
antibodies
antioxidants
antagonists
aqueous
atomic mass
atomic mass units or amu
atomic number
atoms
ATP (adenosine triphosphate)
bases
basic

beta-strand
binding proteins
binding site
biomolecules
bonds
buffer
carbohydrates
carriers
cations
cellulose
chemical modulators
chitin
chromium picolinate
chymotrypsinogen
cofactor
collagen
competitive inhibitors
competitors
compounds
concentration
conformation
conjugated proteins
covalent bonds
covalent modulators
creatine
cyclic AMP (cyclic adenosine
     monophosphate, or cAMP)
cytosine

dalton
deciliter
degrade
denatured
deoxyribose
dextran
dextrose
disaccharides
dissociation
dissociation constant
disulfide bond
DNA (deoxyribonucleic acid)
double bond
down-regulation
eicosanoids
electron-dot shorthand
electrons
element
enzymes
equilibrium
equilibrium constant
equivalent
essential amino acids
FAD (flavin adenine
     dinucleotide)
fatty acids
fibrous proteins
free radicals

functional groups
gamma-amino butyric acid or
  GABA
globular proteins
glucose
glycerol
glycogen
glycolipids
glycoprotein
glycosylated
gram molecular mass
guanine
helium
hemoglobin
high-energy electrons
homocysteine
hydrogen
hydrogenated
hydrophilic
hydrophobic
immunoglobulin
induced-fit model
inhibitors
ion
ionic bonds
irreversible antagonists
isoforms
isotopes
keratin
ligand
lipids
lipoproteins
membrane transporters
milliequivalent
millimoles
modulator
molarity

molecular complementarity
molecules
monosaccharides
monounsaturated
NAD (nicotinamide adenine
  dinucleotide)
natrium
neutrons
nicotine
nitrogenous base
noncovalent interactions
nonpolar molecule
nuclear medicine
nucleic acids
nucleotide
nucleus
oligopeptide
organic molecules
penicillium
peptidases
peptide
peptide bond
percent solutions
periodic table of the elements
pH
phosphorylation
pleated sheets
polar molecules
polymers
polypeptide
polysaccharides
polyunsaturated
primary structure
proteins
proteolytic activation
proteome
protons

purines
pyrimidines
quaternary structure
radiation
radioisotopes
receptors
ribose
ring structure
RNA (ribonucleic acid)
saturated
saturation
secondary structure
shells
signal molecules
single covalent bond
solubility
solutes
solution
solvents
specificity
starch
steroids
substrates
sulfhydryl group
surface tension
tertiary structure
tetramer
thymine
double helix
transcription factors
triacylglycerols
triglycerides
up-regulation
uracil
van der Waals forces

QUANTITATIVE THINKING

In the laboratory, you may be asked to make up or mix solutions. The first example below shows you how to make up a molar solution. The second example shows how to calculate concentration when you mix two different solutions.

Task 1: Make 500 mL of a 150 millimolar (mM) NaCl solution.

Step 1: Calculate the molecular mass of NaCl. One mole of a substance contains its molecular mass in grams; this is known as its gram molecular mass.

Atomic mass of Na is 23 × 1 atom of Na = 23
Atomic mass of Cl is 35.5 × 1 atom of Cl = 35.5
Sum   58.5

One mole of NaCl therefore weighs 58.5 grams.

Step 2: Molar solutions are expressed in moles per liter of solution. You want to make a solution with 0.150 mole per liter (= 150 mM). However, you cannot weigh out moles of a compound. So you must calculate how many grams of NaCl are equal to 0.150 mole. You know that one mole weighs 58.5 grams.

Set up a ratio as shown:

a. $\dfrac{58.5\ \text{g NaCl}}{1\ \text{mole}} = \dfrac{?\ \text{g NaCl}}{0.15\ \text{mole}}$

b. $\dfrac{58.5\ \text{g NaCl} \bullet 0.15\ \text{mole}}{1\ \text{mole}} = ?\ \text{g NaCl}$

c. $? = 8.8\ \text{g NaCl}$

0.15 mole of NaCl therefore weighs 8.8 grams, and a 0.15 M solution has 8.8 g/liter of solution.

Step 3: You know from step 2 how much NaCl to use to make one liter of 0.15 M NaCl. But you have been asked to make up 500 mL, not one liter. Again, set up a ratio:

$\dfrac{8.8\ \text{g NaCl}}{1000\ \text{mL}} = \dfrac{?\ \text{g}}{500\ \text{mL}}$

4.4 grams of NaCl is needed to make up 500 mL of a 150 mM solution.

The next example shows you how to calculate concentration when you mix two different solutions.

Task 2: Mix 2 liters of 3 M NaCl with 1 liter of 6 M glucose. What is the concentration of glucose in the mixed solution? What is the concentration of NaCl?

Step 1: Figure out the *amount* of solute that you have in the two starting solutions:

3 moles NaCl/L × 2 L   = 6 moles NaCl
6 moles glucose/L × 1 L = 6 moles glucose

Step 2: Now put those amounts in the total volume formed when the solutions are added to each other:

$\dfrac{6\ \text{moles NaCl} + 6\ \text{moles glucose}}{3\ \text{L total volume}}$

6 moles/3 liters = 2 moles/liter for NaCl and 2 moles/liter for glucose

The NaCl concentration is therefore 2 M and the glucose concentration is 2 M.

What is the *total* concentration of the mixed NaCl/glucose solution? Show your work.

Task 3: Make 600 mL of a 300 mM glucose solution. (Mol. mass of glucose = 180 Da)

PRACTICE MAKES PERFECT

1. Draw a carbon atom and an oxygen atom with the appropriate number of protons, neutrons, and electrons.

2. A carbon dioxide molecule consists of covalent bonds between one carbon and two oxygen atoms. Draw how their valence (outer shell) electrons are arranged using an electron-dot model.

3. Match the following:

    _____ atomic mass   a. atoms with different numbers of neutrons

    _____ atomic number   b. protons plus neutrons

    _____ isotopes   c. capture and transfer energy

    _____ electrons   d. number of protons

    _____ radioisotopes   e. atoms with different numbers of protons

    _____ radiation   f. unstable isotopes

    _____ atomic mass unit   g. energy emitted by radioisotope

                                       h. dalton

4. Using what you know about atomic number, atomic mass, protons, neutrons, and electrons in atoms, fill in the blank cells in the table below. There is no way to predict the number of neutrons in the different isotopes of an element, but you can guess the number of neutrons in the most common isotope by using the atomic mass. Check your answers against the periodic table.

Element	Symbol	Atomic Number	Protons	Electrons	Neutrons*	Atomic Mass
Calcium		20			20	40.1
Carbon			6	6		12.0
Chlorine			17		18	35.5
	Co		27	27		58.9
Hydrogen				1	0	
Iodine	I	53			74	
Magnesium			12	12	12	
	N	7		7		14
Oxygen	O		8			16.0
Sodium			11			23
Zinc			30		35	
Copper				29	35	
	Fe	26				55.8
Potassium		19	19		20	

* Number of neutrons in the most common isotope.

5. What is the molecular mass of water? Indicate the proper units.

6. Match the following:

    _____ anion   a.   two or more atoms that share electrons

    _____ cation   b.   a pair of electrons shared by two atoms

    _____ covalent bond   c.   molecule that contains more than one element

    _____ ion   d.   atom that gains or loses electrons

    _____ molecule   e.   positively charged ion

    _____ compound   f.   negatively charged ion

a.

b.

c.

d.

7. Which of the above structural formulas for the amino acid leucine ($C_6H_{13}O_2N$) is (are) correct? Be able to explain your reasoning.

8. Answer the following questions about the properties of water.

    a. What type of bond holds a single water molecule together?

    b. What type of bond holds many individual water molecules together?

    c. Why is water called the universal solvent?

    d. When a salt crystal of sodium chloride is dropped into water, how do the molecular properties of water and the ionic properties of the salt interact?

9. What happens to an oxygen atom if it gains a proton?

10. When potassium and chloride form an ionic bond, which ion gains an electron and which ion loses an electron? Which ion(s) becomes stable? Explain your reasoning.

11. Why are polar molecules hydrophilic and nonpolar molecules hydrophobic?

12. What is the gram molecular mass of a half mole of sodium chloride?

13. How would you make a 500 mL NaCl solution that has a molarity of 0.5 M?

14. A 0.1 M solution is equal to how many millimoles per liter?

15. How many grams of glucose are in 100 mL of a 50 mM glucose solution?

16. How would you make 100 mL of a 3% glucose solution? What would be the molarity of this solution? (The molecular mass of glucose is 180.)

17. If you mix 1 liter of 0.4 M glucose with 1 liter of 0.8 M NaCl,

    a.   what is the total concentration of the mixed solution?

    b.   what is the NaCl concentration in the mixed solution?

    c.   what is the glucose concentration in the mixed solution?

18. The plasma concentration of Na^+ is 142 mEq/L. What is the concentration of Na^+ in millimoles per liter?

19. The plasma concentration of Ca^{2+} is 5 mEq/L. What is the concentration of Ca^{2+} in millimoles per liter?

20. Match the following:

    _____ acid   a.   concentration of hydrogen ions

    _____ base   b.   moderates pH changes

    _____ pH   c.   molecule that ionizes and donates H^+

    _____ alkaline   d.   molecule that combines free H^+

    _____ buffer   e.   molecule that produces hydroxide ions (OH^-)

                             f.   solution with a low concentration of H^+

Note: Blanks may have more than one correct answer.

21. Amines are organic compounds that act as acids and bases. Identify the acids and bases in the reaction below.

$$CH_3 - \underset{\underset{\displaystyle H}{|}}{N} - H \;+\; H - OH \;\longrightarrow\; CH_3 - \underset{\underset{\displaystyle H}{|}}{\overset{\overset{\displaystyle CH_3}{|}}{N^+}} - H \;+\; OH^-$$

22. The reaction below occurs in red blood cells. Identify the acids and bases in this reaction.

$$CO_2 \;+\; H_2O \;\longleftrightarrow\; H_2CO_3 \;\longleftrightarrow\; H^+ \;+\; HCO_3^-$$

23. Which HCl solution is more acidic, a 50 mM or a 0.5 M? _____

24. Match the following:

_____ triglycerides   a.   large molecules with repeating units

_____ fatty acids   b.   long carbon chains with terminal carboxyl groups

_____ carbohydrates   c.   fatty acid chains linked to a glycerol

_____ polymers   d.   polar, hydrophilic molecules

_____ steroids   e. $(CH_2O)_n$

                               f.   lipid-related molecules with four carbon rings

25. Match the following:

_____ essential amino acid   a.   made from amino acids

_____ protein   b.   a molecule that has proteins plus lipids

_____ conjugated protein   c.   a molecule that has proteins plus carbohydrates

_____ glycoprotein   d.   amino acid not made by the body

_____ phospholipid   e.   may act as a membrane receptor

                               f.   major component of membranes

Note: Blanks may have more than one answer, and answers may be used more than once.

26. Use the following terms to create a map of biomolecules. Be sure to put labels on your linking arrows. The term biomolecule is the most general term and should appear at the top of your map.

amino acids	lipids	proteins
carbohydrates	lipoproteins	saturated fatty acids
cellulose	maltose	starch
disaccharides	monosaccharides	steroids
fats	peptide	sucrose
fatty acids	phospholipids	triglycerides
glucose	polymer	unsaturated fatty acid
glycogen	polypeptide	
glycolipids	polysaccharides	

3 Compartmentation: Cells and Tissues

LEARNING OBJECTIVES

When you complete this chapter, you should be able to:

- Diagram the body fluid compartments and name the major body cavities.

- Draw the fluid mosaic model of a typical cell membrane and show how the membrane structure relates to its four general functions.

- Differentiate between different categories of membrane proteins according to structure and function.

- Draw, name, and list the functions of organelles found in our cells.

- Predict physiological functions of tissues based on their anatomical properties, including the types of cell junctions present.

- Map the similarities and differences between the four different types of tissue found in our bodies. In each type, what is the composition/function of the extracellular matrix?

- Using structural and functional differences, distinguish between the five types of epithelial tissue. Practice using tissue structure to predict tissue function.

- Using structural and functional differences, distinguish between the seven types of connective tissue.

- Using structural and functional differences, distinguish between the three types of muscle tissue.

- Using structural and functional differences, distinguish between the two types of neural tissue.

SUMMARY

The study of cytology has greatly benefited from the use of sophisticated light and electron microscopes. These instruments, combined with other techniques, have illuminated the structure and function of cellular organelles. This chapter discusses the basic components of a cell, how cells are held together, and the different types of tissues formed by highly differentiated cells.

Cell membranes act as a selectively permeable barrier between the cells and their external environment, mediating transport and communication between the two compartments. In addition, membranes provide structural support. A cell membrane consists of a phospholipid bilayer with proteins that are inserted partially or entirely through the membrane. Many of these proteins are mobile, while others have restricted movement. Membrane proteins can play a structural role; they can form channels, carriers, receptors, or enzymes; or, they might have multiple duties. Carbohydrates attach to some membrane proteins and lipids, serving as cellular markers for recognition by other cells (like cells of the immune system) and giving the membrane a "sugar coating."

An extracelluar matrix, composed mostly of glycoproteins and protein fibers, helps hold cells together and provides a structural base for cell growth and migration during development. Various types of cell junctions, formed by proteins and glycoproteins, also hold cells together and participate in cell-to-cell communication. Anchoring junctions, such as desmosomes, resist physical stresses, while tight junctions prevent the leakage of material between cells. Gap junctions are pores or channels between cells that allow substances, such as ions, to pass directly between cells.

The four basic tissue types are epithelial, connective, muscle, and neural. Epithelia can be functionally divided into five types: exchange, transporting, ciliated, protective, and secretory. The seven types of connective tissue are loose connective tissue, dense regular and dense irregular connective tissue, adipose, blood, cartilage, and bone. These connective tissues consist of an extracellular matrix that contains one or more of the following protein fibers: collagen, elastin, fibrillin, and fibronectin. There are three types of muscle tissue: skeletal, cardiac, and smooth. Neural tissue and muscle are considered to be excitable tissues. Organs are composed of collections of tissues that perform certain functions.

TEACH YOURSELF THE BASICS
FUNCTIONAL COMPARTMENTS OF THE BODY

IP *Fluids & Electrolytes: Introduction to Body Fluids*

1. What are the three major body cavities? What major organs do you find in each? (Fig. 3-1)

2. What separates the body cavities?

The Lumens of Some Organs Are Outside the Body

3. Define and describe the term lumen.

4. What do we mean when we say that, for some organs, the lumen is essentially an extension of the external environment? (Fig. 3-24)

Functionally, the Body Has Three Fluid Compartments

5. Name the two main compartments in the body. (Fig. 1-3)

6. The extracellular fluid can be subdivided further into: (Fig. 3-2)

BIOLOGICAL MEMBRANES

7. Briefly identify the different meanings of the word *membrane* in physiology. (Fig. 3-3)

The Cell Membrane Separates the Cell from Its Environment

8. What are the two synonyms for plasma membrane?

9. List and briefly describe the general functions of the cell membrane.

Membranes Are Mostly Lipid and Protein

10. How does the ratio of lipid to protein in a cell membrane translate into differential cell functions? (Table 3-1)

11. Describe the fluid mosaic model of a cell membrane. Draw a picture to show the orientation of the phosphate heads and the lipid tails. Include in your illustration examples of membrane proteins. (Use a separate sheet of paper if necessary.) (Fig. 3-4)

Membrane Lipids Form a Barrier Between the Cytoplasm and Extracellular Fluid

12. What are the three main types of lipids that make up the cell membrane? (Fig. 3-5a; also see structure of a phospholipid in Fig. 2-8)

13. Define and illustrate a micelle. (Fig. 3-5b) Give an example of the role micelles play.

14. Define and illustrate a liposome. (Fig. 3-5b) Give an example of the role liposomes play.

15. What are sphingolipids?

16. Describe the role cholesterol plays in cell membranes.

Membrane Proteins May Be Loosely or Tightly Bound to the Membrane

17. Anatomically, membrane proteins are classified in three categories. Name and define those three categories. Draw illustrations to help you visualize protein placement in membranes. (Fig. 3-6)

18. Transmembrane proteins are grouped into families according to which classification system? (Fig. 3-7)

  ▶ Revisit phosphorlyation and noncovalent interactions in Ch. 2.

19. What is a lipid raft? (Fig. 3-8)

20. How do cells develop polarity? (Fig. 3-6)

Membrane Carbohydrates Attach to Both Lipids and Proteins

21. Most membrane carbohydrates are either _____ or _____.

22. What is the glycocalyx, and what is its function? Give one specific example.

  ▶ Cell membrane structure is summarized in the map in Fig. 3-9.

INTRACELLULAR COMPARTMENTS

23. What is differentiation? (Fig. 3-10)

Cells Are Divided into Compartments

  ▶ Fig. 3-11 is an overview map of cell structure. Representative organelles are shown in Fig. 3-12.

24. Describe in words the organization of the cell.

The Cytoplasm Includes the Cytosol, Inclusions, and Organelles

25. Name and describe the three components of the cytoplasm.

Inclusions Are in Direct Contact with the Cytosol

26. Why are inclusions in direct contact with the cytosol?

27. How are nutrients stored?

28. Most inclusions with functions other than nutrient storage are made from what?

29. Describe a ribosome. What is the function of a ribosome? (Fig. 3-12)

30. Distinguish between the following terms: fixed ribosome, free ribosome, and polyribosome.

31. Describe proteasomes. What is their function?

32. What are vaults?

Cytoplasmic Protein Fibers Come in Three Sizes
33. Describe the three families of cytoplasmic protein fibers. (Table 3-2) Give examples of each.

34. What are the two general purposes of the insoluble protein fibers of the cell?

Microtubules Form Centrioles, Cilia, and Flagella
35. What is the cell's microtubule organizing center? Describe its function.

36. What are centrioles? What role do they play? They have a specific structure—describe it. (Fig. 3-13a)

37. What are cilia? What is their function? How do they move? Cilia have a specific structure—describe it. (Fig. 3-13c)

38. What are flagella? What is their function? Compare and contrast cilia and flagella. (Fig. 3-10)

The Cytoskeleton Is a Changeable Scaffold

39. Describe the cytoskeleton of a typical cell.

40. List and describe at least five important functions of the cytoskelton. (Fig. 3-14a, b)

Motor Proteins Create Movement

41. Define motor proteins and describe how they work. (Fig. 3-15)

42. Specifically, compare and contrast the following families of motor proteins: myosins, kinesins, dyneins.

Organelles Create Compartments for Specialized Functions

43. What functional advantage do membranous organelles have over nonmembranous organelles?

44. What are the major groups of organelles? (Fig. 3-12)

Mitochondria

45. Draw and label the structure of a mitochondrion. Be sure to include the terms mitochondrial matrix, cristae, outer membrane, inner membrane, and intermembrane space. (Fig. 3-16)

46. What is the primary function of a mitochondrion?

47. What two unusual characteristics do mitochondria possess?

The Endoplasmic Reticulum

48. Distinguish the anatomical and functional differences between rough ER and smooth ER. (Fig. 3-17)

The Golgi Complex

49. Describe the structure and function of the Golgi complex. (Fig. 3-18)

Cytoplasmic Vesicles

50. What are the two types of cytoplasmic vesicles? Describe the differences that differentiate the two.

51. Describe lysosomes. What role do they play? (Fig. 3-19) Why must lysosomal enzymes be activated?

52. Describe peroxisomes. What role do they play? (Fig. 3-19)

The Nucleus Is the Cell's Control Center

53. What is contained in the cell nucleus?

54. Describe or draw the structure of a typical nucleus. Include the terms nuclear envelope, nucleolus, chromatin, and nuclear pores. (Fig. 3-20)

55. Generally speaking, how does a nucleus communicate with the rest of the cell?

TISSUES OF THE BODY

IP   ***Muscular: Anatomy Review—Skeletal Muscle Tissue***

56. Why do cells organize into tissues? Give examples of some specific body tissues.

57. What is histology? What physical features do histologists use to describe tissues?

58. Name the four primary tissue types.

Extracellular Matrix Has Many Functions

59.   Describe the composition of extracellular matrix. Include in your description: proteoglycans and insoluble protein fibers. Is the matrix inert or does it participate in physiological processes?

60.   Give examples of and describe how matrix differs between tissue types. Why is this significant?

Cell Junctions Hold Cells Together to Form Tissues

61.   How are individual cells within tissues connected to each other? What special class of molecules are involved? (Fig. 3-21; Table 3-3)

62.   Describe the structure and function of gap junctions. Where in the body are you likely to find gap junctions? (Fig. 3-21c)

63.   Describe the structure and function of tight junctions. Where in the body are you likely to find tight junctions? (Fig. 3-21a)

64.   Describe the structure and function of anchoring junctions. Where in the body are you likely to find anchoring junctions? (Fig. 3-21b)

65.   Describe different types of cell-cell anchoring junctions. Where in the body are you likely to find these different junctions? (Fig. 3-21b)

66.   Describe the different types of cell-matrix anchoring junctions. Where in the body are you likely to find these?

67.   What are some of the disease states that can arise from cell junction abnormalities?

▶   A summary map of cell junctions is shown in Fig. 3-22.

Epithelia Provide Protection and Regulate Exchange

68.   TRUE/FALSE? Defend your answer.

Any substance that enters or leaves the internal environment of the body must cross an epithelium.

69. What are some of the different types of epithelia and their functions?

Structure of Epithelia

70. What is the basal lamina, and what is its function? (Fig. 3-23)

71. Describe or draw the general structural composition of epithelia.

72. Physiologists characterize epithelia as either _____ or _____. Upon what do they base this characterization?

73. The tightness of an epithelium is directly related to:

Types of Epithelia

74. Name two general types of epithelia.

75. Name the two types of layering and the three cell shapes found in sheet epithelia.

Exchange Epithelia

76. Draw the structure and describe the function of exchange epithelia. (Figs. 3-24, 3-25a; Table 3-4)

Transporting Epithelia

77. Draw the structure and describe the function of transporting epithelia. What are the defining characteristics of transporting epithelia? (Figs. 3-24, 3-25b; Table 3-4)

Ciliated Epithelia

78. Draw the structure and describe the function of ciliated epithelia. (Figs. 3-24, 3-26; Table 3-4)

Protective Epithelia

79. Draw the structure and describe the function of protective epithelia. (Fig. 3-24; Table 3-4)

Secretory Epithelia

80. Draw the structure and describe the function of secretory epithelia. Include a comparison of the different types of secretory epithelia. (Figs. 3-24, 3-27, 3-28; Table 3-4)

81. Compare and contrast the different types of secretions. (Figs. 3-27, 3-28)

Connective Tissues Provide Support and Barriers

82. What is the distinguishing characteristic of connective tissue?

Structure of Connective Tissue

83. Describe the basic components of connective tissue matrix.

84. Compare and contrast fixed cells and mobile cells of connective tissue.

85. What is implied by each of the following suffixes?

    -blast

    -clast

86. What do connective tissue cells secrete?

87. Compare and contrast the different types of protein fibers secreted by connective tissue cells.

Types of Connective Tissue

88. Compare and contrast the different types of connective tissues. Be sure to identify the differences in the various ground substances of each type. (Use a separate sheet of paper.) (Figs. 3-29, 3-30, 3-31, 3-32; Table 3-5)

▶ Fig. 3-32 is a study map showing the components of connective tissues.

Muscle and Neural Tissues Are Excitable

89. What is meant by "excitable" when referring to tissues? What are action potentials?

90. How does the extracellular matrix of muscle and neural tissues compare to that of previous tissue types?

91. List the three types of muscle tissues and give examples for each.

92. What are the two types of neural tissues, and what are their primary functions?

▶ A summary of the characteristics of the four tissue types can be found in Table 3-6.

TISSUE REMODELING

Apoptosis Is a Tidy Form of Cell Death

93. Briefly describe the two forms of cell death.

Stem Cells Can Create New Specialized Cells

94. If cells in the adult body are constantly dying, where do their replacements come from? (Appendix C)

95. Distinguish between the following terms: totipotent, pluripotent, and multipotent.

96. What are some current issues surrounding stem cell research? Include examples of diseases that researchers hope to cure as a result of stem cell research.

ORGANS

97.   What is an organ?

▷   Look in Ch. 1 to review a quick list of the 10 physiological organ systems.

TALK THE TALK

abdomen
abdominopelvic cavity
actin fibers or microfilaments
action potentials
adherin junctions
adipocytes
adipose tissue
anchoring junctions
apical membrane
apocrine glands
apoptosis
atrophy
basal body
basal lamina
basement membrane
basolateral membrane
bilayers
-*blast*
blebs
blister
blood
blood plasma
blood vessels
blood-brain barrier
bone
brown fat
cadherins
calcified
cartilage
catalase
cell adhesion molecules or
    CAMs
cell junctions
cell membrane
cell organelles
cell shape
cell-cell adhesions
centrioles
centrosome
chromatin
cilia
ciliated epithelia
-*clast*

claudins
collagen
columnar
compartmentation
connective tissues
connexins
cranial cavity
cristae*
cuboidal
-*cyte*
cytoplasm
cytoskeleton
cytosol
degenerative diseases
dense connective tissues
dermis
desmosomes
diaphragm
differentiation
ducts
dyneins
elastance
elastin
endocrine glands
endoplasmic reticulum or ER
epidermis
epithelial tissues or epithelia
Escherichia coli or *E. coli*
exchange epithelia
excitable tissues
external lamina
extracellular fluid (ECF)
extracellular matrix
fibroblasts
fibronectin
fixed cells
fixed ribosomes
flagella
fluid mosaic model
focal adhesions
free ribosomes
gap junctions
gland

glial cells
glycocalyx
goblet cells
Golgi complex
ground substance
hair follicles
hemidesmosomes
histology
hormones
hypodermis
immunoliposome
inclusions
integral proteins or transmem-
    brane proteins
integrins
intermediate filaments
intermembrane space
internal organization
interstitial fluid
intracellular fluid (ICF)
intracellular transport
keratin
kinesins
laminin
ligaments
lipid rafts
lipid-anchored proteins
liposomes
loose connective tissues
lumen
lysosomal storage diseases
lysosomes
matrix
matrix metalloproteinases
    (MMPs)
mechanical properties
membrane
membrane modifications
metastasis
micelles
microtubules
microvilli*
mitochondria*

mitochondrial DNA
mitochondrial matrix
mitosis
mobile cells
motor proteins
mucous membranes
mucous secretions
multipotent
muscle tissue
myosins
necrosis
nerve-cell adhesion molecules
   or NCAMs
neural
neural tissue
neurofilament
neurons
nonmembranous organelles
nuclear envelope
nuclear pore complexes
nucleoli
nucleus
occluding
occludins
organelles
organs
paracellular
pelvis
pemphigus

pericardial membrane
pericardial sac
peripheral proteins
peritoneal membrane
peritoneum
peroxisomes
phospholipid matrix
plaques
plasma
plasma membrane
plasmalemma
plasticity
pleural membrane
pleural sac
pluripotent
polarity
polyribosomes
pores
prokaryotic endosymbiont
   theory
proteases
proteasomes
protective epithelia
proteoglycans
reticulum
rheumatoid arthritis
ribosomes
rough endoplasmic reticulum
   (rER)

sebaceous glands
secretion
secretory epithelia
secretory vesicles
sensory receptors
serous secretions
simple squamous epithelium
smooth endoplasmic
   reticulum (sER)
sphingolipids
squamous
stem cells
storage vesicles
stratified
surface keratinocytes
sweat glands
Tay-Sachs disease
tendons
thoracic cavity or thorax
tight junctions
tissue membranes
totipotent
transporting epithelia
tubulin
vaults
vesicles
white fat

* These words are in the plural form. What are their singular forms?

PRACTICE MAKES PERFECT

1. Which organelle, if separated from its cell, would have the highest probability of existing and evolving into a functional life form? Explain your reasoning.

2. What is the advantage of having the nucleus partially separated from the cytoplasm?

3. If intestinal cells lacked a cytoskeleton, how would their structure and function be affected?

4. Why are epithelia more susceptible to developing cancer than most other tissues?

5. A histological examination of a tissue shows that it has tight junctions. Would you expect the solutions normally found on either side of the epithelium to be the same or different? Explain.

6. What anatomical and functional qualities of the epidermis (skin) help regulate body temperature and prevent dehydration?

7. Match the following:

    _____ ground substance   a.   exceptionally strong, inelastic protein fiber

    _____ fibroblast   b.   cell that breaks down extracellular matrix

    _____ collagen   c.   matrix of glycoproteins, water, protein fibers

    _____ elastin   d.   cell that secretes extracellular matrix

    _____ fibrillin   e.   a coiled protein fiber with elasticity

    _____ fibronectin   f.   protein fiber that connects cells to extracellular matrix

                                        g.   a very thin fiber that combines with elastin

8. Match the type of protein fiber(s), if any, that is associated with a particular type of connective tissue.

    a. collagen   b. elastin   c. fibrillin   d. fibronectin   e. none

    _____ under the skin   _____ tendons and ligaments

    _____ sheaths that surround nerves and muscles _____ blood

    _____ cartilage   _____ lungs and blood vessels

    _____ adipose   _____ bones

9. Place the letter of the correct answer in front of the phrase below:

    a. rough endoplasmic reticulum   b. Golgi apparatus   c. both   d. neither

    _____ packages proteins into vesicles

    _____ modifies proteins by adding or subtracting fragments

    _____ a series of interconnected hollow tubes or sacs

    _____ proteins are synthesized in this location

10. Be sure you can distinguish between the following pairs or groups. Give examples when appropriate.

    a. desmosome, intermediate junction, gap junction, tight junction, junctional complex

    b. cilia and flagella

    c. microfilaments, microtubules, microvilli, and intermediate filaments

    d. exchange and transporting epithelia

    e. intracellular, extracellular, and intercellular compartments

    f. cristae and matrix of mitochondria

    g. lysosomes and peroxisomes

    h. rough and smooth endoplasmic reticulum

    i. epithelium and endothelium

    j. apical and basolateral sides of an epithelium

    k. movement of substances across a tight epithelium and a leaky epithelium

    l. endocrine and exocrine glands

    m. serous and mucous secretions

    n. fibroblasts, melanocytes, adipocytes, macrophages, and mast cells

    o. collagen and elastin

    p. tendons and ligaments

BEYOND THE PAGES

Lorenzo's Oil: Rent this movie about a family whose child has an inherited peroxisomal disorder.

The Visible Human project: The Visible Man was a 39-year-old prisoner who was executed and donated his body to science. It was embedded in gelatin, frozen, and sliced crosswise into 1878 slices 1 mm thick. The slices were photographed, digitized, and colored. The entire set occupies about 14 gigabytes, so don't even think about downloading it! To view it, go to: http://www.nlm.nih.gov and search *visible human*.

Cell Biology: A wonderful collection of links to resources on many aspects of cell biology. Check out http://www.cellbio.com.

4 Energy and Cellular Metabolism

LEARNING OBJECTIVES

When you complete this chapter, you should be able to:

- Explain the relationships between the following terms: potential energy, kinetic energy, endergonic reactions, exergonic reactions, coupled reactions, net free energy, reversible reactions, irreversible reactions, and activation energy.

- Explain why it is important to understand the foundations of biochemistry before you study physiology.

- Show how enzymes facilitate biochemical reactions. Why are enzymes necessary? How do they work? How are they regulated?

- Categorize and describe the key chemical reactions that make use of enzymes.

- Create a map of the general processes involved in human metabolism and ATP production. Determine the key substrates, products, and net energy yields from each process. How are the processes related? What role does oxygen play?

- Sketch a diagram to show how ATP is synthesized in the mitochondrion. What is oxidative phosphorylation? Also, how is H^+ movement related to ATP production in the mitochondrial matrix?

- Map the processes for synthesis of glucose and glycogen.

- Map the processes for lipid synthesis.

- Map the processes for protein synthesis. If you're pressed for time in this chapter, don't neglect protein synthesis. This topic is key to understanding many of our later physiology topics.

SUMMARY

This chapter discusses how cells obtain and store energy in the form of chemical bonds. The potential energy stored in chemical bonds is later used as kinetic energy to perform work. In biological systems, this work can be chemical, transport, or mechanical work. The types of work performed by chemical bond energy include protein synthesis, transport of substances across membranes, and muscle contraction.

Chemical reactions are of two general types, exergonic or endergonic. During exergonic reactions, the free energy of the products is less than that of the reactants. In endergonic reactions, the opposite occurs. Many exergonic and endergonic biological reactions are coupled and use nucleotides (e.g., ATP) to capture, store, and transfer energy. Some of these reactions are irreversible, but many are reversible. Most reversible reactions require the aid of an enzyme to overcome their activation energy.

Enzymes are involved in many biological reactions. Enzymatic activity depends in part upon the ability of a substrate to bind to the active site of the enzyme. Numerous other factors affect enzyme activity including proteolytic activation, cofactors or coenzymes, modulation, and the concentration of enzyme, substrate, or product. The reaction rate of enzymes is measured by how fast products are synthesized. This rate is determined by modulators, enzyme and substrate concentrations, and the ratio of substrates to products (law of mass action). Types of biological reactions catalyzed by enzymes include oxidation-reduction, hydrolysis and dehydration, and addition-subtraction-exchange reactions.

Metabolic reactions are either catabolic (net breakdown) or anabolic (net synthesis). The activity of metabolic pathways can be altered in a variety of ways, including mechanisms that regulate enzyme activity and ATP concentrations.

Glucose, fatty acids, glycerol, and amino acids are used as energy sources to produce ATP. During glycolysis (an anaerobic pathway), one glucose molecule yields 2 pyruvate molecules, 2 ATP, 2 NADH, 2 H^+, and 2 H_2O molecules. In the absence of sufficient oxygen, pyruvate is converted to lactate. However, during aerobic metabolism pyruvate is converted to acetyl CoA, which enters the citric acid cycle. The yield from the citric acid cycle metabolism of two pyruvate molecules is 8 NADH, 2 $FADH_2$, 2 ATP, and 2 CO_2 molecules. NADH and $FADH_2$ enter the electron transport system and donate their electrons and H^+ to produce up to 38 ATP, H_2O, and heat.

Glucose is not the only fuel source for making ATP during anaerobic and aerobic respiration. Glycogen can be converted to glucose-6-phosphate. Amino acids can be converted (via deamination) to pyruvate, acetyl CoA, or other intermediates of the citric acid cycle. Glycerol can enter glycolysis, and fatty acids can be converted to acetyl CoA to enter the citric acid cycle.

Many fuel sources can be synthesized by the body. For example, glucose can be synthesized from glycogen, glycerol, or amino acids whenever there is not enough glucose for ATP production. Most glucose is synthesized in the liver and released into the bloodstream for transport to other cells. Nerve cells normally use only glucose for fuel, so the body always tries to maintain adequate glucose concentrations in the blood. Lipids can be synthesized from glucose, acetyl CoA, glycerol, and fatty acids.

Proteins are synthesized by elaborate processes that include transcription and translation. Genes are selected for transcription by initiation factors that bind to specific promoter regions on DNA. DNA is transcribed into mRNA, and mRNA is translated into a protein. These proteins are sorted, modified, packaged, and directed to a specific destination.

TEACH YOURSELF THE BASICS

ENERGY IN BIOLOGICAL SYSTEMS

1. What are the properties of living organisms? (Table 4-1)

2. Where do plants acquire energy? In what forms do they store excess energy? (Fig. 4-1)

3. Where do animals acquire energy? In what forms do they store excess energy?

Energy Is Used to Perform Work
4. Define energy.

5. List three kinds of work in biological systems and give an example of each.

6. What is a concentration gradient?

Energy Comes in Two Forms: Kinetic and Potential

7. Describe the difference between kinetic and potential energy. Give an example of each. (Fig. 4-2)

Energy Can Be Converted from One Form to Another

8. Kinetic energy converted from potential energy is used to do work. However, some of that energy is lost to the environment. The amount of energy lost in the transformation depends on the _____ of the process.

9. How is potential energy stored in biological systems?

Thermodynamics Is the Study of Energy Use

10. State the first law of thermodynamics.

11. State the second law of thermodynamics.

12. What is entropy?

CHEMICAL REACTIONS

13. Define bioenergetics.

Energy Is Transferred Between Molecules During Reactions

14. Combine these words into a sentence that explains their relationship: reaction, product, molecule(s), substrate, reactant. (Table 4-2)

15. How do we measure the rate of a chemical reaction?

16. Define and give an example of free energy.

17.   What is the relationship between free energy and chemical bonds?

Activation Energy Gets Reactions Started

18.   Define and give an example of activation energy. (Fig. 4-3)

Energy Is Trapped or Released During Reactions

19.   Compare and contrast endergonic reactions and exergonic reactions. Describe the free energy of the products in each reaction type. (Fig. 4-4)

20.   An important biological example of an exergonic reaction is the hydrolysis of ATP to yield ADP. Draw out this chemical reaction. From which bond is the most energy released? (Use a separate sheet of paper if necessary.)

Coupling Endergonic and Exergonic Reactions

21.   Where does the activation energy for metabolic reactions come from?

22.   In what ways do our cells utilize energy from coupled reactions? Are reactions always directly coupled? What indirect ways are used to take advantage of highly exergonic reactions? (Fig. 4-5)

Net Free Energy Change Determines Reaction Reversibility

23.   Compare and contrast reversible reactions with irreversible reactions. How does the net free energy determine the reversibility of a reaction? (Use a separate sheet of paper if necessary.) (Figs. 4-4a, 4-6)

24.   Are most biological reactions reversible or irreversible?

ENZYMES

25.   What role do enzymes play in a chemical reaction? How do they accomplish this?

26.   What are substrates?

Enzymes Take Part in Typical Protein Interactions

27.   Like other proteins that bind to substrates, which three qualities do enzymes exhibit? (Also refer to Protein Interactions in Ch. 2.)

28.   What are isozymes? How are they useful in medical diagnoses? (Table 4-3)

Enzymes May Be Activated, Inactivated, or Modulated

29.   Describe different mechanisms of enzyme activation.

30.   What are coenzymes? What is an example of a coenzyme?

31.   How are enzymes inactivated? (Fig. 4-7)

32.   What is the ultimate purpose of regulating enzyme activity?

Enzymes Lower the Activation Energy of Reactions

33.   In thermodynamic terms, how does an enzyme increase the rate of a reaction? (Fig. 4-8)

Reaction Rates Are Variable

34.   How do we measure the rate of an enzymatic reaction?

35.   What factors can alter the rate of an enzymatic reaction? In mammals, which two factors are of primary significance?

36.   In the absence of an enzyme, describe the reaction rate:

37. If an enzyme is present, describe the reaction rate:

38. When does saturation occur? (See Reaction Rates in Ch. 2.)

39. What is one effective way our cells control rates of enzymatic reactions?

Reversible Reactions Obey the Law of Mass Action

40. In the case of a reversible reaction, what determines in which direction the reaction is to go? (Fig. 4-9a)

41. What happens when some factor disturbs the equilibrium of a reversible reaction? (Fig. 4-9b, c)

42. How is equilibrium restored? (Fig. 4-9d)

43. Describe the law of mass action. (Also refer to discussion of equilibrium in Ch. 2.)

Enzymatic Reactions Can Be Categorized

▶ Most enzymes are instantly recognized by the suffix –*ase*. The first part of the enzyme's name refers to the type of reaction, to the substrate upon which the enzyme acts, or both.

44. What is the term for the process of adding a phosphate group? What type of enzyme carries out this function?

45. Briefly describe the reactions in the general categories of enzymatically catalyzed reactions listed below. Include specific examples and representative enzymes when possible. (Table 4-4)

    oxidation-reduction reactions

    hydrolysis-dehydration reactions

addition-subtraction-exchange reactions

ligation reactions

METABOLISM

46. Define metabolism.

47. Distinguish between catabolic and anabolic reactions.

48. Define kilocalorie (kcal).

49. Of the energy released during catabolism, where is most of it stored?

50. What is a metabolic pathway? What are intermediates? (Fig. 4-10)

Cells Regulate Their Metabolic Pathways
51. List the ways in which cells regulate metabolism.

Modulation of Enzymes
52. What role do hormones play in regulation of enzymes?

53. What is feedback inhibition? What role does it play in modulation of enzymes? (Fig. 4-11)

Enzymes and Reversible Reactions
54. What is the advantage of having a reaction that is regulated by two enzymes (one for the forward direction, one for the reverse direction)? (Fig. 4-12a, b, c)

Compartmentation of Enzymes in the Cell

55. What advantage does a cell gain by isolating some enzymes within specific intracellular compartments? (Refer back to discussion of organelles in Ch. 3.)

Ratio of ATP to ADP

56. What role does the ratio of ATP to ADP play in the regulation of metabolic reactions?

ATP Transfers Energy Between Reactions

57. Revisit the structure of ATP from Ch. 2. Sketch an ATP molecule here. (Use a separate sheet of paper.)

58. Identify the high-energy phosphate bond in the ATP molecule.

59. What is the biological significance of ATP?

60. What is the difference between aerobic and anaerobic pathways?

61. Compare ATP production by aerobic and anaerobic pathways.

ATP PRODUCTION

IP *Muscular: Muscle Metabolism*

62. Name two pathways that produce compounds carrying high-energy electrons. What are those high-energy compounds? (Fig. 4-13)

63. Which pathway transfers electrons from the high-energy molecules to ATP?

64. What are the by-products of these collective reactions?

65. Describe how glucose, lipids, and proteins enter glycolysis.

66. How is ATP production an example of compartmentation?

Glycolysis Converts Glucose and Glycogen into Pyruvate

67. Describe glycolysis in terms of the end products produced from one molecule of glucose. What are the fates of these end products? (Fig. 4-14)

68. Which steps are the endergonic steps? What form of energy is consumed in these steps? Does this make glycolysis an endergonic or exergonic pathway?

69. Is glycolysis an aerobic or anaerobic pathway? Why? What is the significance of this?

Anaerobic Metabolism Converts Pyruvate into Lactate

70. Under what conditions is pyruvate converted into lactate? (Fig. 4-15)

71. What is the net energy yield from the conversion of one glucose to two lactate? What high-energy molecule is used in the process?

Pyruvate Enters the Citric Acid Cycle in Aerobic Metabolism

72. In aerobic metabolism, pyruvate is converted into _____. (Fig. 4-15)

73. Where in the cell does this reaction take place? (Fig. 4-15)

74. What are the end products of the citric acid cycle, and what are their fates? (Fig. 4-16)

75. What is the net energy yield for two pyruvate molecules completing the citric acid cycle? (Fig. 4-16)

High-Energy Electrons Enter the Electron Transport System

76. Relate the terms chemiosmotic theory, electron transport system, cytochromes, and oxidative phosphorylation. (Use a separate sheet of paper if necessary.) (Fig. 4-17)

77. How do NADH and $FADH_2$ participate in the electron transport system?

ATP Synthesis Is Coupled to Hydrogen Ion Movement

78. How is potential energy created across the inner mitochondrial membrane? Where in the mitochondrion is this potential energy stored? (Use a separate sheet of paper if necessary.) (Fig. 4-17)

79. Draw and label a diagram to explain how the electron transport system makes ATP and water. (Fig. 4-17)

One Glucose Molecule Can Yield 30–32 ATP

80. The total energy production from 1 glucose molecule in aerobic metabolism = _____ ATP. (Fig. 4-18)

81. Explain how the net energy yield can vary.

Large Biomolecules Can Be Used to Make ATP

▶ Review the overview of the metabolic pathways in Fig. 4-13.

Glycogen Converts to Glucose

82. What is the metabolic role of glycogen in animals?

83. When glycogen is broken down for energy, it converts to _____ or _____. (Include approximate percentages.) (Fig. 4-19)

84. Which product above is a more efficient option for the cell? Why?

85. What is the name of the process by which glycogen is broken down for metabolism?

Proteins Can Be Catabolized to Produce ATP

86. What is the first step in protein catabolism? (Fig. 4-20a)

87. Organic acids derived from catabolized amino acids can enter which of the metabolic pathways? (Fig. 4-20b)

88. What are the toxic by-products of protein catabolism? How are they handled? (Fig. 4-20c)

Lipids Yield More Energy per Unit Weight than Glucose or Proteins

89. Why are lipids the primary fuel-storage molecule for the body?

90. Describe the process of lipolysis. (Fig. 4-21)

91. At which stage of the metabolic pathways does glycerol enter?

92. Describe the process of beta-oxidation. (Fig. 4-21, step 3)

93. What happens to the products of beta-oxidation?

94. What happens to the excess acetyl CoA? What are the health implications?

SYNTHETIC PATHWAYS

Glycogen Can Be Made from Glucose

95. What two locations have the largest glycogen stores?

96. Describe the process by which glycogen is synthesized. (Fig. 4-19)

Glucose Can Be Made from Glycerol or Amino Acids

97. What is gluconeogenesis? (Fig. 4-22) When is it employed?

98. What organ is the primary site of gluconeogenesis during periods of fasting?

Acetyl CoA Is an Important Precursor for Lipid Synthesis

99. What two-carbon precursor can be used to synthesize fatty acids? (Fig. 4-23, step 1)

100. Where in a cell does triglyceride formation take place?

101. Can the body make cholesterol or must it come from the diet? (Review the structure of cholesterol in Fig. 2-8.)

Proteins Are the Key to Cell Function

102. What makes proteins highly variable and highly specific?

The Protein "Alphabet" Begins with DNA

103. What is a codon? (Fig. 4-24)

104. What is the start codon?

105. Which codons are recognized as stop codons?

106. Relate the terms mRNA, transcription, and translation.

Translating DNA's Cells to Protein Is a Complex Process

107. Define the concept of a gene, as we currently understand it.

108. What is the difference between constitutively active genes and regulated genes? (Fig. 4-25)

109. Briefly describe what happens to a newly transcribed piece of mRNA before it leaves the nucleus and enters the cytoplasm. (Fig. 4-25)

110. How is mRNA used to create proteins?

111. What is post-translational modification?

In Transcription, DNA Guides the Synthesis of RNA

112. Why is transcription and mRNA processing compartmentalized in the nucleus?

113. Describe the process by which transcription is initiated. (Fig. 4-26)

114. What is the role of RNA polymerase?

Alternative Splicing Creates Multiple Proteins from One DNA Sequence

115. Explain the relationship between introns, exons, mRNA processing, RNA interference, and alternative splicing. (Fig. 4-27)

116. What are the advantages of alternative splicing?

Translation of mRNA Produces a String of Amino Acids

117. Differentiate between mRNA, tRNA, and rRNA. (Fig. 4-28)

118. Outline or map the process of translation. Include the following terms: mRNA, rRNA, tRNA, ribosomes, ribosomal subunits, ribosome-mRNA complex, codon, anticodon, dehydration synthesis, peptide bond, termination, and ribonuclease. (Use a separate sheet of paper.)

Protein Sorting Directs Proteins to Their Destination

119. What is a signal sequence or targeting sequence? What happens to proteins bearing a signal/targeting sequence? (Fig. 4-29)

120. What happens to proteins without a targeting sequence? (Fig. 4-29)

Proteins Undergo Post-Translational Modification

121. What are the possible post-translational modifications a newly synthesized protein could experience?

Protein Folding

122. Describe the process of protein folding. What bonds and or forces come into play?

123. What are molecular chaperones?

124. How does the cell deal with misfolded proteins?

Cross-Linkage

125. What do we mean by cross-linkage? Give some examples.

Cleavage

126. What is accomplished by post-translational cleavage?

Addition of Other Molecules or Groups

127. List examples of molecules that might be added to a protein.

Assembly into Polymeric Proteins

128. What are polymeric proteins? Give an example.

Proteins Are Modified and Packaged in the Golgi Complex

129. Trace the path of a newly synthesized protein from the ribosome to a secretory vesicle. (Fig. 4-29, steps 5–10)

—————————————————————————————

TALK THE TALK

acetyl CoA
activation energy
acyl unit
addition reaction
aerobic or oxidative
alternative splicing
amination
anabolism
anaerobic
anticodon
antisense strand
-ase
ATP synthase
base pairs
beta-oxidation
bioenergetics
carbonic anhydrase
catabolism
catalysts
chemical reaction
chemical work
chemiosmotic theory
chymotrypsin
cis-face
cisternae
citric acid cycle
cleavage
closed system
codons
coenzymes
compartmentation
concentration gradients
constitutively
cross-linkage
cytochromes
deamination
dehydratase
dehydration reactions
dehydration synthesis
disulfide bonds
efficiency
electron transport system
emergent properties
endergonic reaction
end-product inhibition
energy
entropy
enzymes
equilibrium
equilibrium constant

exchange reaction
exergonic reaction
exons
fatty acid synthetase
feedback inhibition
first law of thermodynamics
free energy
gene
glucokinase
gluconeogenesis
glucose 6-phosphatase
glycogenolysis
glycolysis
guanosine triphosphate (GTP)
high-energy phosphate bond
hydrolysis reaction
induced
intermediary metabolism
intermediates
introns
irreversible reaction
isozymes
ketone bodies
key intermediates
kilocalorie
kinase
kinetic energy
Krebs cycle
lactate dehydrogenase
law of conservation of energy
law of mass action
LeChâtelier's principle
ligase
lipase
lipolysis
lysase
mechanical work
messenger RNA (mRNA)
metabolism
microRNA
molecular chaperones
molecular interactions
mRNA processing
net free energy change of the
    reaction
open system
oxidation-reduction
oxidative phosphorylation
oxidized
oxidoreductase

panothenic acid
peptidase
peptide bond
phosphorylation
polymeric proteins
polymers
post-translational
    modification
potential energy
products
proenzymes or zymogens
promoter
proteasomes
protein sorting
reactants
reaction rate
reduced
regulated
repressed
respiration
retrograde
reversible reaction
ribonucleases
ribosomal RNA (rRNA)
RNA interference
RNA polymerase
saccharidase
second law of thermodynamics
sense strand
signal sequence or targeting
    sequence
start codon
stop codon
subtraction reaction
synthetases
termination
tetramers
transamination
transcription
transcription factors
trans-face
transfer RNA (tRNA)
transferase
translation
transport vesicles
transport work
triplets of bases
ubiquitin
vitamins

FOCUS ON PHYSIOLOGY:
THE LAW OF MASS ACTION

The Law of Mass Action is a simple relationship that holds for chemical reactions ranging from those in a test tube to those in the blood. This law says that the ratio of the concentrations of the molecules is a constant number, known as the *equilibrium constant*, K.

$$K = \frac{[\text{substrates}]}{[\text{products}]}$$

Mathematically:

$$K = \frac{[A]^m \cdot [B]^n}{[C]^p \cdot [D]^q}$$

Where m, n, p, and q represent the respective number of molecules of A, B, C, and D in the balanced equation.

In very general terms, the Law of Mass Action says that when a reaction is at equilibrium, the ratio of the substrates to the products is always the same. If the concentration of a substrate (A or B) increases, the equilibrium will be disturbed. To restore the proper substrate/product ratio, some of the added substrates will convert into product. Conversely, if the amount of product (C or D) decreases, such as would occur when they are used up in a different reaction, some substrate will convert into product to replace that which was lost. This has the effect of decreasing the amount of A and B but keeping the ratio at the value needed for K.

The equilibrium constant is important in physiology because as the concentrations of substrates change, the change will be reflected in the concentrations of the products. This is most easily illustrated by equating the chemical reaction to a scale, with the number of blocks on each side representing the concentrations of the substrates and products. The reaction is reversible. The equilibrium constant, K, is indicated by the position of the triangular base of the scale. For simplicity, our example has an equilibrium constant of 1, so that the reaction is in equilibrium when there are equal numbers of blocks on each side: 1 = [S]/[P]. The figure below shows the reaction at equilibrium, with substrate and product concentrations balanced at the equilibrium constant (K) for the reaction.

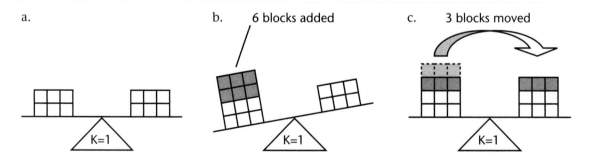

In part b. of the figure, some outside force has added substrate to the left side of the reaction sending the reaction out of equilibrium. Now, without removing any blocks from the scale, how can you get the reaction back to equilibrium? Easy ... move some of the added blocks over to the product side. Then each side still has an equal number of blocks (the 1:1 ratio that we set up as our equilibrium constant) and the reaction is at equilibrium. Notice that *both* product and substrate have increased, but by an equal amount. You cannot change the K since this is a constant for any given reaction.

The next figure shows a different reaction that is at equilibrium when the product is twice the substrate d. That means the K for this reaction is 1/2. Notice that the balance point has moved to compensate for the K value: 1/2 = [S]/[P]. In part e. of this figure, nine units have been added to the substrate side. How can you rearrange the blocks so that you balance the scale at the 1/2 ratio? Try this on your own, using part f.

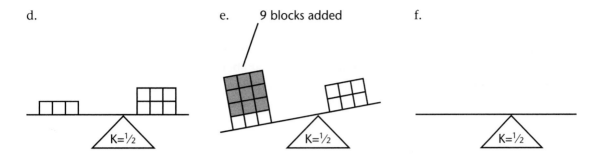

In the figures below, you again have the reaction with a K value of 1/2 g. In part h., three units of product are removed by another metabolic reaction. Without adding or subtracting any new units to the scales, move the units between the scales until the reaction is at equilibrium again, as in part i.

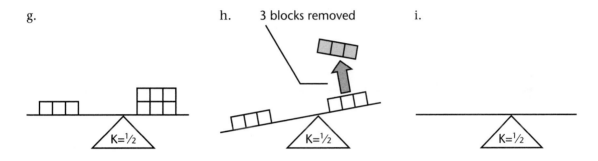

Now, instead of blocks on a scale, assume that the reaction is: $A \rightarrow C + D \rightarrow E$

If more substrate A is added to the reaction, what happens to the amounts of the following when the reaction reaches equilibrium?

A _____   C _____   D _____

If C and D are converted into E, when the system again reaches equilibrium, what has happened to the amount of A in the system?

PRACTICE MAKES PERFECT

1. Match the following:

_____entropy

_____potential energy

_____kinetic energy

_____exergonic

_____endergonic

a. ATP

b. a muscle contracting

c. energy in chemical bonds

d. second law of thermodynamics

e. free energy

f. reaction that releases a lot of energy

g. reaction that requires a lot of energy

Note: More than one letter may apply to each blank.

2. Most breakfast cereals would not be very nutritious unless they were supplemented with vita-
    mins and ions such as Mg^{2+}. How do some of these ions and vitamins affect enzyme activity?

3. In the human body, carbonic anhydrase (CA) has maximum activity at pH 7.4. Some other ani-
    mals have an isozyme that has maximum activity at about 6.4. Draw a curve of the activity of
    human carbonic anhydrase and its isozyme as a function of pH, using the graph below.

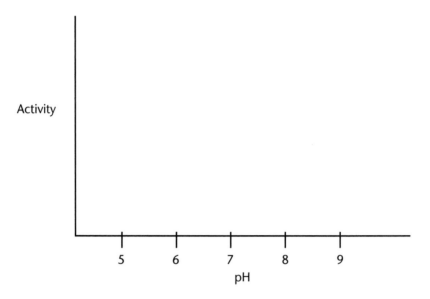

4. Answer the questions using the two graphs on the next page.

    a.   Which graph represents an exergonic and which graph represents an endergonic reaction?

    b.   Which reaction is more likely to proceed in the forward direction? Why?

    c.   Which reaction products yield more free net energy?

    d.   Which reactions below belong with graph A and which belong with graph B? (Reactions
         2, 3, and 4 are from the glycolysis pathway.)

    1.   _____ $6 CO_2 + 6 H_2O$ + sunlight $\rightarrow C_6H_{12}O_6 + 6 O_2$

    2.   _____ glucose + ATP $\rightarrow$ glucose-6-phosphate + ADP

    3.   _____ 3-phosphoglyceraldehyde + P_i + $NAD^+ \rightarrow$ 1,3 diphosphoglycerate + NADH

    4.   _____ 1,3 diphosphoglycerate + ADP $\rightarrow$ 3-phosphoglycerate + ATP

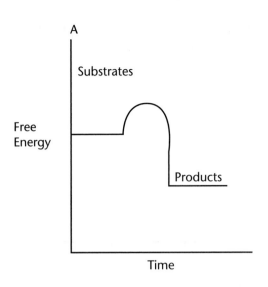

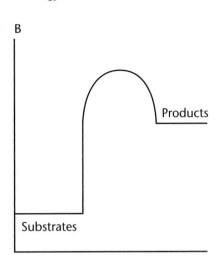

5. Match the following:

_____hydrolysis   a.   lose electron

_____dehydration   b.   removal of water

_____oxidation   c.   –OH is added to one molecule and –H to another

_____reduction   d.   gain electron

_____kinase   e.   loss of activity

_____deamination   f.   transfer of an amino group

_____transamination   g.   enzyme that transfers a phosphate group

                                     h.   removal of an amino group

6. Ribonuclease is a digestive enzyme secreted by the pancreas. This enzyme catalyzes the hydrolysis
    of RNA.

    a.   A 100 mM solution of ribonuclease is added to a series of test tubes with RNA concentra-
         tions ranging from 10% to 70%. Plot the reaction rate as a function of substrate concen-
         tration on the graph in part c. Assume that the reaction rate is maximal at a 50%
         substrate concentration.

    b.   What is the relationship between the substrate and the enzyme when the reaction rate is
         maximal?

c.   If the enzyme concentration was only 50 mM, indicate on the graph below how the reaction rate would change as a function of substrate concentration.

Reaction
Rate

| 10 | 20 | 30 | 40 | 50 | 60 | 70 |

RNA Concentration (%)

7.   During anaerobic metabolism, the reaction shown below occurs. Which molecules have been reduced and which have been oxidized?

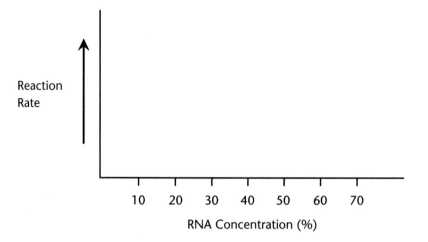

8.   During glycolysis, the reaction shown below occurs. What type of reaction is this?

OH
|
CH_2 O
| ‖
CH — O — P — O⁻   ⇌   H₂O
| |
COO⁻   O⁻

2-phosphoglycerate

CH_2 O
‖   ‖
C — O — P — O⁻
| |
COO⁻   O⁻

phosphoenolpyruvate

9. Match the following:

_____chemiosmotic theory

_____Sir Hans A. Krebs

_____NADH and FADH$_2$

_____glycolysis

_____beta-oxidation

_____glycogen

_____gluconeogenesis

_____glycogenolysis

_____acetyl coenzyme A

_____deamination

a.   described citric acid cycle

b.   conversion of proteins or fats to glucose

c.   converted to glucose-6-phosphate

d.   using H$^+$ gradient to make ATP

e.   derived from pyruvate and the vitamin pantothenic acid

f.   pathway that converts glucose to pyruvate

g.   donates electrons and H$^+$

h.   glycogen breakdown

i.   removes NH$_3$ from amino acid to make organic acid that enters aerobic metabolic pathway

j.   converts fatty acid or glycerol into pyruvate

k.   converts fatty acids into acetyl CoA

10. Explain the function of the following in the electron transport system:

a.   NADH

b.   FADH$_2$

c.   oxygen

d.   ATP synthase

e.   H$^+$

f.   complex inner membrane proteins

MAPS

1. The three major food biomolecules—proteins, carbohydrates, and fats—are used as fuel sources to produce ATP. Create a map using the list of terms (and any terms you wish to add) to show:

   a.   the important or key intermediates

   b.   where the breakdown products of these food groups enter the pathway

   c.   the major products for each part of the pathway (identify the fates of these products)

   d.   both aerobic and anaerobic pathways

   e.   where these reactions occur in the cell

acetyl CoA	electrons	H^+
ADP + P_i	fats	H_2O
aerobic	fatty acids	lactic acid
anaerobic	gluconeogenesis	liver
ATP	glucose	mitochondria
carbohydrates	glycerol	NADH
citric acid cycle	glycogen	oxygen
cytoplasm	glycogenesis	proteins
electron transport system	glycolysis	pyruvate

2. Design a map that outlines the events of translation and protein sorting. Include in your answer the following terms (plus any terms you wish to add):

amino acid

anticodon

codon

Golgi complex

mRNA

peptide bond

polypeptide

ribosomes

rough endoplasmic

   reticulum

signal sequence

transport vesicles

tRNA

BEYOND THE PAGES

Suppose that both Sarah and David were carriers of the Tay-Sachs gene. What parenting options are available for such a couple?

A couple where both partners are carriers has a 1-in-4 chance of conceiving a child who will be born with the disease. These are the basic options that the genetic counselor presents to them:

1. The couple can decide not to have children.

2. They can decide to adopt children.

3. They can elect to use an egg or sperm donor who does not carry the gene.

4. If they decide to have their own children, they can take the 1-in-4 chance that their child will be born with the disease. Screening techniques are available to test the genetic makeup of a fetus as early as 6 weeks after conception to determine whether it will be born with Tay-Sachs disease. If the couple's personal convictions allow, they can screen for the disease in the fetus and elect to abort a fetus that would be born with the disease.

5 Membrane Dynamics

LEARNING OBJECTIVES

When you complete this chapter, you should be able to:

- Draw a model of the body fluid compartments, including typical concentrations of key substances (ions, and so on) in each compartment for a cell in a homeostatic "steady state."

- Predict whether a molecule will cross a cell membrane by diffusion, protein-mediated transport, or vesicular transport.

- Summarize the properties of diffusion as well as the factors that affect the rate of diffusion, and describe changes in the rate of diffusion that take place when these factors change.

- Differentiate between the different types of protein-mediated transport and diagram the mechanisms of each, including the general energy requirements.

- Apply the principles of specificity, competition, and saturation to solute movement by transport proteins.

- Differentiate between phagocytosis, endocytosis, and exocytosis and diagram the mechanisms of each process.

- Draw and label a polarized transporting epithelial cell, and model the transcellular transport of three representative molecules (glucose, Na^+, water), including direction of movement and relevant membrane proteins, ions, and concentration gradients.

- Describe solutions in terms of comparative osmolarities (iso-, hypo-, and hyper-).

- Distinguish between osmolarity and tonicity, and predict the tonicity (iso-, hypo-, hyper-) of a solution relative to a cell. Describe osmotic water movement across a cell membrane in the presence of nonpenetrating and penetrating solutes.

- Define the resting membrane potential of a cell and relate it to the homeostatic steady state created by the selective permeability of the cell membrane to specific ions.

- Describe how scientists measure the resting membrane potential of a cell.

- Diagram the mechanism of insulin secretion as a representative model of integrated membrane processes. Include relevant transported substances, membrane proteins, ions, membrane potentials, and transport processes.

SUMMARY

Be careful not to confuse the terms homeostasis and equilibrium—they are not synonymous. Homeostasis more accurately refers to the **dynamic** (constantly moving) **steady states** (no *net* change) our cells maintain between the intracellular and extracellular compartments. Taking a closer look, we see that our cells maintain osmotic equilibrium—but chemical and electrical *dis*equilibrium—across the cell membrane. Other processes then leverage these dynamic steady states to influence water movement and cellular communication.

   The law of mass balance provides the foundation of homeostasis—what comes in must be offset by what goes out; and, conversely, if something goes out, it must be replaced in order to restore mass balance. Put in more specific terms: The total amount of a substance in the body = intake + production – excretion – metabolism.

Cell membranes act as a selectively permeable barrier between the cells and their external environment, mediating transport and communication between the two compartments. In addition, membranes provide structural support. A cell membrane consists of a phospholipid bilayer with proteins that are inserted partially or entirely through the membrane. Many of these proteins are mobile, while others have restricted movement. Membrane proteins can play a structural role; they can form channels, carriers, receptors, or enzymes; or, they might have multiple duties. Carbohydrates attach to some membrane proteins and lipids, serving as cellular markers for recognition by other cells (like cells of the immune system) and giving the membrane a "sugar coating." Review membrane structure in Ch. 3.

The ability of a substance to cross a cell membrane depends upon the properties of the membrane and upon the size and lipid solubility of the substance. Small, lipid soluble substances can cross a membrane by passive diffusion. Channels formed by transmembrane proteins allow the passage of water and smaller ions. Larger molecules and ions are transported by protein carriers. One form of carrier-mediated transport, primary active transport, uses ATP while other forms of carrier-mediated transport use the energy stored in concentration gradients. Large molecules and particles move into and out of cells by vesicles that fuse with the membrane. Epithelial cells are polarized, with the apical and basolateral membranes containing different types of protein-mediated transporters. This allows the one-way transport of certain molecules.

Water is able to move freely across most membranes, moving primarily through open water channels. The movement of water from a region of lower solute concentration to one of higher solute concentration is called osmosis. Osmolarity describes the number of particles in a solution; tonicity is a comparative term that describes how the volume of a cell changes by osmosis when exposed to a solution.

Ions and most solutes do not diffuse freely across cell membranes, and therefore, a state of chemical disequilibrium exists between the cell and its environment. All cells have a resting membrane potential resulting from chemical and electrical gradients of ions. Depending upon the type of cell, resting membrane potentials are usually between -50 to -90 mV, relative to the extracellular fluid. The resting potential in human cells is established by a membrane that is more permeable to K^+ than to Na^+ or other ions. Active transport maintains this potential by transporting Na^+ and K^+ ions across the membrane. Controlled changes in membrane potential, by means of ion movement, provide a means for intracellular and intercellular communication.

TEACH YOURSELF THE BASICS

MASS BALANCE AND HOMEOSTASIS

IP *Fluids & Electrolytes: Introduction to Body Fluids*

1. Define the law of mass balance. What equation is used to summarize it? How is it used to maintain homeostasis? (Figs. 5-1, 5-2)

Excretion Clears Substances from the Body

2. What is clearance? What organs are involved?

3. Why is clearance an indirect measure of the movement of a substance? What is the mathematical definition of mass flow, and how is it a more direct measure of a substance's movement?

4. Review from Ch. 3 (see Functional Compartments of the Body): What are the two major compartments of the body? Do these break down into further subcompartments? List all fluid compartments.

Homeostasis Does Not Mean Equilibrium

5. What is implied by homeostasis?

6. Describe what we mean when we say that the extracellular and intracellular fluid compartments usually exist in a state of dynamic disequilibrium.

7. What is the only molecule that moves freely between most cells and the extracellular fluid?

8. What is osmotic equilibrium? How does it arise?

9. What is chemical disequilibrium? Give some examples of specific solutes that exist in a state of chemical disequilibrium in our body. (Fig. 5-3)

10. How is chemical disequilibrium a hallmark of a living organism? How is chemical disequilibrium related to the second law of thermodynamics? (Review the second law of thermodynamics in Ch. 4.)

11. How does the chemical disequilibrium of ions create an electrical disequilibrium? What is the result of electrical disequilibrium?

Transport Occurs Within and Between Compartments

12. Compare and contrast movement of substances within compartments and between compartments. In which of these scenarios is a substance most likely to cross a membrane?

13. Identify and briefly describe some of the general processes involved in physiological transport of materials. How do these general transport processes differ from more specific transport processes?

DIFFUSION

14. Define and relate the following terms: selectively permeable, permeable, impermeable.

15. Which two properties of a molecule determine whether it can diffuse across a membrane?

16. What is the difference between active transport and passive transport?

17. What are the different ways a molecule can move across a membrane? (Fig. 5-4)

Diffusion Uses Only the Energy of Molecular Movement

18. Define diffusion. (Fig. 5-5)

19. List seven properties of diffusion. (Fig. 5-5; Table 5-1)

Lipophilic Molecules Can Diffuse Through the Phospholipid Bilayer

20. What three additional factors influence the rate of simple diffusion of a molecule across the phospholipid bilayer? For each factor, if it were to increase in magnitude, would the rate of diffusion increase or decrease? (Fig. 5-6)

21. Fick's law of diffusion describes simple diffusion. (Fig. 5-6; Table 5-1) Write in words the mathematical expression for Fick's law:

    Rate of diffusion ∝

22. What factors influence membrane permeability?

23. Membrane thickness is a constant in most physiological situations; therefore we are able to simplify Fick's law to derive the equation for flux. What is flux? What is the equation for flux?

PROTEIN-MEDIATED TRANSPORT

24. Why is simple diffusion not an option for most molecules in our body?

25. Differentiate between mediated transport, facilitated diffusion, and active transport.

Membrane Proteins Function as Structural Proteins, Enzymes, Receptors, and Transporters

▷ Describe and give examples of each of the following functional categories of membrane proteins. Wherever possible, give specific examples.

Structural Proteins
26. What are the three major roles of structural proteins? (Review transporting epithelia, gap junctions, and the cytoskeleton in Ch. 3.)

Enzymes
27. What are membrane enzymes? How are they different in function from intracellular enzymes?

Receptors

28. Describe membrane receptor proteins. (Fig. 5-8)

Transporters

29. What do membrane transporters do? What are the different types of transporters? How do they differ? (Fig. 5-9)

Channel Proteins Form Open, Water-Filled Passageways

30. Describe and draw a typical channel protein. (Fig. 5-10)

31. What types of molecules pass through channel proteins? Give examples of channel proteins.

32. What factors determine a channel protein's specificity?

33. Distinguish between an open channel and a gated channel. (Fig. 5-9)

34. Describe or draw the anatomy of a gated channel. (Fig. 5-9)

35. List three types of gated channels and indicate the stimulus that opens each type of gate.

Carrier Proteins Change Conformation to Move Molecules

36. What molecules can cross a membrane by using a carrier protein?

37. Distinguish between uniport carriers, symport carriers, and antiport carriers. (Fig. 5-9)

38. What is a cotransporter?

39. Describe the structure of a carrier protein.

40. Describe or diagram the molecular mechanism by which carrier proteins move molecules. (Fig. 5-11)

41. How do carrier proteins differ from channel proteins? Compare and contrast the two types of transporters.

Facilitated Diffusion Uses Carrier Proteins

42. Compare and contrast facilitated diffusion with simple diffusion.

43. Compare and contrast facilitated diffusion with active transport.

44. What types of molecules enter and leave our cells by facilitated diffusion?

45. How are facilitated diffusion and equilibrium related? (Fig. 5-12; Table 5-1)

Active Transport Moves Substances Against Their Concentration Gradients

46. What is active transport? What does it accomplish? Why does it require the input of energy? (Review potential energy, Fig. 4-2.)

47. Distinguish between primary (direct) active transport and secondary (indirect) active transport.

48. How does the mechanism for both types of active transport compare to facilitated diffusion? What accounts for the similarities? The differences?

Primary Active Transport

49. Diagram the structure and mechanism of the Na^+-K^+-ATPase as an example of primary active transport. (Figs. 5-13, 5-14; Table 5-2)

Secondary Active Transport

50. How is the relatively high extracellular $[Na^+]$ used to drive transport of other molecules against their concentration gradient across a membrane? Give some specific examples. (Table 5-3)

51. Describe and diagram the structure and mechanism of the Na^+-glucose transporter (SGLT) as a representative example of secondary active transport. (Fig. 5-15)

52. Why does the body have both a facilitated diffusion GLUT carrier and a SGLT Na⁺-glucose symporter?

▶ The use of energy to drive active transport is summarized in Fig. 5-16.

Carrier-Mediated Transport Exhibits Specificity, Competition, and Saturation

▶ Review our earlier discussion of specificity, competition, and saturation in Ch. 2.

Specificity
53. Give specific examples of how specificity applies to carrier-mediated transport.

Competition
54. How does competition relate to specificity?

55. Use the experiment represented in the graph of Fig. 5-17 to demonstrate the principle of competition.

56. Give an example of competitive inhibition. (Fig. 5-18; also see competitive inhibition in Ch. 2)

57. Give an example of a medicinal application of competition.

Saturation
58. Describe how the principle of saturation applies to carrier-mediated transport. (Fig. 5-19) Include a description of transport maximum.

59. How can cells increase their transport capacity and avoid saturation?

VESICULAR TRANSPORT

▶ Macromolecules that are too large to enter or leave cells through protein channels or carriers are moved with the aid of vesicles.

60. The two primary modes of vesicular transport are _____ and _____

Phagocytosis Creates Vesicles Using the Cytoskeleton

61. Describe the process of phagocytosis. (Fig. 5-20) What is a phagosome?

Endocytosis Creates Smaller Vesicles

62. How does endocytosis differ from phagocytosis?

63. What is pinocytosis?

Receptor-Mediated Endocytosis

64. Diagram the process of receptor-mediated endocytosis. (Fig. 5-21) What role does clathrin play? What is an endosome?

65. What is membrane recycling? How does this work?

66. What types of substances are transported via receptor-mediated endocytosis?

67. What disease states can arise from abnormalities in receptor-mediated endocytosis?

Potocytosis and Caveolae

68. What distinguishes potocytosis from receptor-mediated endocytosis?

69. Diagram the process of potocytosis.

70. What functions do caveolae carry out? What disease states can arise from abnormal caveolae? (Fig. 5-24)

Exocytosis Releases Molecules Too Large for Transport Proteins

71. How is exocytosis related to endocytosis? (Fig. 5-21, steps 8–9)

72. What types of molecules are exported by means of exocytosis?

73. What are the roles of Rabs and SNAREs?

74. Compare and contrast regulated exocytosis, constitutive exocytosis, and intermittent exocytosis.

75. Is exocytosis an active or passive process?

EPITHELIAL TRANSPORT

76. What is transepithelial transport? Why is it important?

77. Compare and contrast the apical membrane and basolateral membrane of a transporting epithelial cell. How did this polarization arise? (Fig. 5-22) What are alternate names for the apical membrane and basolateral membrane?

78. What is the physiological significance of transporting epithelial cell polarization?

79. Contrast the terms absorption and secretion relative to epithelial cell membrane polarization.

Epithelial Transport May be Paracellular or Transcellular

80. Define and contrast paracellular transport and transcellular transport.

81. Describe the energy requirements for protein-mediated transcellular transport. How do the energy requirements of protein-mediated transcellular transport relate back to the concept of dynamic steady states you learned about earlier in the chapter?

82. Can transporting epithelial cells manipulate their permeability? Please explain.

Transcellular Transport of Glucose Uses Membrane Proteins

83. Diagram and describe the movement of glucose across a transporting epithelial cell. Include in your diagram all membrane proteins, ions, and directionality of transport. (Fig. 5-23)

Transcytosis Uses Vesicles to Cross an Epithelium

84. Describe the process of transcytosis. When is it used? (Fig. 5-24) What role does the cytoskeleton play in transcytosis?

OSMOSIS AND TONICITY

85. Why is water movement such an important aspect of physiology?

The Body Is Mostly Water

86. What is the physiological significance of the "70-kg man"? What is his total body water?

87. How does a woman's total body water compare to a man's? (Table 5-4)

88. Describe the distribution of water among the body compartments. (Fig. 5-25; review fluid compartmentation in Ch. 3)

The Body Is in Osmotic Equilibrium

89. Define and describe osmosis. (Fig. 5-26)

90. What is osmotic pressure?

Osmolarity Describes the Number of Particles in Solution

91. How does osmolarity differ from molarity?

92. What is the equation used to determine osmolarity?

93. HCl, as a strong acid, completely dissociates into two ions in aqueous solution. What is the osmolarity of a 1 mM solution of HCl?

94. Contrast osmolarity with osmolality.

Comparing Osmolarities of Two Solutions

95. Distinguish between isosmotic, hyposmotic, and hyperosmotic.

96. If solution A has more particles per liter than solution B, then solution A is said to be _____

osmotic to solution B while solution B is _____ osmotic to solution A. (Table 5-5)

97. How does this discussion of relative osmolarities relate to water movement? Can you predict water movement based on relative osmolarities? Why or why not?

Tonicity of a Solution Describes the Volume Change of a Cell Placed in That Solution

98. Contrast osmolarity with tonicity.

99. Distinguish between isotonic, hypotonic, and hypertonic. (Table 5-6)

▶   Hint: To remember what happens to a cell in hypotonic solutions versus hypertonic solutions, look at the letter directly preceding the "t" in each word. For hypotonic, the "o" can be a visual reminder that that cell swells in a hypotonic solution. The "r" in hypertonic can be a visual reminder that cells shrink in hypertonic solutions.

100. Contrast penetrating solutes with nonpenetrating solutes. How do they each affect water movement across a membrane?

101. Solution C has 100 mosmoles/L and solution D has 200 mosmoles/L. Can we describe the tonicities of these two solutions relative to each other? Explain your reasoning.

102. What determines the tonicity of a solution relative to a cell? (Figs. 5-27, 5-28)

103. What are the rules for predicting tonicity?

104. How can a solution be isosmotic to a cell and also hypotonic?

105. Rules for osmolarity and tonicity are given in Table 5-7. Spend some time going over the examples in the book until you are comfortable with osmolarity and tonicity. These will be important topics to understand as we move forward. It might help you to draw the tonicity examples given in the book and either rework the math or write out the rules until you have a firm grasp on what's happening in each scenario and why.

▸ Table 5-8 lists the most common IV solutions and their approximate osmolarity and tonicity relative to the normal human cell.

THE RESTING MEMBRANE POTENTIAL

IP *Nervous I: The Membrane Potential*

106. List some of the chemical players that contribute to the electrical makeup of our bodies. In which fluid compartments are they typically more concentrated?

107. Are our fluid compartments in electrical equilibrium? Why or why not? What is the significance of this arrangement? Do intracellular and extracellular compartments have a neutral net electrical charge? Explain.

Electricity Review

108. List the highlights of the electrical principles important for our discussion of electricity in physiological systems:

The Cell Membrane Enables Separation of Electrical Charge in the Body

109. Describe how the cell membrane accomplishes the separation of electrical charge. (Fig. 5-29)

110. How does the cell create an electrical gradient? An electrochemical gradient?

111. Define and describe the resting membrane potential difference (membrane potential).

▶ When measuring membrane potentials in living systems, it's convention to assign a charge of 0 mV to the extracellular fluid. (Fig. 5-30) However, remember that the extracellular fluid is not really electrically neutral. (Fig. 5-29c)

The Resting Membrane Potential Is Due Mostly to Potassium

112. What is the equilibrium potential? Write the equation for determining the equilibrium potential of a single ion. What equation must be used when multiple ions are involved?

113. Describe the electrochemical forces at play as ions come to their equilibrium potentials across a cell membrane. (Figs. 5-31, 5-32, 5-33)

114. What role do transport proteins play in creating and maintaining electrochemical gradients? Give some examples of specific transporters.

Changes in Ion Permeability Change the Membrane Potential

115. What two factors influence a cell's membrane potential?

116. Compare and contrast the terms depolarization, repolarization, and hyperpolarization. (Fig. 5-34)

117. Which four ions contribute the most to changes in membrane potential? In which fluid compartment is each ion most concentrated?

▶ It's important to learn that a significant change in membrane potential requires the movement of very few ions.

INTEGRATED MEMBRANE PROCESSES: INSULIN SECRETION

118. Do only nerve and muscle cells use changes in membrane potential as signals for physiological responses?

119. Diagram and describe the process of insulin secretion in pancreatic beta cells as a key example of integrated membrane function. (Fig. 5-35) Include examples of membrane proteins and demonstrate the role of membrane potential in bringing about a physiological response to a specific parameter change.

TALK THE TALK

absorption
active transport
amoeba
antiport carriers
apical
aquaporin
atmospheres (atm)
ATPases
ATP-binding cassette (ABC) transporter superfamily
ATP-gated K^+ channel or K_{ATP} channel
basolateral
bulk flow
carrier proteins
caveolae
caveolins
cellular homeostasis
channel proteins
chemical disequilibrium
chemical gradient
chemically-gated channels
clathrin
clathrin-coated pits
clearance
colligative
competition
competitive inhibitor
concentration gradient
conductor
constitutive
cortisol
cotransporter
depolarized
difference
diffusion
dynamic
dynamic disequilibrium
electrical disequilibrium
electrical gradient
electricity
electrochemical gradient
electrodes
electrogenic
electron
endosome
epithelial transport
equilibrium potential
excretion
exocytosis
facilitated diffusion
Fick's law of diffusion

fluid
flux
functionally
gated channels
glucose symporter
GLUT transporter
ground
hexoses
hyperosmotic
hyperpolarized
hypertonic
hyposmotic
hypotonic
impermeable
insulator
internal environment
invaginates
ion channels
isosmotic
isotonic
law of conservation of electrical charge
law of mass balance
leak channels or pores
lipophobic
load
mass flow
mediated transport
membrane enzymes
membrane potential
membrane receptor proteins
membrane recycling
membrane transporters
metabolite
micropipets
millimeters of mercury (mm Hg)
molarity (M)
mole
monovalent
mucosal
muscular dystrophy
Na^+-K^+-ATPase
net movement
nonpenetrating solutes
open channels
osmolality
osmolarity
osmoles
osmometer
osmosis
osmotic equilibrium
osmotic pressure

ouabain
particles
passive transport
penetrating solutes
permeable
phagocytosis
phagocytes
phagosome
pinocytosis
polarized
potential
potocytosis
primary (direct) active transport
rab
receptor-mediated endocytosis
recording electrode
reference electrode
repolarization
resting membrane potential difference or membrane potential
saturation
secondary (indirect) active transport
secretion
selectively permeable
serosal
SGLT (sodium glucose transporter)
simple diffusion
SNAREs
solute carrier (SLC) superfamily
specificity
steady states
structural proteins
symport carriers
tonicity
total body water
transcellular transport
transcytosis
transepithelial transport
transport maximum
uniport carriers
vesicular transport
voltage-gated Ca^{2+} channel
voltage-gated channels
voltmeter
water channels
xenobiotics

QUANTITATIVE PHYSIOLOGY

VOLUMES OF DISTRIBUTION

The techniques by which the volumes of the body compartments are determined provides us with interesting insight into experimental design. The determination of a volume of liquid can be estimated using a *dilution technique*. There are several requirements for this experiment. 1. The volume to be measured must be well-mixed, so that any marker that is given will distribute evenly throughout the compartment. 2. The marker must be able to be measured. If the system is a living organism, the marker must also be nontoxic, not metabolized or excreted, and must distribute only into the compartment being measured.

To carry out a dilution experiment, a known amount of marker in a known volume is administered. It is allowed to distribute; then a sample of the compartment liquid is removed and analyzed for its marker content. Because the concentration of marker in the sample is the same as the concentration throughout the compartment, the following ratio is used to calculate the compartment volume:

$$\frac{\text{amount marker in sample (known)}}{\text{volume of sample (known)}} = \frac{\text{total amount of marker (known)}}{\text{total volume of compartment (unknown)}}$$

For example, 1 gram of dye is put into a giant vat of water. The dye is stirred until it is evenly distributed throughout the vat. A 1 mL sample of water is taken out, analyzed, and found to contain 0.02 mg of dye. What is the volume of the vat?

0.02 mg dye/1 mL = 1000 mg dye/χ mL

χ = 1000 mg dye • 1 mL/0.02 mg dye

χ = 50,000 mL or 50 L

The dye distributed into a volume of 50 L. For this reason, the volumes calculated by this method are known as *volumes of distribution*.

The determination of body compartment volumes in humans has been done using a variety of markers. The marker must be restricted to the compartment being measured in order to give accurate results. *Total body water* has been estimated using water with radioactive isotopes of hydrogen. Both deuterium oxide (D_2O or heavy water) and tritium oxide can be used. Calculating the volume of distribution for the *extracellular fluid (ECF) volume* requires a molecule that can move freely between the plasma and the interstitial fluid but that cannot enter the cells. Sucrose, the disaccharide commonly known as table sugar, fits this requirement as does a molecule called *inulin*. Inulin is a plant polysaccharide extracted from the roots of dahlias. It is not metabolized by humans but is excreted in the urine, so this must be taken into account when estimating ECF volume. The final compartment that can be directly measured is the *plasma volume*. This measurement requires a large molecule that distributes in the plasma but cannot cross the leaky epithelium to the interstitial fluid. Since endogenous plasma proteins meet this requirement, researchers found a dye, Evans blue, that binds to plasma proteins and therefore distributes only in the plasma.

There are no markers for the interstitial fluid and the intracellular compartment, but we have been able to accurately estimate those volumes as well. Using the information in the paragraph above, can you explain how to calculate interstitial fluid and the intracellular volumes? (See answer in workbook appendix.)

PRACTICE MAKES PERFECT

1. According to Einstein, the time required for a molecule to diffuse from point A to point B is proportional to the square of the distance. Fill in the table below and graph the relationship:

Distance	Time
1 mm	
2 mm	
3 mm	
4 mm	

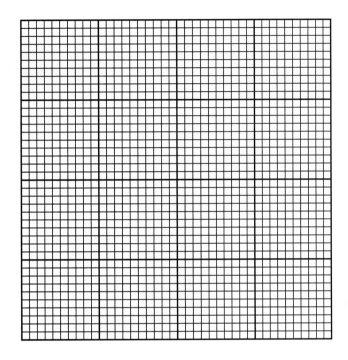

2. A person with pneumonia has an accumulation of fluid in her lungs. This has an effect similar to that of increasing membrane thickness by a factor of 2. How does this affect the rate of diffusion? Give a quantitative and qualitative answer. Assume that all other parameters do not change.

3. Complete the table below.

Method	Movement Relative to Concentration Gradient	Energy Source	What Affects Rate?	Through Membrane Bilayer or Through Protein Transporter?	Exhibits Specificity?	Exhibits Competition?	Exhibits Saturation?	Examples
Simple diffusion								
Facilitated diffusion								
Primary active transport								
Secondary active transport								

4. Match the transporter with all the descriptions that apply.

_____Na$^+$-K$^+$-ATPase   a.   symport

_____Na$^+$-glucose transporter   b.   antiport

_____Ca^{2+}-ATPase   c.   active transport

_____Na$^+$-K$^+$-2 Cl$^-$-transporter   d.   cotransport

_____Na$^+$-H$^+$-transporter   e.   secondary active transport

5. After completing physiology, you become so fascinated by biology that you go on to become a world-famous zoologist. In the year 2045 you are part of a scientific expedition to investigate newly discovered life forms on the planet Zwxik in another solar system. You are assigned to study a single-cell organism found in the aqueous swamps of Zwxik. In one of your first studies using a radioisotope of calcium, you observe that calcium moves freely in and out of the cell during daylight hours but is unable to get into or out of the cell in the dark. What is your hypothesis about how calcium is crossing the cell membrane in this organism? Be as specific and as scientific as possible.

6. In the compartments below, diagram the osmotic, chemical, and electrical equilibrium/disequi-librium that exists in the living body. For ions, use large symbols where concentrations are high and small symbols where concentrations are low.

    Key: Na^+, K^+, Ca^{2+}, Cl^-, proteins, osmolarity (mOsm), plasma, interstitial fluid, cells

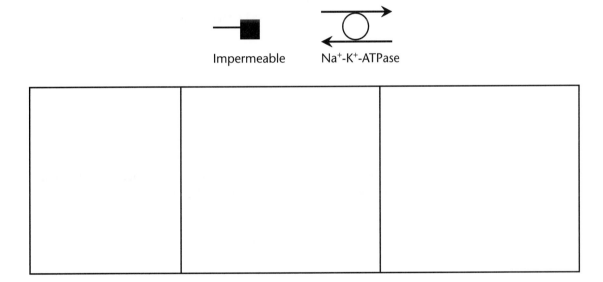

7. The addition of solute to water disrupts the hydrogen bonds of water and interferes with the crystalline lattice formation of ice. As a result, the water will not freeze until cooled below the freezing point for pure water, a phenomenon known as freezing point depression. For an ideal solution, its freezing point will drop 1.86° Celsius for each osmole of solute/liter of water. A plasma sample from a patient shows a freezing point depression of 0.55° C. What is the osmo-larity of this patient's plasma? What is the intracellular osmolarity in this patient?

8. You put 5 grams of glucose into a giant beaker of water and stir. You then take a 1 mL sample and analyze it for glucose. If the 1 mL sample contains 1 mg of glucose, how much liquid is in the beaker?

9. You are monitoring the absorption of a new drug, Curesall, across the intestine of a rat *in vitro*. After numerous experiments, you get data that give you the graph below. The only difference between experiments A and B is that in experiment A the apical solution contains 150 mM Na^+ and in experiment B the apical solution contains 50 mM Na^+. What conclusion(s) can you draw about Curesall absorption based on the graph? Based on what you have learned about different kinds of transport, hypothesize about the type of transporter that carries Curesall.

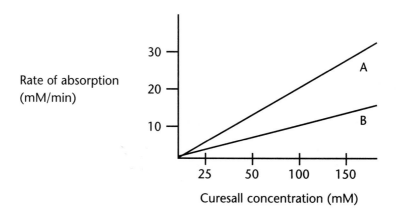

10. What is the osmolarity of a 0.9% NaCl solution? Assume complete dissociation of NaCl. (To review molar solutions, percent solutions, and atomic weights, see Ch. 2 of the text.)

11. What is the osmolarity of a 5% dextrose (= glucose, $C_6H_{12}O_6$) solution?

12. An artificial cell with zero glucose inside is placed in a solution of glucose. The amount of glucose inside the cell is measured at different times. The two graphs below show the results of the experiments.

Graph 1:

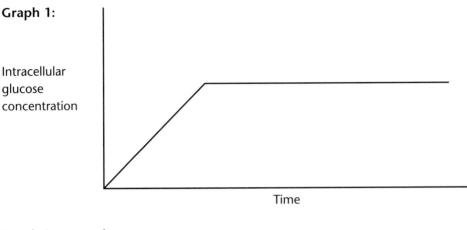

Graph 2:

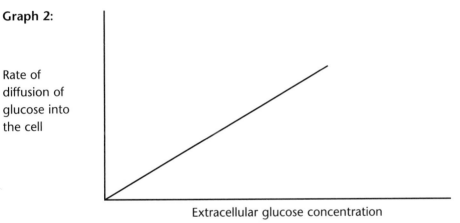

Extracellular glucose concentration

a. In what units might the rate of diffusion be measured?

b. Is glucose moving into the artificial cell by diffusion or by protein-mediated transport, or can you tell from these graphs? If it is mediated transport, is it active or passive?

c. Why does the line in graph 1 level off?

13. For each row in the table below, assume that the row represents two compartments separated by a membrane that is freely permeable to water but impermeable to solutes. Show which way water will move by osmosis for each row by drawing in an arrow in the direction of water movement in the column labeled "Membrane." In the last column, tell the osmolarity of solution A relative to solution B.

Solution A	Membrane	Solution B	Osmolarity of A Relative to B
100 mM glucose		100 mM urea	
200 mM glucose		100 mM NaCl	
300 mOsM NaCl		300 mOsM glucose	
300 mM glucose		200 mM CaCl$_2$	

14. A one-liter intravenous infusion of 300 mOsM NaCl is administered during a 15-minute period to a person whose starting osmolarity was 300 mOsM. What effect will this infusion have on the volume and osmolarity of the extracellular and intracellular fluid compartments? Explain your answer.

15. A red blood cell whose internal osmolarity is 300 osmol/L is placed into a solution that has a composition of 200 osmol/L NaCl and 100 osmol/L of urea.

    This solution is _____ osmotic to the cell. This solution is _____ tonic to the cell.

    Draw a fully labeled graph showing the volume change of the cell from the time it is placed into the solution (at the arrow) until it reaches equilibrium.

16. A person comes into the emergency room dehydrated after working out in the summer sun. His ECF volume is 13 liters and his ECF osmolarity is 340 MOsM. (Ignore ICF in this problem.)

    a. How many milliosmoles of solute are in his ECF when he comes in?

    b. An intravenous (IV) infusion of 1 liter of 160 mOsM NaCl is given. How many millimoles of solute are in the IV?

    c. Assuming that no water or solute moves out of the ECF, calculate the new ECF volume and osmolarity after the man has been given the IV.

17. Assume that a membrane that is permeable to Cl$^-$ but not to Na$^+$ separates two compartments. Two different sodium chloride solutions are placed in the compartments. The concentration of NaCl on side 1 is 0.3 M, while the NaCl solution on side 2 is 2 M. In which direction will ion movement occur? Explain. Will a membrane potential difference develop across the membrane? Explain.

18. If you are told that the extracellular K^+ concentration has increased from 2 mEq/L to 3 mEq/L, does this also mean that the extracellular fluid has become more positive? What happens to the resting membrane potential of a liver cell when the extracellular K^+ concentration increases? Explain.

19. You are doing an experiment and have an intracellular recording electrode stuck into a liver cell. The resting membrane potential is –70 mV relative to the outside of the cell. At the arrow-head you add ouabain to the extracellular fluid bathing the cell. On the graph below draw what will happen to the resting membrane potential of the cell over time. Write an explanation of the forces at work in this situation.

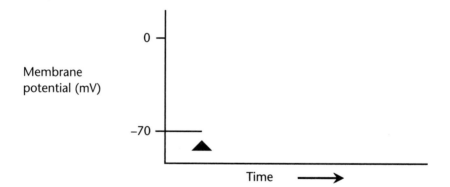

BEYOND THE PAGES

Interactive animations of osmosis, diffusion, and the Nernst equation can be found at http://cweb.middlebury.edu/cr/labbook/.

Try Google searches (www.google.com) for the following topics:

Nernst—This brings up pages with the biography of Walther Nernst as well as pages with problems, simulators, and other interactive Nernst equation goodies.

IV fluids—This brings up pages with clinical information about the use of IV fluids.

6 Communication, Integration, and Homeostasis

LEARNING OBJECTIVES

When you complete this chapter, you should be able to:

- List cell-to-cell communication methods and distinguish when one communication method might be used instead of another.

- Contrast paracrine signals, autocrine signals, neural signals, hormones, neurohormones, and cytokines.

- Draw examples of intracellular receptors (cytosolic and nuclear) and membrane receptors (ligand-gated ion channel, receptor-enzyme, G protein-coupled receptors, and integrin) and describe their cellular signaling mechanisms.

- Describe the processes of biological signal transduction and signal amplification. Diagram the mechanisms of the G protein-coupled adenylyl cyclase-cAMP system and the phospholipase C system as key representatives of cellular signal transduction.

- Apply the concepts of specificity, saturation, and competition to signal pathway modulation. Describe specific ways signal pathways are modulated.

- List Cannon's postulates and use them to describe homeostatic control systems.

- Contrast local control and reflex control.

- Diagram the seven steps of a reflex pathway, describe each component, and distinguish between the response loop and the feedback loop.

- Diagram examples of negative and positive feedback loops and how they modulate the response loop.

- Apply your understanding of specificity, nature of a signal, speed of a signal, duration of action, and coding for stimulus intensity to describe neural and endocrine regulation of control systems.

- Diagram control pathways for neural, neuroendocrine, and endocrine reflexes.

SUMMARY

For a society of distinct individuals to exist, its members must learn to communicate effectively. The cells in a multicellular organism comprise such a society, and they have evolved effective and efficient means of communication so that each component can do its job correctly and respond to changes in the environment. This chapter introduces the concepts of cellular communication and provides patterns by which you can organize more detailed information that will be presented later for each specific physiological system. In other words, learn the patterns presented in this chapter, and use them to make files in your mental filing cabinet under which you can store specific information later.

Communication in the body is either electrical or chemical. Electrical communication is limited mostly to nerves and muscles, while chemical communication is an activity in which all cells participate to some extent. Local communication is accomplished by paracrines, autocrines, and cytokines that work close to the cells that secrete them. Long-distance communication is accomplished by any combination of the following: hormones, neurohormones, cytokines, or electrical impulses. Chemical communication is limited by diffusion, as increasing distance increases the time required for action. Hormones (and sometimes cytokines) are secreted into the blood and carried to all parts of the body but ultimately have to diffuse to their target cells. Therefore, electrical impulses are a faster form of long-distance communication (milliseconds-minutes) than hormones (minutes-hours). However, electrical communication is short-lived, while chemical communication can have long-lasting effects.

A cellular response to a signal depends more on the receptor than on the signal. For instance, different cells can have different responses to the same signal depending on the receptor type (isoform) present. There are four major categories of membrane receptors: ligand-gated ion channels, integrins linked to the cytoskeleton, receptor-enzymes, G protein-coupled receptors. In the latter three categories, ligand binding causes receptors to turn on enzymatic machinery that ultimately directs the cellular response. If the receptor is on the cell membrane, it turns on signal transduction machinery to amplify the signal and relay it to the appropriate destinations via second messengers. Signal ligands are called first messengers, and products of the transduction machinery are called second messengers.

If the receptor is in the cytoplasm, then no signal transduction machinery is employed, but enzymes are still activated to direct the cellular response—often transcription of new gene products. In addition, receptors can be down-regulated or up-regulated depending on a given situation.

Homeostasis means maintaining a "similar condition" despite changes in physiological parameters. Walter B. Cannon outlined four postulates regarding the properties of homeostasis. The body uses a system of local control and reflex control pathways to maintain homeostasis. Local control depends on paracrines, autocrines, and cytokines. Reflex control pathways handle widespread or systemic challenges, and these reflex pathways are composed of nervous, endocrine, and cytokine components (starting to look familiar?).

Reflex pathways are further broken down into response loops and feedback loops. Response loops contain the following steps: stimulus, receptor, afferent pathway, integrating center, efferent pathway, effector, response (cellular and systemic). This is an important pattern to learn, and the various endocrine and/or nervous system combinations involved serve as good file headings for future information (Fig. 6-31). Feedback loops can be either negative feedback loops or positive feedback loops. Negative feedback loops turn off the response that created them, and are therefore homeostatic. Positive feedback loops continue the response that created them, are therefore not homeostatic, and must be controlled from outside the response loop. Feedforward control responses can anticipate changes. Biological rhythms often coincide with a predictable environmental change and reflect changes in the setpoint of the cycling function.

TEACH YOURSELF THE BASICS

CELL-TO-CELL COMMUNICATION

1. Distinguish between chemical communication methods and electrical communication methods.

2. What are targets?

3. List the basic methods of cell-to-cell communication.

Gap Junctions Create Cytoplasmic Bridges

4. What kinds of signals pass through gap junctions? (Fig. 6-1a; review cell junctions in Ch. 3)

5. What proteins create gap junctions?

6. Where do you find gap junctions?

Contact-Dependent Signals Require Cell-to-Cell Contact

7. Give examples of contact-dependent signaling. What are some of the proteins involved? (Fig. 6-1b)

Paracrine and Autocrine Signals Carry Out Local Communication

8. Distinguish between paracrine and autocrine signals.

9. How do chemical signals secreted by cells spread to adjacent cells? How does this limit the effectiveness of this paracrine and autocrine signaling? (Fig. 6-1c)

10. Give examples of some important local signal molecules.

Long-Distance Communication May Be Electrical or Chemical

11. What are hormones? How do they reach their targets?

12. Why don't all cells react to hormones circulating in the body? (Fig. 6-2a)

13. In the nervous system, what kinds of signals are used to transmit information?

14. Distinguish between a neurotransmitter (Fig. 6-2b), neuromodulator, and neurohormone. (Figs. 6-2c, 6-31)

Cytokines May Act as Local and Long-Distance Signals

15. What are cytokines? Give examples of some of the cellular responses they mediate.

16.   How are cytokines different from hormones?

SIGNAL PATHWAYS

17.   Why do some cells respond to a chemical signal while other cells ignore it? (Fig. 6-2a)

18.   What features do all signal pathways share in common? (Fig. 6-3)

Receptor Proteins Are Located Inside the Cell or on the Cell Membrane

19.   How do chemical signals interact with receptors? What factors govern cellular placement of receptor proteins? (Fig. 6-4)

20.   List differences between lipophobic and lipophilic signal molecules. How does the molecular anatomy of a signal molecule affect its functionality in the cellular communication machinery?

21.   What are the major categories of membrane receptors? (Fig. 6-5)

Membrane Proteins Facilitate Signal Transduction

22.   What is signal transduction? (Fig. 6-6) What is a transducer?

23.   Describe signal amplification. (Fig. 6-7; Table 6-1)

24. Briefly diagram the basic steps in signal transduction. (Fig. 6-8)

25. What is a second messenger? List some examples. (Table 6-2)

26. How do signal transduction pathways constitute a cascade? (Fig. 6-9)

▶ Take some time to familiarize yourself with the common signal amplification pathways presented in Table 6-1 and the second messengers in Table 6-2.

Receptor-Enzymes Have Protein Kinase or Guanylyl Cyclase Activity

27. Describe the action of receptor-enzymes. (Figs. 6-5, 6-10)

28. Name two enzymes that are found to participate in receptor-enzyme complexes.

Most Signal Transduction Uses G Proteins

29. Briefly describe the G protein-coupled receptors (GPCR). (Fig. 6-5)

30. What types of ligands bind with GPCRs? Give some examples.

31. Describe how G proteins are activated and what happens upon activation.

32. What are the two most common amplifier enzymes for G protein-coupled receptors?

Many Lipophobic Hormones Use GPCR-cAMP Pathways

33. Diagram the G protein-coupled adenylyl cyclase-cAMP system. (Fig. 6-11) Identify the amplifier enzyme and the second messenger. (Use a separate sheet of paper.)

G Protein-Coupled Receptors Also Use Lipid-Derived Second Messengers

34. Diagram the G protein-coupled phospholipase C (PL-C) signal transduction system. (Fig. 6-12) Identfiy the amplifier enzyme and the second messenger. (Use a separate sheet of paper.)

35. What cellular actions do the lipid-derived second messengers initiate?

Integrin Receptors Transfer Information from the Extracellular Matrix

36. Describe the structure and function of integrins. (Fig. 6-5)

The Most Rapid Signal Pathways Change Ion Flow Through Channels

37. What are the simplest receptors? Describe how these receptors produce a cellular response. (Fig. 6-13; review cell membrane potential in Ch. 5)

▶ Spend time reviewing and re-creating the map in Fig. 6-14, which is a summary of signal transduction systems.

NOVEL SIGNAL MOLECULES

Calcium Is An Important Intracellular Signal (Fig. 6-15)

38. How does extracellular calcium enter the cell?

39. Where and how is intracellular calcium stored? (Review organelle structure and function in Ch. 3.)

40. Diagram a Ca^{2+}-based signal pathway. What effects can a calcium spark initiate? (Use a separate sheet of paper.) (Fig. 6-15)

Gases Are Ephemeral Signal Molecules

41. What is nitric oxide (NO), and how does it bring about a cellular response?

42. How is NO synthesized in tissues?

43. What are some of the effects brought about by NO?

44. How does carbon monoxide (CO) exert signal effects?

Some Lipids Are Important Paracrine Signals

45. What are orphan receptors?

46. What are eicosanoids, and why are they important?

47. How are eicosanoid signal molecules made? (Fig. 6-16)

48. How is arachidonic acid made?

49. Name two major groups of paracrines derived from arachidonic acid. Give examples of their actions.

50. How are NSAIDs related to lipid-derived paracrines?

MODULATION OF SIGNAL PATHWAYS

51. For most signal molecules, the target cell response is determined by _____ rather than _____.

Receptors Exhibit Saturation, Specificity, and Competition

52. Why do receptors exhibit characteristics of specificity, competition, and saturation? (Review these principles in Ch. 2, protein reaction rates; Ch. 4, enzymes; and Ch. 5, membrane protein-mediated transport.)

Specificity and Competition: Multiple Ligands for One Receptor

53. Use the given example of norepinephrine and epinephrine to describe specificity and competition for membrane receptors.

Agonists and Antagonists (Fig. 6-17)

54. (Agonists/Antagonists?) are:

   • Molecules or drugs that mimic normal ligand

   • Bind to and activate receptors

55. (Agonists/Antagonists?) are:

   • Molecules or drugs that mimic normal ligand

   • Bind and block ligand from binding so that the receptor never gets turned on

56. How are agonists and antagonists utilized in the development of pharmaceutics?

Multiple Receptors for One Ligand

57. What are isoforms? What is their significance?

58. Give an example of isoforms found in the body. (Fig. 6-18)

Up-Regulation and Down-Regulation Enable Cells to Modulate Responses

59. What is saturation? Why does it happen?

60. Describe how a cell accomplishes down-regulation. Why is it necessary? Under what conditions does a cell initiate down-regulation? Give an example.

61. When a cell down-regulates a receptor, what happens to its responsiveness to the ligand that binds to that receptor?

62. Describe how a cell accomplishes up-regulation. Why is it necessary? Under what conditions does a cell initiate up-regulation?

63. When a cell up-regulates a receptor, what happens to its responsiveness to the ligand that binds to that receptor?

Cells Must Be Able to Terminate Signal Pathways

64. What are some ways in which a cell is able to terminate signal pathways? Give examples.

Many Diseases and Drugs Target the Proteins of Signal Transduction

65. Give an example of a disease involving abnormal receptors. Identify the abnormal receptor. (Table 6-3)

CONTROL PATHWAYS: RESPONSE AND FEEDBACK LOOPS

▶ Review homeostasis. (Homeostasis was first discussed in Ch. 1, but discussed in more depth in Ch. 5.)

66. What are the three components of a physiological control system in its simplest form? (Fig. 6-19)

Cannon's Postulates Describe Regulated Variables and Control Systems

67. List Walter B. Cannon's four postulates. (Figs. 6-20, 6-21, 6-18)

Homeostasis May Be Maintained by Local or Long-Distance Pathways

68. Describe what is meant by local control.

69. What is a reflex control pathway?

70. What role do cytokines play in long-distance reflex pathways?

71. What are two other control systems involved with long-distance pathways?

Local Control

▶ Note: Local control involves paracrines or autocrines, and its response is restricted to where the change took place.

72. Give an example of local control.

Reflex Control

73. A reflex pathway can be broken into two parts. Name them:

74. List the three main components of a response loop.

75. Fill in the missing terms:

   stimulus ➔ _____ ➔ _____ pathway ➔ _____

   center ➔ _____ pathway ➔ _____ ➔ response

Receptor

▶ Note: This is a new and different application of the word *receptor*.

76. Give two different meanings for the word *receptor*.

77. What is a threshold for a receptor? What role does the threshold play in the reflex?

Afferent Pathway

78. What is an afferent pathway? In other words, from where to where does it relay information?

Integrating Center

79. What is the role of the integrating center?

80. Where are the integrating centers for neural reflexes?

81. Where are the integrating centers for endocrine reflexes?

Efferent Pathway

82. From where to where does an efferent pathway carry information?

83. In the nervous system, what is the distinguishing characteristic of an efferent signal? In other words, what distinguishes one efferent neural signal from the next?

84. How are efferent signals named or described in nervous system pathways?

85. In the endocrine system, what is the distinguishing characteristic of the efferent signal?

Effectors

86. Define and give examples of effectors.

Responses

87. Distinguish between and give examples of cellular and systemic responses.

Response Loops Begin with a Stimulus and End with a Response

88. Use the example of the aquarium heater to understand the different components of a simple response loop. (Fig. 6-25)

89. Compare and contrast tonic and antagonistic control.

Setpoints Can Be Varied

90. Distinguish between acclimatization and acclimation.

91. What are some ways that setpoints can vary in individuals?

Feedback Loops Modulate the Response Loop

92. What is the purpose of a feedback loop?

93. What completes a reflex response?

Negative Feedback Loops Are Homeostatic

94. How does negative feedback interact with the original stimulus? (Figs. 6-26, 6-27a)

95. Negative feedback loops _____ the original physiological parameter being regulated.

Positive Feedback Loops Are Not Homeostatic

96. How does positive feedback interact with the original stimulus? (Fig. 6-27b)

97. Give an example of a positive feedback loop. (Fig. 6-28)

98. What shuts off a positive feedback loop?

Feedforward Control Allows the Body to Anticipate Change and Maintain Stability

99. Give an example of feedforward control. How does feedforward control differ from negative feedback and positive feedback?

Biological Rhythms Result from Changes in a Setpoint

100. What is a circadian rhythm? (Fig. 6-29)

101. What is the adaptive significance of functions that vary with a circadian rhythm?

102. Name some body functions that exhibit circadian rhythms.

Control Systems Vary in Their Speed and Specificity

103. Why does the body need different types of control systems? (Fig. 6-30)

104. Compare and contrast neural and endocrine control systems on the basis of five key areas that follow (Table 6-4):

     specificity

     nature of the signal (Fig. 6-31)

     speed

     duration of action

     coding for stimulus intensity

Complex Reflex Control Pathways Have Several Integrating Centers (Fig. 6-31; Table 6-5)

105. For neural reflexes, endocrine reflexes, and neuroendocrine reflexes, identify the afferent pathway, the integrating center, and the efferent pathway.

106. Take some time now to draw out the different reflexes from Fig. 6-31 for yourself. Color-code the components and learn the distinctions between the reflex types.

TALK THE TALK

acclimation
acclimatization
acetylcholine
adenylyl cyclase
adrenergic
adrenergic receptors
afferent pathway
agonists
amplifier enzyme
anaphylaxis
anchor proteins
antagonistic control
antagonists
arachidonic acid
autocrine signal
calmodulin
CAMs
cascade
cellular response
centimeters
central receptors
chemical signals
circadian rhythm
connexins
contact-dependent signals
cyclic AMP (cAMP)
cycloxygenase (COX)
cytokines
cytosolic receptors or nuclear
   receptors
dephosphorylation
desensitization
diacylglycerol (DAG)
distal
drug tolerance
effectors
efferent pathway
eicosanoids
electrical signals
endogenous
endothelial-derived relaxing
   factor (EDRF)

epinephrine or adrenaline
feedback loop
feedforward control
first messenger
fitness
G protein
G protein-coupled adenylyl
   cyclase-cAMP system
G protein-coupled receptors
   (GPCR)
gap junctions
guanylyl cyclase
half-life
histamine
hormones
hydrogen sulfide
inositol trisphosphate (IP_3)
input signal
integrating center
integrin receptors
Janus kinase (JAK)
leukotrienes
ligand
ligand-gated
lipophilic signal molecules
lipophobic signal molecules
local communication
local control
long-distance communication
negative feedback
neurocrine
neurohormone
neuromodulator
neurotransmitter
norepinephrine
orphan receptors
oscillates
output signal
oxytocin
paracrine signal
peripheral receptors
phospholipase A_2 (PLA2)

phospholipase C (PL-C)
physiological control system
positive feedback loop
prostaglandins and throm-
   boxanes
prostanoids
protein kinase A (PK-A)
protein kinases
protein phosphates
receptor
receptor-channel
receptor-enzymes
reflex
reflex control pathways
regulated variables
reinforces
response loop
responses
second messenger system
second messengers
selective serotonin reuptake
   inhibitors (SSRIs)
sensitivity
sensor
setpoint
signal amplification
signal transduction
specificity
stimulus
syncytium
systemic
systemic response
target cells or targets
threshold
tonic control
transduce
transducer
troponin
tyrosine kinase
wheal

PRACTICE MAKES PERFECT

1. Describe the basic ways cells communicate with each other. Which modes are faster?

2. Why are there different isoforms and subtypes of receptors? Is the number of receptors on a cell constant?

3. What is the advantage of having a cascade of events as opposed to having just a single event drive a response?

4. Would you expect the secretion of growth hormone to be regulated through negative feedback? Explain.

MAPS

1. Use the following terms to design a map or flowchart describing the events that are associated with signal transduction and the activation of second messenger systems. You may add any terms that you deem necessary. You may draw a cell and associate the terms with different components of the cell (a structure/function map).

adenylyl cyclase	G protein-coupled receptors	nitric oxide (NO)
amplifier enzyme	GDP	phospholipase C
calcium	GTP	protein kinase
calmodulin	guanylyl cyclase	receptor-enzymes
cAMP	hormone	second messenger
cGMP	integrin receptors	target cell
DAG	IP3	tyrosine kinase
first messenger	membrane receptor	

2. Use the following terms to design a map or flowchart describing the events that are associated with reflex control pathways. You may add any terms you feel are necessary to complete the map.

afferent pathway	integrating center	sensory receptor
antagonistic control	negative feedback	stimulus
effector	receptor	threshold
efferent pathway	response	tonic control

Introduction to the Endocrine System

LEARNING OBJECTIVES

When you complete this chapter, you should be able to:

- Create a chart that distinguishes the three classes of hormones (peptides, steroids, and amino acid derivatives) according to how they are synthesized, stored, and released; how they are transported in the blood; and their cellular mechanisms of action. You should also be able to work backwards and predict the classification of an unknown hormone from knowledge of its synthesis, transport, cellular mechanism, and so on.

- List the similarities and differences between the anterior pituitary and the posterior pituitary.

- List (full spellings and abbreviations) the hormones secreted by the anterior pituitary and posterior pituitary. Which ones have trophic effects?

- Diagram the reflex pathways of specific hormones in order to demonstrate simple endocrine reflexes and neuroendocrine reflexes. Include at least one example that contains multiple integrating centers and at least one example that demonstrates how an integrating center must "decide" what to do with multiple input signals.

- Distinguish between short-loop feedback and long-loop feedback in a diagram of the cortisol control pathway.

- Predict the expected response from a target cell in the presence of multiple hormones based on an application of your understanding of hormone interactions (synergism, permissiveness, antagonism).

- Diagnose primary and secondary endocrine pathologies based on knowledge of feedback loops and an understanding of hormone hypersecretion, hyposecretion, and abnormal tissue responsiveness.

SUMMARY

A hormone is traditionally defined as a chemical secreted by a cell or group of cells into the blood for transport to a distant target, where it exerts its effect at very low concentrations. Chemically, hormones fall into one of three classes: peptide, steroid, or amino acid derivative. Generally, the three groups can be distinguished by how they are synthesized, stored, and released; how they are transported in the blood; and the mechanisms by which they cause a cellular response. Practice classifying hormones according to these aspects in order to integrate hormones presented in future chapters.

Hormones can interact with other hormones, and in doing so, alter cellular response. Three types of hormone interaction are discussed in this chapter: synergism, permissiveness, and antagonism. Familiarize yourself with these distinctions so you will be able to apply them to hormone interactions you encounter in later chapters.

The release of hormones can be under multiple levels of control: classic hormones are under direct control of the parameter they regulate, but other hormones can have a multilevel control pathway. Revisit reflex control pathways in Ch. 6 (Fig. 6-31) and apply those basic concepts to hormone control pathways. Hormones can be classified according to their reflex pathways, so an understanding of the basics presented in this chapter will help you understand other hormone control pathways you encounter.

If one hormone causes the release of another hormone, the first hormone is called a trophic hormone. Most of the six hormones of the anterior pituitary are trophic hormones, and they are in turn controlled by trophic neurohormones of the hypothalamus. The hypothalamus and the pituitary communicate via the hypothalamic-hypophyseal portal system. The posterior pituitary is derived from neural tissue, which is a characteristic that distinguishes it from the anterior pituitary. The posterior pituitary only releases two hormones, which are actually neurohormones synthesized in the hypothalamus. Become well acquainted with the distinctions between the anterior pituitary and posterior pituitary and spend some time learning the hormones secreted by each.

Endocrine pathologies can be classified as primary or secondary, depending on where in the reflex pathway the problem occurs. Endocrine pathologies can arise in a number of ways (e.g., hypersecretion, hyposecretion, abnormal tissue responsiveness). Apply what you've learned about negative feedback in hormone reflex pathways to develop your understanding of endocrine pathology diagnosis. Your efforts here will help you to integrate future cases of hormone physiology and pathophysiology.

TEACH YOURSELF THE BASICS
HORMONES

IP *Endocrine System: Endocrine System Review*
1.   Hormones are responsible for:

2.   What are the ways hormones act on their target cells?

Hormones Have Been Known Since Ancient Times
3.   List the four classic steps for identifying an endocrine gland.

What Makes a Chemical a Hormone?
4.   Define hormone and describe its generalized actions.

▶   The Anatomy Summary in Fig. 7-2 lists the major hormones of the body and the glands or cells that secrete them, along with the major properties of each hormone.

Hormones Are Secreted by a Cell or Group of Cells
5.   What types of cells and tissues secrete hormones?

Hormones Are Secreted in the Blood
6.   Define secretion. (Review secretion in Ch. 3.)

7. What is an ectohormone?

8. What are pheromones? What do they do?

Hormones Are Transported to a Distant Target

9. Compare and contrast the following terms: hormone, factor, paracrine, autocrine, candidate hormone. Give examples of when the distinctions become blurry.

Hormones Exert Their Effect at Very Low Concentrations

10. Define nanomolar and picomolar.

11. What distinguishes hormones from cytokines?

Hormones Act by Binding to Receptors

12. What is meant by cellular mechanism of action?

13. What determines the variable responsiveness of a cell to a particular hormone?

Hormone Action Must Be Terminated

14. Why is it necessary for hormone action to be terminated? (There is more than one possible answer.)

15. What generally happens to hormones in the bloodstream?

16. What is the half-life of a hormone?

17. How are hormones that have bound to target membrane receptors terminated?

THE CLASSIFICATION OF HORMONES

IP *Endocrine System: Biochemistry, Secretion and Transport of Hormones, and the Actions of Hormones on Target Cells*

18. Review organelle function. Can you describe all the steps of protein synthesis, starting with transcription? (See organelles in Ch. 3; protein synthesis in Ch. 4.)

19. List three chemical classes of hormones. (Table 7-1)

Most Hormones Are Peptides or Proteins

20. What are the monomer molecules for peptides? (Review protein structure in Ch. 2, Fig. 2-9.)

▶   Exclusion rule: If a hormone isn't a steroid or an amine, then it must be a peptide.

Peptide Hormone Synthesis, Storage, and Release

21. Where in the cell are peptide hormones synthesized? (Fig. 7-3)

22. Explain the differences between the preprohormone, the prohormone, and the hormone. Where is each made?

23. How are peptide hormones released from endocrine cells?

Post-Translational Modification of Prohormones

24. Name some of the post-translational modificiations a prohormone can undergo. (Fig. 7-4; review post-translational modification in Ch. 4)

Transport in the Blood and Half-Life of Peptide Hormones

25. Peptide hormones are (lipophobic/lipophilic?) and therefore (will/will not?) dissolve in plasma.

26.   Describe the half-life of most peptide hormones.

Cellular Mechanism of Action of Peptide Hormones

27.   Do peptide hormones have membrane receptors or intracellular receptors? Why?

28.   What is the target cell's response to hormone-receptor binding? (Fig. 7-5; review membrane receptors, second messengers, and signal transduction systems in Ch. 6)

29.   Is the target response quick or slow? Explain.

Steroid Hormones Are Derived from Cholesterol (Fig. 7-6)

30.   List the tissues/organs from which steroid hormones are secreted.

Steroid Hormone Synthesis and Release

31.   Where in the cell are steroid hormones sythesized?

32.   Can steroid hormones be stored? Explain. How does this affect steroid hormone synthesis?

33.   Describe how steroid hormones are released from endocrine cells.

Transport in the Blood and Half-Life of Steroid Hormones

34.   Why are carrier proteins required to transport steroid hormones in the bloodstream?

35.   Describe the half-life of steroid hormones. How do carrier proteins affect half-life?

36. Describe how steroid hormones release from their carrier proteins and leave the plasma to interact with the target cell. (Fig. 7-7; review dissociation constants [K_d] in Ch. 2; review law of mass balance in Ch. 5)

Cellular Mechanism of Action of Steroid Hormones

37. Do most steroid hormones have membrane receptors or intracellular receptors? Why?

38. What is a target cell's response to hormone-receptor binding? (Fig. 7-7)

39. Is the target response quick or slow? Explain.

40. Explain nongenomic responses to steroid hormones. How do these types of responses differ from traditional steroid responses?

Some Hormones Are Derived from Single Amino Acids (Fig. 7-8)

41. From which two amino acids do all hormones in this class derive?

42. Give some examples of amino acid–derived hormones.

▸ Be sure you can list the tissues/glands that secrete steroid and amino acid–derived hormones. If a hormone doesn't come from one of these locations, then it is most likely a peptide hormone.

CONTROL OF HORMONE RELEASE

IP *Endocrine System: The Hypothalamic-Pituitary Axis*

▸ Review local control, reflex control, and reflex pathways in Ch. 6.

Hormones Can Be Classified by Their Reflex Pathways

▸ Reflex pathways are a convenient method by which to classify hormones and simplify your learning of the pathways that regulate their secretion.

43. Compare Fig. 6-27a with Fig. 7-9. How does a hormone create negative feedback that turns off a reflex?

44. Why do endocrine cells need to evaluate signals and "decide" the best response?

The Endocrine Cell Is the Sensor in the Simplest Endocrine Reflexes

45. Describe the components of a simple endocrine reflex. What serves as the sensor? The integrating center?

46. As an example of a simple endocrine reflex, diagram the secretion of parathyroid hormone. (Fig. 7-10) What initiates the reflex? What turns off the reflex? (Use a separate sheet of paper if necessary.)

Many Endocrine Reflexes Involve the Nervous System

47. What are some ways the nervous system and endocrine system overlap in function?

48. Name some ways nervous input can influence endocrine function.

Neurohormones Are Secreted into the Blood by Neurons

49. Name three groups of neurohormones and the specific tissues that synthesize and secrete them.

The Pituitary Gland Is Actually Two Fused Glands

50. Compare and contrast the tissue types that make up the anterior and posterior lobes of the pituitary gland. (Fig. 7-11)

51. What are alternate names for the posterior and anterior lobes?

The Posterior Pituitary Stores and Releases Two Neurohormones

52. What two neurohormones are stored and released by the posterior pituitary?

53. Where are these neurohormones made, and how do they get to the posterior pituitary? How does this process differ from the synthesis, storage, and release pattern followed by traditional peptide hormones? (Fig. 7-12; review axonal transport in Ch. 8)

54. Briefly describe the functions of the two posterior pituitary neurohormones.

The Anterior Pituitary Secretes Six Hormones (Fig. 7-13)

55. List (spell out) the six hormones synthesized by endocrine cells of the anterior pituitary.

56. In the answer above, put a star next to the hormone(s) that do not function as a trophic hormone.

Feedback Loops Are Different in the Hypothalamic-Pituitary Pathway

57. How many integrating centers are involved in the endocrine reflexes of the anterior pituitary hormones? Identify these integrating centers. (Fig. 7-14)

58. How does negative feedback in the hypothalamic-pituitary pathway differ from previously described patterns of negative feedback?

59. Distinguish between long-loop negative feedback and short-loop negative feedback. Give examples of each. (Figs. 7-14, 7-15)

The Hypothalamic-Hypophyseal Portal System Directs Trophic Hormone Delivery

60. How do hypothalamic trophic hormones reach the anterior pituitary? (Fig. 7-16)

61. Describe and diagram the structure of a portal system. Identify the locations of three portal systems in the body.

62. What is the primary advantage of using a portal system?

Anterior Pituitary Hormones Control Growth, Metabolism, and Reproduction
63. List the six anterior pituitary hormones again, but this time, include their target tissues.

64. Re-create and describe what is happening in the complex endocrine pathway illustrated in Fig. 7-17. (Use a separate sheet of paper.)

HORMONE INTERACTIONS
65. Identify three types of hormone interactions.

In Synergism, the Effect of Interacting Hormones Is More Than Additive (Fig. 7-18)
66. Describe synergism (potentiation).

67. Give an example of a synergistic hormone interaction.

A Permissive Hormone Allows Another Hormone to Exert Its Full Effect
68. Describe permissiveness as it pertains to hormone interactions.

69. Give an example of a permissive hormone interaction.

Antagonistic Hormones Have Opposing Effects

70.   Describe antagonism as it pertains to hormone interactions.

71.   Give an example of an antagonistic hormone interaction.

ENDOCRINE PATHOLOGIES

72.   What are three basic patterns of endocrine pathology?

73.   To illustrate endocrine pathologies, we'll use a single example: cortisol production by the adrenal cortex (see Fig. 7-15). Diagram the endocrine reflex for cortisol production, identifying the components, signals, hormones, and feedback.

Hypersecretion Exaggerates a Hormone's Effects

▶   Hypersecretion of a hormone can be due to benign tumors (adenomas), malignant tumors, or iatrogenic causes.

74.   Define exogenous. Can exogenous sources of hormone participate in reflex pathways? (Fig. 7-19)

75.   Define atrophy. How is it a factor in endocrine pathologies?

76.   Why is it important to taper doses of exogenous steroid?

Hyposecretion Diminishes or Eliminates a Hormone's Effects

▶   Hyposecretion of a hormone can be due to atrophy of the endocrine gland or lack of vitamins or minerals.

77. Using our example of the cortisol pathway, identify responses to hyposecretion of CRH, ACTH, or cortisol.

Receptor or Second Messenger Problems Cause Abnormal Tissue Responsiveness

78. Describe hormone levels in conditions of abnormal tissue responsiveness. (Review Modulation of Signal Pathways in Ch. 6.)

Down-Regulation

79. (High/Low?) hormone levels will cause down-regulation of receptors. Why?

Receptor and Signal Transduction Abnormalities

80. Describe abnormalities in hormone receptors. Give examples of pathologies that can result.

81. Describe abnormalities in signal transduction pathways. Give examples of pathologies that can result.

Diagnosis of Endocrine Pathologies Depends on the Complexity of the Reflex

82. What is the difference between a primary endocrine pathology and a secondary endocrine pathology? Give an example of each.

83. Explain why the concentrations of trophic hormones vary in cases of primary and secondary pathologies. (Fig. 7-20)

84. For the following scenarios for our cortisol reflex, diagnose the condition as a primary or secondary pathology and identify the tissue location of the abnormality (or alternate causes).

    high cortisol, high ACTH, low CRH

    high cortisol, low ACTH, low CRH

85. Fig. 7-21 shows patterns of hormone secretion in hypocortisolism. Answer the Figure Question to see if you understand the basics of diagnosing endocrine pathologies.

HORMONE EVOLUTION

86. Why does insulin from cows, pigs, or sheep work in humans? Where do we now get the insulin we use for exogenous administration?

87. Speculate as to why growth hormone from cows, pigs, or sheep doesn't work in humans.

88. Define vestigial. Name some vestigial structures in our endocrine system.

89. How has the evolutionary conservation of hormone activity aided in the study of human endocrine function? (Fig. 7-22)

 ## FOCUS ON:
THE PINEAL GLAND (FIG. 7-22)

90. Once thought to have no function, the pineal gland is now known to secrete _____.

91. What are the functions of melatonin in humans?

TALK THE TALK

adenohypophyseal

adenohypophysis

adiponectin

adrenocorticotrophic hormone (ACTH, also called corticotropin)

albumin

amino acid–derived hormone

antagonism

anterior pituitary

atrophy

autism

autocrines

axillary

calcitonin gene-related peptide (CGRP)

candidate hormones

catecholamine

cellular mechanism of action

cholecystokinin (CCK)

classic hormones

comparative endocrinology

corticosteroid-binding globulin

cortisol

co-secretion

C-peptide

cytokines

degraded

down-regulate

-drome

ectohormone

eicosanoids

endocrinology

endogenous

erythropoietin

etiology

evolutionary conservation

exogenous

factor

follicle-stimulating hormone (FSH)

genomic effect

glands

goiter

gonadotropin

growth factor

growth hormone (GH, also called somatotropin)

growth hormone-inhibiting hormone (GHIH)

half-life

hormone

hormone deficiency

hormone excess

hyperinsulinemia

hypersecretion

hyposecretion

hypothalamic-hypophyseal portal system

iatrogenic

inhibiting hormones

long-loop negative feedback

luteinizing hormone (LH)

melanocyte-stimulating hormone (MSH)

melatonin

negative feedback

nongenomic responses

organotherapy

oxytocin

paracrines

parathyroid hormone (PTH)

permissiveness

pheromones

pituitary gland

portal system

posterior pituitary or neurohypophysis

post-translational modification

potentiation

preprohormones

primary hypersecretion

primary pathology

prohormone

prolactin (PRL)

pro-opiomelanocortin

proteolytic

pseudohypoparathyroidism

releasing hormones

replacement therapy

resistin

secondary hyposecretion

secondary pathology

short-loop negative feedback

signal sequence

signal transduction

somatostatin

tamoxifen

testicular feminizing syndrome

thyroid hormones

thyroid-stimulating hormone (TSH, also called thyrotropin)

thyrotropin-releasing hormone (TRH)

transcription factor

trophic

trophic hormone

-tropin

vasopressin

vestigial

PRACTICE MAKES PERFECT

1. Fill in the chart below for the two primary types of hormones.

	Peptide	Steroid
Transport in plasma		
Synthesis site in endocrine cell		
Method of release from endocrine cell		
General response of endocrine cell to hormone stimulation (i.e., mechanism of action)		

2. Write "A" next to the statements that apply to the anterior pituitary, and "P" next to the statements that apply to the posterior pituitary.

_____ connected to hypothalamus by nerve fibers

_____ connected to hypothalamus by blood vessels

_____ secretes hormones produced by hypothalamus

_____ controlled by releasing hormones from hypothalamus

_____ secretes peptide hormones

3.   Which of the following seems to be a major function of the pineal gland?

    a.   sense of appetite
    b.   sense of thirst
    c.   regulation of behavioral responses
    d.   regulation of temperature responses
    e.   acts as a biological clock

4.   Match the following:

    _____ cortisol

    _____ aldosterone

    _____ growth hormone

    _____ vasopressin

    _____ prolactin

    _____ thyroxine

    a.   hormone(s) produced by the adrenal cortex

    b.   hormone(s) produced by the anterior pituitary

    c.   the two hormones most important for normal growth and development

    d.   hormone whose target is the kidney

Note: Answers may be used once, more than once, or not at all.

5.   TRUE/FALSE? Defend your answers.

    a.   Endocrine gland cells that synthesize steroid hormones have lots of hormone stored in vesicles in the cytoplasm.

    b.   These same endocrine cells have lots of smooth endoplasmic reticulum and Golgi complexes.

6.   Hormone XTC causes decreased neural transmission of pain, inducing a pleasant but misleading lack of pain. Another hormone, Y-ME, acts on XTC-producing cells to inhibit transcription of XTC-mRNA. The ultimate result of combining XTC and Y-ME is normal pain sensation.

    a.   Would the interaction between XTC and Y-ME be best described as:

        1. synergism   2. antagonism   3. permissiveness   4. none of the above

    b.   From the information given, would you say that hormone XTC is probably:

        1. a peptide hormone   2. a steroid hormone   3. insufficient information to decide

7. You have been doing research on the pancreatic endocrine cells that secrete insulin and the adrenal cortex cells that secrete corticosteroids. You prepared tissue for examination under the electron microscope, but the labels fell off the jars when the fixative dissolved the glue. You sent the tissue off anyway and got back the following description for one of the tissues. Which tissue is being described? Defend your answer.

   "...cells are close to blood capillaries. Numerous dense, membrane-bounded granules throughout the cytoplasm with reduced rough endoplasmic reticulum and free ribosomes. Cells with fewer secretory granules show an increase in rough ER and ribosomes."

8. A woman has secondary hypocortisolism due to a pituitary problem. If you give her ACTH, what happens to her cortisol secretion? Draw the complete control pathway with hormones and the glands as part of your answer.

9. Graves' disease is caused by the production of auto-antibodies to the TSH receptor. These antibodies interact with the TSH receptor to stimulate the thyroid gland in a manner similar to TSH. The antibodies are not subject to negative feedback. Which of the sets of lab values below would indicate Graves' disease? *Briefly* defend your choice. (*Hint:* On which tissue would you find TSH receptors?)

	Serum thyroxine	Serum TSH
Patient A	6 µg/100 mL	1.5 µIU*/mL
Patient B	16 µg/100 mL	0.75 µIU/mL
Patient C	2.5 µg/100 mL	20 µIU/mL
Patient D	12 µg/100 mL	10 µIU/mL
Normal	4–11 µg/100 mL	1.5–6 µIU/mL

*IU = international units, a standard way of quantifying TSH amounts.

Graph the data for the Graves' disease patients above. What kind of graph is most appropriate: a scatter plot? a bar graph? a line graph?

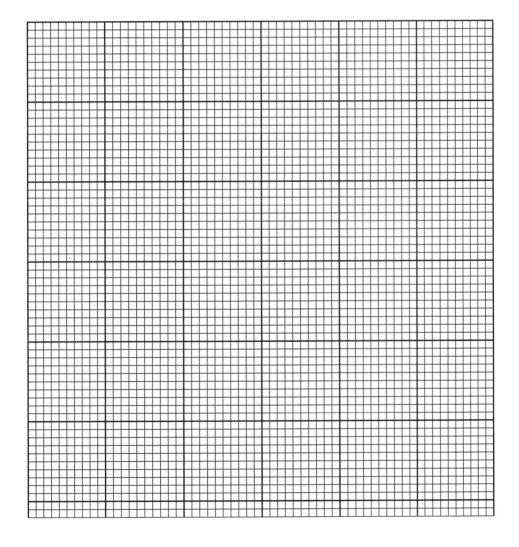

8 Neurons: Cellular and Network Properties

LEARNING OBJECTIVES

When you complete this chapter, you should be able to:

- ▨ Map the divisions of the nervous system, including representative neural cells and tissues.

- ▨ Use the Nernst and GHK equations to show: (1) how permeable ions contribute to the resting membrane potential, and (2) how ion movement affects the resting membrane potential.

- ▨ Diagram graded potentials and describe how they can either inhibit or initiate an action potential.

- ▨ Diagram an action potential and describe the roles of ions and ion channels (and their gating mechanisms) during the rising phase, the falling phase, and refractory periods.

- ▨ Diagram the mechanism by which an action potential is conducted within a neuron and explain the factors contributing to signal directionality and speed of conduction.

- ▨ Diagram mechanisms of electrical and chemical synaptic transmission.

- ▨ Explain the role of neurocrines and differentiate between the terms neurotransmitter, neuromodulator, and neurohormone (Ch. 7).

- ▨ Create a map using adrenergic, cholinergic, and glutaminergic receptors as examples to show how multiple receptor types can amplify the effects of neurotransmitters. In your map, identify which receptors are ion channel-receptors and which ones are G protein-coupled receptors. Sketch simple representative mechanisms for each. (Review membrane receptors in Ch. 5 and signal transduction in Ch. 6.)

- ▨ Distinguish between and draw diagrams or maps to demonstrate the following concepts: convergence and divergence; presynaptic and postsynaptic modulation; spatial summation and temporal summation; and long-term potentiation and long-term depression.

- ▨ Outline key processes of nervous system development and neuron repair.

SUMMARY

This chapter contains information that forms the foundation of our understanding of cell excitability and nerve cell function. Be warned! It is very heavy with terminology and contains some of the most difficult information you will need to master in physiology. If you did not understand the concept of electrochemical gradients and the movement of ions along those gradients, go back and learn this before you begin this chapter (discussion of membrane potentials in Ch. 5 and signaling and other communication principles in Ch. 6).

The nervous system is a complex arrangement of branching divisions, each with specific functions. Take some time to map out the divisions of the nervous system and become familiar with their duties. The nervous system is composed of two basic cell types: neurons and glial cells. The neurons are the functional units of the system, producing electrical and chemical signals to communicate with other cells. Although glial cells do not participate directly in the transmission of electrical signals over long distances, they do communicate with and provide important biochemical and structural support to neurons.

A neuron has three parts: the cell body, the dendrites, and the axon. Axons lack organelles for protein synthesis, so the cell body ships materials to the axon terminals via axonal transport. Neurons synapse on their target cells. Information passes from the presynaptic cell, across the synapse, to receptors on the postsynaptic target cell(s).

Electrical signaling in neurons is a result of ion movement across the membrane. Compare creation of membrane potential with electrical signaling. What specific cellular components make electrical excitability possible? It is very important that you take the time to learn the specifics of how ion movement creates graded potentials and action potentials (APs). Graded potentials are summed, and if they surpass the threshold, an AP is created. APs cannot be summed: they are all or none. Conduction mechanisms allow APs to travel the axon at high speeds and at full strength. Conduction is affected by axon diameter and resistance. Saltatory conduction takes advantage of myelin sheaths and nodes of Ranvier to give high-speed conduction at biologically reasonable axon diameters. Take time to learn how conduction works and the factors that can alter it.

Cells communicate by electrical or chemical means (review Ch. 6). At electrical synapses, electrical signals pass directly from cell to cell. At chemical synapses, neurotransmitters are released onto the postsynaptic cell, where they bind specific receptors to elicit a specific response. Neurotransmitters are made in the cell body or axon terminal and stored in synaptic vesicles until signaled for release. When signaled, synaptic vesicles are released by calcium-dependent exocytosis (Ch. 5). If these details are fuzzy, go back through this chapter and Chs. 5 and 6 to review.

Neurons must "decide" how to handle multiple neural signals. This might be handled through convergence or divergence of the presynaptic neurons. It could also be accomplished by spatial summation and/or temporal summation on the postsynaptic neuron. There are other forms of postsynaptic and presynaptic modulation that might be used, such as long-term potentiation and long-term depression.

TEACH YOURSELF THE BASICS

Spend some time studying neuroscience terminology and synonyms in Table 8-1.

ORGANIZATION OF THE NERVOUS SYSTEM

(Try putting this material into a map or flowchart.)

1. What are the two major divisions of the nervous system? (Fig. 8-1)

2. The peripheral nervous system can be divided further into:

3. Afferent neurons link sensory receptors to _____. (Discussed further in Ch. 10.)

4. Efferent neurons carry information (to/from?) the CNS. Efferent neurons can be classified into two groups:

   somatic motor neurons (Ch. 12), which control:

   autonomic neurons (Ch. 11), which control:

5. Autonomic neurons can be divided further into these two groups:

▶ The enteric nervous system is a network of neurons in the walls of the digestive tract (Ch. 21).

CELLS OF THE NERVOUS SYSTEM

IP *Nervous System I: Anatomy Review*
6. Name the two primary cell types found in the nervous system.

Neurons Are Excitable Cells That Generate and Carry Electrical Signals
7. Define neuron. Describe and illustrate the generalized structure of a model neuron. (Fig. 8-2)

▶ Neurons can be grouped according to structure or function. See Fig. 8-3 for anatomical and functional groupings of neurons.

8. The functional classification separates neurons into three groups: sensory (afferent) neurons, interneurons, and efferent neurons. Compare and contrast these functional groups.

The Cell Body Is the Control Center of the Neuron
9. Describe a neural cell body (soma). What would the extensive cytoskeleton suggest about functions of neurons?

10. What happens if a neuron is cut apart?

Dendrites Receive Incoming Signals
11. Describe the structure and function of dendrites. (Figs. 8-3 and 8-30)

12. How is dendritic function different in the CNS and the PNS?

Axons Carry Outgoing Signals to the Target

13. Describe and illustrate the structure of an axon and associated structures. (Fig. 8-2) Include in your description/illustration: collaterals, axon terminal, axon hillock, synapse, presynaptic cell, postsynaptic cell, and synaptic cleft.

14. Describe the generalized function of an axon. How do axons translate an electrical signal in order to communicate with an adjacent target cell?

15. What is axonal transport? What is its purpose?

16. Describe the mechanism behind slow axonal transport.

17. Describe the mechanism behind fast axonal transport. (Fig. 8-4)

18. Distinguish between forward (anterograde) axonal transport and backward (retrograde) axonal transport.

Glial Cells Provide Chemical and Physical Support for Neurons (Fig. 8-5)

19. Are there more neurons or glial cells in the nervous system?

20. Does neural tissue have extensive or scant extracellular matrix?

21. What are some of the functions carried out by glial cells?

22. How do glial cells communicate with each other and with neurons?

23. What are the two main types of glial cells in the PNS? What are the four types of glial cells in the CNS?

24. What is myelin? Why is it so important? (Fig. 8-6)

25. Compare and contrast Schwann cells and oligodendrocytes. (Figs. 8-5a, 8-6)

26. What are nodes of Ranvier? What is their functional significance?

27. Describe the role of satellite cells in the PNS.

28. What is a ganglion? What is the equivalent structure in the CNS?

29. Describe the role of the following CNS glial cells: microglia, astrocytes.

Adults Have Neural Stem Cells

30. What are ependymal cells? What is the ependyma, and why is it significant?

31. What are neural stem cells? Why are researchers actively trying to develop clinical applications of neural stem cells?

32. Describe the development of neurons and glial cells from neural stem cells.

ELECTRICAL SIGNALS IN NEURONS

IP *Nervous System I: The Membrane Potential; Ion Channels; The Action Potential*

33. Why are neurons and muscle cells considered excitable tissues?

The Nernst Equation Predicts Membrane Potential for a Single Ion

34. What two factors influence membrane potential?

35. Why is the Nernst equation ineffective for determining the resting membrane potential of neurons?

The GHK Equation Predicts Membrane Potential Using Multiple Ions

36. Why is the GHK equation more effective at determining the resting membrane potential of neurons?

37. Which three ions are most influential in creating the membrane potential in resting cells? Write out the GHK equation for these three ions. (Table 8-2)

Ion Movement Across the Cell Membrane Creates Electrical Signals

38. A sudden (increase/decrease?) in Na^+ permeability allows Na^+ to (enter/leave?) the cell. When Na^+ moves, it is moving (down/against?) its concentration gradient and (down/against?) its electrical gradient.

39. The (influx/efflux?) of Na^+ ions (depolarizes/hyperpolarizes/repolarizes?) the membrane potential, creating an electrical signal. (Review the Resting Membrane Potential in Ch. 5.)

40. Describe two examples of how ion movement can hyperpolarize a cell.

   ▶ A significant change in membrane potential occurs with the movement of very few ions.

Gated Ion Channels Control the Ion Permeability of the Neuron

41. How does a cell change its ion permeability? (Review ion channels in Ch. 5.)

42. How are ion channels classified?

43. Name four types of selective ion channels in the neuron:

44. Mechanically gated ion channels open in response to:

45. Chemically gated ion channels open in response to:

46. Voltage-gated ion channels open in response to:

47. Describe variations in ion channel behavior.

Changes in Channel Permeability Create Electrical Signals

48. Define what is meant by an ion's current. What determines the current direction?

49. Distinguish between the two basic types of electrical signals: graded potentials and action potentials. (Table 8-3)

Graded Potentials Reflect the Strength of the Stimulus That Initiates Them

50. Define graded potential. Why are these signals called "graded"? (Fig. 8-7)

51. What initiates a graded potential? This can vary between different types of neurons.

52. What is local current flow?

53. What determines the strength of the initial depolarization?

54. Why do graded potentials lose strength as they move through the cytoplasm?

55. What are trigger zones? Where are they located in sensory neurons? In efferent neurons and interneurons? (Fig. 8-2)

56. What is the functional significance of a trigger zone? What happens when a graded potential reaches a trigger zone? (Fig. 8-8)

57. What is the distinction between excitatory and inhibitory graded potentials?

Action Potentials Travel Long Distances Without Losing Strength

58. Define action potential (AP).

59. How do APs differ from graded potentials?

60. Does the strength of the graded potential that initiates the AP have any influence on the amplitude of the AP? What is the all-or-none phenomenon?

61. Describe some of the more complex patterns of APs observed in brain neurons.

Action Potentials Represent Movement of Na⁺ and K⁺ Across the Membrane

▶ Voltage-gated ion channels are responsible for the ion movement that creates an action potential. (Fig. 8-9)

62. Describe the role of Na⁺, K⁺, and voltage-gated ion channels in each of the following phases of an action potential:

    the rising phase

    the falling phase

63. How is the membrane potential returned to its resting permeability?

Na⁺ Channels in the Axon Have Two Gates

64. In the resting neuron, the activation gate is (open/closed?) and the inactivation gate is (open/closed?) (Fig. 8-10a)

65. Still looking at the resting neuron, describe Na⁺ movement through the channel with its resting gate configuration.

66. Depolarization opens the _____ gate. (Fig. 8-10b)

67. Describe Na⁺ movement through the channel after a depolarization event. What type of feedback loop is involved, if any? (Figs. 8-10c, 8-11; review feedback loops in Ch. 6)

68. Explain how the inactivation gate closes. How is the movement of Na⁺ affected by the closure of the inactivation gate? (Fig. 8-10d)

69. What must happen to the gates before the next action potential can take place? (Fig. 8-10e)

Action Potentials Will Not Fire During the Absolute Refractory Period

70. Define refractory period.

71. Distinguish between absolute refractory period and relative refractory period. Describe ion movement and gate configurations during each. (Fig. 8-12)

72. Why can a greater than normal stimulus trigger an AP during the relative refractory period but not during the absolute refractory period?

73. In what ways do refractory periods affect signal transmission in neurons?

Stimulus Intensity Is Coded by the Frequency of Action Potentials

74. How does the neuron transmit information about the strength and duration of the stimulus that started an action potential? (Fig. 8-13)

One Action Potential Does Not Alter Ion Concentration Gradients

75. TRUE/FALSE? Defend your answer.

Following an action potential, the Na^+ concentration gradient reverses. Explain.

76. TRUE/FALSE? Defend your answer.

If the Na^+-K^+-ATPases in a neuron membrane are poisoned and nonfunctional, the neuron will immediately become unable to fire action potentials. Explain. (Review Na^+-K^+-ATPase in Ch. 5.)

Action Potentials Are Conducted from the Trigger Zone to the Axon Terminal

77. Describe and diagram the conduction of action potentials. Why is it that the signal doesn't lose strength over distance like in a graded potential? (Figs. 8-14, 8-15)

78. How do the mechanisms involved in conduction compare to the mechanisms involved in the initiation of an AP? (Figs. 8-11, 8-14, 8-15)

▶ There is no single action potential. An AP is simply a representation of membrane potential in a membrane segment at a given period in time. A series of recording electrodes along an axon will measure a series of identical APs in different stages, just like falling dominos caught frozen in the act. (Fig. 8-16)

79. When Na^+ channels in the middle of an axon open, depolarizing local current flow will spread in both directions along the axon. What happens to the current that flows backward from the trigger zone into the cell body?

Larger Neurons Conduct Action Potentials Faster

80. List two physical parameters (called cable properties) that affect the speed of action potential conduction.

81. Explain how axon diameter affects AP conduction speed. (Fig. 8-17)

Conduction Is Faster in Myelinated Axons

82. How does the axon prevent against current leakage? In other words, how is membrane resistance to current flow created in an axon?

83. Describe how saltatory conduction allows some animals to have high-resistance, small diameter axons that rapidly transmit signals. (Fig. 8-18) Include in your description a discussion of the nodes of Ranvier.

84. What makes conduction more rapid in myelinated axons?

85. What happens to AP conduction in axons that have lost their myelin? Name a disease resulting from demyelination of axons.

Electrical Activity Can Be Altered by a Variety of Chemicals

86. What happens to AP conduction in a neuron whose Na^+ channels have been blocked?

87. What happens to the likelihood of firing an action potential when the extracellular K^+ increases? Why? (Fig. 8-19)

88. Explain how hypokalemia decreases neuronal excitability. What happens to the membrane potential difference in this instance? (Fig. 8-19d)

▶ Because of the importance of K^+ in nervous system function, the body regulates K^+ levels tightly. We will discuss K^+ regulation in later chapters.

CELL-TO-CELL COMMUNICATION IN THE NERVOUS SYSTEM

IP Nervous System II: Anatomy Review; Synaptic Transmission; Ion Channels

89. What factors affect the specificity of neural communication?

Information Passes from Cell to Cell at the Synapse

90. Describe and diagram the anatomy of a synapse. (Fig. 8-20)

Electrical Synapses

91. Explain how information is transmitted at electrical synapses.

92. What is the main advantage of an electrical synapse?

Chemical Synapses

93. Explain and diagram how information is transmitted at a chemical synapse. (Figs. 8-20, 8-21)

94. Where are polypeptide neurotransmitters and protein enzymes for axon metabolism made, and how do they get to the axon?

Calcium Is the Signal for Neurotransmitter Release at the Synapse

95. Which cellular process is utilized to release neurotransmitters from the axon terminal into the synaptic cleft? (Fig. 8-21)

96. Describe and diagram the events at the synapse that result in neurotransmitter release. (Fig. 8-21)

97. Describe the "kiss-and-run" model of neurotransmitter release.

Neurocrines Convey Information from Neurons to Other Cells

98. Distinguish between neurocrines, neurotransmitters, neuromodulators, neurohormones, paracrines, and autocrines. (Review autocrines and paracrines in Ch. 6, and neurohormones in Ch. 7.)

The Nervous System Secretes a Variety of Neurocrines (Table 8-4)

99. List the seven structural classes of neurocrines.

100. Contrast CNS and PNS neurocrines.

Acetylcholine

101. How is acetylcholine (ACh) made? (Review glycolysis and the citric acid cycle in Ch. 4.)

102. Where is ACh synthesized? (Fig. 8-22)

103. Neurons that secrete ACh and receptors that bind ACh are described as _____.

Amines

104. From where are amine neurotransmitters derived? (Review amino acid–derived hormones in Ch. 7.)

105. List examples of important amine neurocrines and include their functions.

Amino Acids

106. Give examples of four amino acids that behave as neurotransmitters in the CNS. What are their functions?

Peptides

107. List examples and include the functions of some peptides that act as neurocrines.

Purines

108. Name some purine neurotransmitters.

Gases

109. Nitric oxide (NO) is an unstable gas synthesized from _____ and _____. What neurocrine role does it play?

110. What other gases might also be produced in small amounts to serve as a neurocrine?

Lipids

111. Name some lipid neurocrines.

Multiple Receptor Types Amplify the Effects of Neurotransmitters

▶ All neurotransmitters except nitric oxide have one or more receptor types with which they bind. Each receptor type might also have multiple subtypes. (Table 8-4)

112. What are the two categories of membrane receptors into which neurotransmitters fall? What are the descriptive terms given to the function of these receptor types? (Review membrane receptors in Ch. 5.)

Cholinergic Receptors

113. What are the two main subtypes of cholinergic receptors? By which cellular mechanism does each exert its effect?

114. Where in the body are each of the subtypes commonly found?

Adrenergic Receptors

115. Name the classes of adrenergic receptors found in our bodies. How do they initiate cellular response?

Glutaminergic Receptors

116. What is significant about glutamate in the CNS? What are the two receptor types?

117. Describe the mechanism of action for AMPA receptors.

118. Describe the mechanism of action for NMDA receptors. What is unusual about these receptors? (Fig. 8-32)

Not All Postsynaptic Responses Are Rapid and of Short Duration

119. In a fast postsynaptic response, creating a fast synaptic potential, how does the neurotransmitter exert its effect on the postsynaptic cell? (Fig. 8-23)

120. Distinguish between excitatory postsynaptic potential (EPSP) and inhibitory postsynaptic potential (IPSP).

121. In a slow postsynaptic response, creating a slow synaptic potential, how does the neurotransmitter exert its effect on the postsynaptic cell? (Fig. 8-23)

Neurotransmitter Activity Is Rapidly Terminated

122. How is the short duration of nervous signaling achieved? (Fig. 8-24; review ligand binding in Ch. 2)

123. Distinguish between the methods used to inactivate acetylcholine and norepinephrine.

INTEGRATION OF NEURAL INFORMATION TRANSFER

IP *Nervous System II: Synaptic Potentials & Cellular Integration*

124. Contrast divergence and convergence. (Figs. 8-25, 8-26, 8-27) What are the advantages of each?

Neural Pathways May Involve Many Neurons Simultaneously

125. Define and describe spatial summation. (Fig. 8-28a) How does it create an opportunity for integration ("decision making" by a cell or group of cells)?

126. Define and describe postsynaptic inhibition. (Fig. 8-28b) How does it create an opportunity for integration ("decision making" by a cell or group of cells)?

127. Define and describe temporal summation. (Fig. 8-29) How does it create an opportunity for integration ("decision making" by a cell or group of cells)?

▶ Figure 8-30 shows a 3-D distribution of synapses on dendritic spines. The summed input from these synsapses determines the activity of the postsynaptic neuron.

Synaptic Activity Can Also Be Modulated at the Axon Terminal

128. How is presynaptic modulation accomplished? (Fig. 8-31a)

129. Distinguish between presynaptic inhibition and presynaptic facilitation.

130. Why is presynaptic modulation a more precise means of control than postsynaptic modulation? (Fig. 8-31b)

131. What are some ways changes in the responsiveness of the target cell can alter synaptic activity? (Fig. 8-23)

Long-Term Potentiation Alters Synaptic Communication

132. What are long-term potentiation and long-term depression? What's the significance of these processes?

133. Describe and diagram the role of glutamate and its receptors (AMPA and NMDA) in long-term potentiation. (Fig. 8-32)

Disorders of Synaptic Transmission Are Responsible for Many Diseases

134. Some nervous system disorders are caused by synaptic transmission abnormalities. Give some examples of these disorders.

135. List some pharmacological agents that can alter synaptic transmission.

Development of the Nervous System Depends on Chemical Signals

136. How do embryonic nerve cells find their correct targets and make synapses? (Fig. 8-33)

137. In order for a synapse to form correctly and survive, what must take place?

138. Define and describe plasticity.

When Neurons Are Injured, Segments Separated from the Cell Body Die

139. If a neuron is damaged at the cell body, what happens to the neuron?

140. Describe what happens to a neuron if the cell body survives but the axon is severed. (Fig. 8-34)

TALK THE TALK

5-hydroxytryptamine (or 5-HT)
absolute refractory period
acetycholine (ACh)
acetylcholinesterase (AChE)
action potentials (APs)
activation
adenosine
adenosine monophosphate
  (AMP)

adenosine triphosphate (ATP)
adrenaline
adrenergic
adrenergic receptors
afferent (or sensory) neurons
all-or-none phenomenon
amines
amino acids
AMPA receptors

amplitude
amyotrophic lateral sclerosis
  (ALS, Lou Gehrig's disease)
anaglesia
anaxomic
antagonistic control
aspartate
astrocytes
atrial natriuretic peptide (ANP)

autocrine
automatic division
axon hillock
axon terminal
axonal transport
axons
axoplasmic (cytoplasmic) flow
Batten disease
bipolar
brain
bursting
cannabinoid receptors
Cannabis sativa
carbon monoxide (CO)
cell body
central nervous system (CNS)
channel kinetics
chemical synapses
cholecystokinin (CCK)
cholinergic
cholinergic receptors
collaterals
conductance
conduction
convergence
current
current leak
cytoplasmic resistance
demyelinating disease
dendrite spines
dendrites
depolarize
divergence
dopamine
efferent neurons
efflux
electrical synapses
electrochemical
emergent properties
enteric nervous system (ENS)
ependyma
ependymal cells
epinephrine
equilibrium potential
excitability
excitable tissues
excitatory
excitatory postsynaptic potential (EPSP)
fast axonal transport
fast synaptic potential
fusion pore
gamma-aminobutyric acid (GABA)
ganglion

gap junctions
gases
glial cells
glutamate
Goldman-Hodgkin-Katz (GHK) equation
graded potentials
growth cones
hyperkalemia
hyperpolarize
hypokalemia
I_{ion}
inactivation
inactivation gates
influx
inhibitory
inhibitory postsynaptic potentials (IPSP)
initial segment
innervated
interneurons
ionotropic
kiss-and-run pathway
lipids
local current
long-term depression
long-term potentiation (LTP)
metabotropic
microglia
mixed nerves
monoamine oxidase (MAO)
motor nerves
multiple sclerosis
multipolar
muscarine
muscarinic
myelin
Nernst equation
nerves
nervous system
neural stem cells
neurocrine
neuroepithelium
neurons
neurotoxins
neurotransmitters
neurotrophic factors
nicotine
nicotinic
nitric oxide (NO)
NMDA receptors
N-methyl-D-aspartate
nodes of Ranvier
noradrenergic neurons
norepinephrine

nucleus
oligodendrocytes
opioid peptides (enkephalins and endorphins)
overshoot
oxidative stress
paracrine signals
parasympathetic branches
peptides
peripheral nervous system (PNS)
phrenic nerve
plasticity
postsynaptic cell
postsynaptic inhibition
presynaptic cell
presynaptic facilitation
presynaptic inhibition
processes
pseudounipolar
purinergic
purines
reactive oxygen species (ROS)
rectifying synapse
refractory
refractory period
relative refractory period
saltatory conduction
satellite cell
Schwann cells
sensory nerves
serotonin
slow axonal transport
slow synaptic potentials
somatic motor division
spatial summation
spikes
spinal cord
substance P
subthreshold
suprathreshold
sympathetic
synapse
synaptic cleft
synaptic plasticity
temporal summation
threshold voltage
tonically active
trigger zone
trophic
undershoot
varicosities
vasopressin
visceral nervous system

PRACTICE MAKES PERFECT

1. What is probably the biggest advantage of using the nervous system for a homeostatic response rather than the endocrine system?

2. On the figure below, label the boxes with either K^+ or Na^+ to show the ion concentration in the two body compartments.

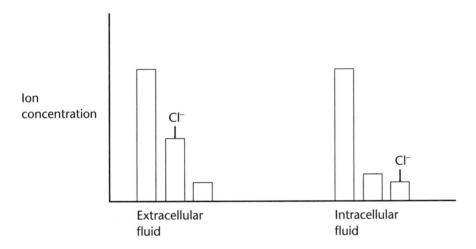

3. Anatomically trace a signal from one end of a neuron to the other, using the correct sequence. Assuming that there is a chemical synapse, describe how a signal passes from one neuron to the next.

4. In the year 2045 you are part of a scientific expedition to investigate newly discovered life forms on the planet Zwxik in another solar system. You are assigned to study a single-cell organism found in the aqueous swamps of Zwxik. In one of your first studies (see Ch. 5), using a radioisotope of calcium, you observe that calcium moves freely in and out of the cell during daylight hours but is unable to get into or out of the cell in the dark.

    You follow up this study with an electrophysiology study. You use an intracellular electrode to measure the electrical charge inside the cell and find that it has a resting membrane potential of +60 mV when the outside fluid is arbitrarily set at 0 mV. From additional studies you find that the ion concentrations in the cell and the surrounding swamp water are as follows:

	Cell	Swamp
K^+	50	50
Na^+	175	15
Cl^-	200	90

All concentrations in millimoles per liter.

You now use contemporary molecular biology techniques to insert some protein channels into the cell membrane. These channels will allow both Na^+ and K^+ to pass; no other molecules can go through them.

a. Predict which ion(s) will move. Tell what direction it/they will move and what force(s) is acting on it/them.

b. The Nernst potential for an ion describes two things: 1. the resting membrane potential if it were determined by only one permeable ion; and 2. the point at which ion movement across a membrane ceases because electrical and osmotic work directly oppose each other. Using the Nernst equation, determine the Nernst potential for Na^+ in the above situation. You have measured the temperature in the swamp to be 37° C.

_____ 5. During repolarization of a nerve fiber:

    a. potassium ion leaves the cell
    b. potassium ion enters the cell
    c. neither occurs, as the fiber is in its refractory period

_____ 6. Treatment of a nerve cell with cyanide, an inhibitor of ATP synthesis, will: (circle all that are correct)

    a. immediately reduce the resting membrane potential to zero
    b. cause a slow increase in the intracellular sodium concentration
    c. immediately prevent the cell from propagating action potentials
    d. inhibit the movement of potassium ions across the membrane

_____ 7. The conduction of an action potential along an axon: (circle all that are correct)

    a. is faster for a strong stimulus than a weak one
    b. occurs at a constant velocity
    c. is faster in unmyelinated nerve fibers than in myelinated nerve fibers
    d. decreases in magnitude as it is propagated along the axon
    e. is slower in a long nerve than in a short nerve of the same diameter

8. Fill in the following table that asks you to list different types of voltage- and chemically gated ion channels in neurons. Tell which ion(s) moves through the channel. For all channels, fill in the blocks showing the type of gating (chemical or electrical) and the physiological process that is linked to the ion movement. For action potentials, state to which phase the ion movement is linked.

Location of channel	Ion(s) that move(s)	Chemical or voltage gating?	Physiological process in which ion participates
dendrite			
axon			
axon			
axon terminal			

9. What is the function of the ground electrode when recording an action potential in a neuron? (See Ch. 5, Fig. 5-30.)

10. What would you guess is the reason(s) that stretching the nerve disrupts its functions?

11. Explain to your parents or your unscientific roommate what an action potential graph looks like, what it represents, and how it is generated.

MAPS

1. Draw a map showing the relationship of the central and peripheral nervous systems, in as much detail as you can. Label the parts with the corresponding parts of a nervous reflex: AP for afferent pathway, IC for integrating center, and EP for efferent path.

BEYOND THE PAGES

The Jackson Laboratory in Bar Harbor, Maine, is a nonprofit research institution that emphasizes genetic research. They develop and keep track of the mutant mouse and rat strains that are being developed and that have yielded so much valuable information for medical research. Each summer they have outstanding undergraduate students come to the lab to conduct biomedical research under the guidance of staff scientists. To learn more about the Jackson Laboratory and its programs, check out its web site at www.jax.org.

The Central Nervous System

LEARNING OBJECTIVES

When you complete this chapter, you should be able to:

- Explain why some neuroscientists would propose changing the functional unit of the nervous system from the individual neuron to neural networks.

- Apply knowledge of nervous system evolution to identify the anatomical similarities humans share with lower-order organisms and demonstrate what makes humans unique.

- Map the anatomical divisions of the CNS.

- Distinguish between white matter and gray matter in the CNS.

- Diagram the three layers of meninges.

- Diagram the secretion and reabsorption of cerebrospinal fluid and describe its purposes and composition (relative to plasma).

- Diagram the blood-brain barrier and explain both how it is formed and how it controls the movement of substances.

- Diagram the spinal cord.

- Map the major anatomical divisions of the brain and identify their primary functions: brain stem (medulla, pons, midbrain), cerebellum, diencephalon (thalamus, hypothalamus, pituitary, pineal gland), and cerebrum (cortex [sensory areas, motor areas, association areas], basal ganglia, limbic system).

- Explain the role of the corpus callosum.

- Describe the distinct layering of neurons in the cerebral cortex.

- Describe cerebral lateralization.

- Distinguish between the sensory system, the cognitive system, and the behavioral state system.

- Describe how the sensory system initiates reflexes and identify the three major types of motor output.

- Describe how the behavioral state system modulates motor output.

- Describe how the reticular activating system influences states of arousal.

- Outline the distinctions between the different stages of sleep and arousal. How are the brainwaves different in each state?

- Describe the role of the suprachiasmatic nucleus in regulating circadian rhythms.

- Explain how emotion and motivation represent a probable overlap of the behavioral state system and the cognitive system.

- Distinguish between emotions and moods.

- Describe the process of learning and differentiate between associative learning and nonassociative learning; habituation and sensitization.

- Describe how memories are formed and differentiate between short-term memory, working memory, and long-term memory; reflexive memory and declarative memory.

- Diagram the cerebral processing of spoken and visual language and describe the role of Wernicke's area and Broca's area—and what happens when either or both areas are damaged.

SUMMARY

The central nervous system (CNS) is made of the brain and spinal cord. Both are encased in membrane and protected by bone. Cerebrospinal fluid (CSF), secreted by the choroid plexus, cushions the CNS and provides a chemically controlled environment. Impermeable brain capillaries create a blood-brain barrier to protect the brain from potentially harmful substances in the blood. Gray matter consists of nerve cell bodies, dendrites, and axon terminals. Cell bodies form layers or cluster into nuclei. White matter, on the other hand, is made primarily of myelinated axons. CNS neurons exhibit plasticity.

The spinal cord is the main pathway between the brain and effectors of the body. The cord is protected by the vertebrae. Severing the spinal cord leads to loss of sensation or paralysis. In the spinal cord, each spinal nerve has a dorsal and ventral root. Dorsal roots carry incoming sensory information, while ventral roots carry information from the CNS to effectors.

The brain is divided into the brain stem, the cerebellum, and the cerebrum. The brain stem plays a major role in the unconscious functions of the body, and it is where many neurons cross sides. The cerebellum processes sensory information and coordinates body movements. The cerebrum, consisting of two hemispheres connected by the corpus callosum, accomplishes higher order brain functions like memory, perception, reasoning, and cognition.

The diencephalon lies between the brain stem and the cerebrum and is composed of the thalamus and hypothalamus. The thalamus is an integration and relay center for sensory information; the hypothalamus plays a key role in homeostasis, autonomic function, and endocrine function. The cerebrum houses the basal ganglia, participating in movement control; and the limbic system, linking higher emotion and brain function.

The cerebral cortex consists of neurons that are highly ordered into vertical columns and horizontal layers. The cortex has sensory areas, motor areas, and association areas. The sensory areas are specialized for specific sensory information: visual cortex, auditory cortex, olfactory cortex. Motor areas direct skeletal muscle movement. Association areas integrate sensory information and create our perception of the sensory world. For example, language is processed in Wernicke's area and Broca's area. Each cerebral hemisphere has become lateralized, not sharing certain functions with the other hemisphere.

The CNS has a variety of neurotransmitters and neuromodulators that create complex pathways for information storage and transmission. Diffuse modulatory systems influence a wide range of body functions. Sleep is a reversible state of inactivity. Sleep consists of REM and slow-wave (non-REM) sleep.

Learning is our acquisition of knowledge. Associative learning occurs when we learn to associate two stimuli. Nonassociative learning includes imitative behaviors like language. Habituation is showing a decreased response to a repeated stimulus, and sensitization is just the opposite. Memory has multiple storage levels: short-term and long-term. Short-term memories can be consolidated into long-term memories.

This chapter presents many details with which you are simply going to have to get familiar.

▶ HINT: Break this information up in a way that makes the most sense to you, and then integrate the details. For example, if it makes more sense for you to divide by function, go that way. If you prefer to move around by anatomy, do that. You'll probably have to change game plans later in the chapter, but do what works best for you.You might find flash cards, maps, and charts particularly useful for this chapter.

TEACH YOURSELF THE BASICS

EMERGENT PROPERTIES OF NEURAL NETWORKS

1. What is plasticity? What role does it play in neural networks?

2. Distinguish between affective and cognitive behaviors.

EVOLUTION OF NERVOUS SYSTEMS

3. Write a brief narrative highlighting the evolution of nervous systems. (Fig. 9-1)

4. What is a ganglion?

5. What is a spinal reflex?

6. More than anything else, which part of the brain makes us human?

ANATOMY OF THE CENTRAL NERVOUS SYSTEM

The CNS Develops from a Hollow Tube

7. Outline the development of the CNS. (Figs. 9-2, 9-3)

8. Which three regions of the brain are obvious by four weeks of human development? By six weeks? (Fig. 9-3)

The CNS Is Divided into Gray Matter and White Matter

9. Describe the composition of the CNS.

10. Describe gray matter. What are nuclei? (Fig. 9-4c)

11. Describe white matter. What are tracts?

12. Describe the consistency of the brain and spinal cord.

Bone and Connective Tissue Support the CNS

13. Why do the brain and spinal cord require protection?

14. Name the bones that protect the brain and spinal cord. (Fig. 9-4a)

15. The stacked vertebrae are separated by tissue disks. What type of tissue makes up these disks? What is the function of these disks?

16. What are meninges?

17. List the three layers of meninges, moving from bone toward the neural tissue. (Fig. 9-4b, c)

18. Fluid within the skull is divided into what two distinct extracellular compartments?

The Brain Floats in Cerebrospinal Fluid

19. What is the choroid plexus? How do choroid plexus cells accomplish secretion of cerebrospinal fluid (CSF) into the ventricles? (Fig. 9-5b)

20. Describe and diagram the anatomical route followed by cerebrospinal fluid from its point of secretion to its point of reabsorption back into the blood. (Use a separate sheet of paper if necessary.) (Fig. 9-5)

21. Name the two purposes of the cerebrospinal fluid.

22. Describe the composition of the cerebrospinal fluid. How does CSF compare with plasma?

23. When physicians need a sample of CSF, how do they acquire it?

The Blood-Brain Barrier Protects the Brain from Harmful Substances in the Blood

24.   What is the blood-brain barrier and what is its function? (Fig. 9-6)

25.   Why are brain capillaries less leaky than other capillaries?

26.   What cells induce formation of tight junctions in brain capillaries and how do they do this?

27.   What kinds of molecules can cross the blood-brain barrier?

28.   Name two brain areas that lack a blood-brain barrier. What are the functions of these areas that require their contact with blood?

Neural Tissue Has Special Metabolic Requirements

29.   Neurons have a high rate of oxygen consumption so that they can produce ATP for what purpose? (Review ATP Synthesis in Ch. 4 and Active Transport in Ch. 5.)

30.   Why is a large percentage of the blood pumped by the heart directed to the brain?

31.   What is the only fuel source for neurons under normal metabolic circumstances?

32.   Define hypoglycemia.

THE SPINAL CORD

33.   The spinal cord is divided into four regions. Name these regions. (Fig. 9-4a)

34. Draw a diagram, map, or outline to explain the relationships between the following terms: axons, brain, cell bodies, columns, dorsal horns, dorsal roots, dorsal root ganglia, efferent signals, gray matter, nuclei, sensory information, spinal reflexes, tracts (ascending, descending, propriospinal), ventral horns, ventral roots, white matter. (Figs. 9-7, 9-8)

THE BRAIN

35. List the six major divisions of the brain. (Fig. 9-9)

The Brain Stem Is the Transition Between Spinal Cord and Midbrain

36. What is the oldest and most primitive region of the brain? What is its structure?

37. What are cranial nerves? How many pairs are there, and in general terms, what are their functions? (Table 9-1)

38. What is the reticular formation?

39. What are some of the processes that involve brain stem nuclei?

The Brain Stem Consists of Medulla, Pons, and Midbrain

40.   Describe and diagram the basic anatomy of the brain stem. (Fig. 9-9d)

41.   Differentiate between corticospinal tracts and somatosensory tracts. Between which regions do they convey information?

42.   Corticospinal tracts cross to the other side of the body in the region called the

_____.

43.   Where is the pons located and what is its primary function?

44.   Where is the midbrain (mesencephalon) located and what is its primary function?

The Cerebellum Coordinates Movement

45.   Where is the cerebellum located? (Fig. 9-9a–c)

46.   From which sensory receptors does the cerebellum receive sensory input?

The Diencephalon Contains the Centers for Homeostasis

47.   Where is the diencephalon located? (Fig. 9-10)

48.   Name the two main sections of the diencephalon.

49.   Which two endocrine structures are located in the diencephalon?

50.   Describe the structure and identify the primary function of the thalamus.

51.   Explain why the thalamus is often described as a relay station. Can it function as an integrating center?

52. Where is the hypothalamus located? From which sources does it receive input?

53. Describe the relationship between the anterior pituitary and the hypothalamus. How does this compare to the relationship between the posterior pituitary and the hypothalamus?

54. Name eight major functions of the hypothalamus. (Table 9-2)

▶ Review hypothalamic control of anterior pituitary hormone release in Ch. 7.

55. What is the pineal gland? (Review the "Focus on . . . the Pineal Gland" box in Ch. 7.)

The Cerebrum Is the Site of Higher Brain Functions

56. The two hemispheres of the cerebrum are connected at the _____ _____. (Figs. 9-9c, 9-10)

57. Name the four lobes found in each hemisphere. (Fig. 9-9b, c, e)

58. What is the adaptive significance of the intricate folding of the surface of the cerebrum?

The Cerebrum Has Distinct Regions of Gray and White Matter

59. What three major regions comprise the cerebral gray matter? (Fig. 9-11)

60. Describe the anatomical arrangement of cortical neurons. (Fig. 9-12)

61. The basal ganglia are involved with:

62. The role of the limbic system is:

63. Name the major components of the limbic system. (Fig. 9-13)

64. Describe white matter in the cerebral cortex.

BRAIN FUNCTION (FIG. 9-14)

65. Identify three systems that influence output by the motor systems of the body:

The Cerebral Cortex Is Organized into Functional Areas

66. Name the following functionally specialized areas of the cerebral cortex. (Fig. 9-15)

_____ translate input into perception

_____ direct movement

_____ integrate information, direct voluntary behaviors

67. Describe cerebral lateralization. (Fig. 9-16)

68. The right side of the brain is associated with which functional skills?

69. The left side of the brain is associated with which functional skills?

70. Why are these generalizations subject to change?

71. What are some noninvasive imaging techniques currently available for evaluating brain function? (Fig. 9-17)

Sensory Information Is Integrated in the Spinal Cord and Brain

72. Describe the general flow of sensory information to the brain. (Figs. 9-14a, 9-8)

73. The primary somatic sensory cortex in the _____ lobe receives information

from _____.

74. What type of information is transmitted?

75. Damage to this region results in:

76. The special senses of vision, hearing, taste, and olfaction have different brain regions devoted to processing their sensory input: (Fig. 9-15)

The visual cortex in the _____ lobe receives information from

_____.

The auditory cortex in the _____ lobe receives information from

_____.

The olfactory cortex, a small region in the _____ lobe, receives information from

_____.

The gustatory cortex in the _____ lobe receives information from the _____.

Sensory Information Is Processed into Perception

77. Define perception. (Fig. 9-18)

78. Which brain areas integrate sensory information into perception?

The Motor System Governs Output from the CNS

79. Identify and characterize the three major types of motor output.

80. Where do voluntary movements initiated by the cognitive system originate? (Fig. 9-15)

81. What roles do the cerebellum, basal ganglia, and pyramidal cells play in voluntary movements?

82. Damage to a motor area will be exhibited as paralysis or loss of function on the _____ side of the body.

83. Where are visceral and neuroendocrine responses coordinated?

84. The brain stem is responsible for maintaining many of the automatic life functions we take for granted, such as:

85. Which aspects of hypothalamic function are part of the motor system?

The Behavioral State System Modulates Motor Output

86. Describe diffuse modulatory systems. How are they generally classified? (Fig. 9-19; Table 9-3)

87. What are some of the brain functions modulated by the diffuse modulatory systems?

The Reticular Activating System Influences States of Arousal

88. Define consciousness.

89. Describe the reticular activation system. How does it influence the "conscious brain"?

90. How do brain wave patterns differ with different states of arousal? How are brain wave patterns measured? (Fig. 9-20)

Why Do We Sleep?

91. Is sleep a passive state or an active state? Why?

92. Name the two major phases of sleep and describe them. (Fig. 9-20)

93. Why do we sleep? What are some of the ideas being explored to answer this question?

94. List some common sleep disorders.

Physiological Functions Exhibit Circadian Rhythms

95. What are circadian rhythms? (Review Biological Rhythms in Ch. 6.)

96. Why is the suprachiasmatic nucleus important to circadian rhythms? Where is it located?

97. How does our circadian "clock" work?

98. What are some examples of biological functions influenced by circadian rhythms?

Emotion and Motivation Involve Complex Neural Pathways (Fig. 9-21)

99. Why do emotion and motivation represent a probable overlap of the behavioral state system and the cognitive system?

100. The amygdala is a center for:

101. Name the brain regions involved in neural pathways for emotions. (Fig. 9-21) What are some of the physical results of emotions?

102. What is motivation?

103. What are three properties shared by motivational drives?

104. Give examples of some motivated behaviors.

Moods Are Long-Lasting Emotional States

105. Define moods.

106. Describe depression. What are some current therapies for depression?

Learning and Memory Change Synaptic Connections in the Brain

107. What have scientists discovered about the underlying basis for cognitive function?

Learning Is the Acquisition of Knowledge

108. What is learning?

109. Distinguish between associative and nonassociative learning.

110. Distinguish between habituation and sensitization.

Memory Is the Ability to Retain and Recall Information

111. What are memory traces? Where are they located?

112. What is parallel processing? What is its role in learning?

113. How do we know the hippocampus plays an important role in learning and memory?

114. Explain anterograde amnesia.

115. Distinguish between short-term and long-term memory. (Fig. 9-22)

116. What is working memory?

117. Describe the process of memory consolidation. (Fig. 9-22)

118. Distinguish between reflexive and declarative memory. Which brain regions are involved in each?

119. What is Alzheimer's disease? How is it diagnosed? How is it treated?

Language Is the Most Elaborate Cognitive Behavior

120. Why is language considered a complex behavior? (Fig. 9-23)

121. Language ability is found primarily in the _____ cerebral hemisphere, even in most left-handed or ambidextrous people.

122. The ability to communicate through speech has been divided into two processes. Name these two processes.

123. Damage to Wernicke's area results in _____ aphasia. Describe this condition.

124. Damage to Broca's area results in _____ aphasia. Describe this condition.

125. Mechanical forms of aphasia occur as a result of damage to which brain region(s)? Describe what the resulting condition(s) might be.

Personality Is a Combination of Experience and Inheritance

126. If we all have similar brain structure, what makes us different?

127. What is schizophrenia? How is it diagnosed? How is it treated?

TALK THE TALK

affective behaviors
Alzheimer's disease
amygdala
anterograde amnesia
Aplysia
arachnoid
artificial intelligence
ascending tracts
association areas
associative learning
auditory cortex
basal ganglia
behavioral state system
blood-brain barrier
brain-derived neurotrophic
    factor (BDNF)
brain stem
Broca's area
central canal
cerebellum
cerebral cortex
cerebral dominance
cerebral lateralization
cerebrum
cholingeric

cingulate gyrus
circuits (pathways)
cognitive behaviors
cognitive system
columns
consciousness
consolidation
corpus callosum
corticospinal tracts
cranial nerves
cranium
declarative (explicit) memory
delta waves
depression
descending tracts
descending ventricles
diencephalon
diffuse modulatory systems
dopaminergic
dorsal horns
dorsal root
dorsal root ganglia
drives
dura mater
electroencephalogram (EEG)

electroencephalography
ependyma
expressive aphasia
forebrain
frontal
ganglia
gray matter
gustatory cortex
gyri
habituation
hindbrain
hippocampus
hypoglycemia
hypothalamic-hypophyseal
    portal system
hypothalamus
insomnia
lateral geniculate nucleus
lateral ventricles
limbic system
long-term memory
magnetic resonance imaging
    (MRI)
medulla oblongata
membrane

memory
memory traces
meninges
mesencephalon
midbrain
moods
motivation
motor areas
motor association area
muscle memory
nerve cords
nerve net
neural crest cells
neural plate
neural stem cells
neural tube
nonassociative learning
noradrenergic
nuclei
occipital
olfactory cortex
paralysis
parietal
Parkinson's disease
pia mater
pons

positron emission tomography
  (PET)
primary motor cortex
primary somatic sensory cortex
procedural memory
propriospinal tracts
pyramidal cells
pyramids
receptive aphasia
reflexive (implicit) memory
rapid eye movement (REM) sleep
reticular activating systems
reticular formation
roots
satiety
schizophrenia
selective serotonin reuptake
  inhibitors (SSRIs)
sensitization
sensory areas
sensory system
serotonergic
short-term memory
sinuses
skull
sleep apnea

slow-wave sleep (deep sleep or
  non-REM sleep)
somatosensory tracts
somnambulism
spinal nerves
spinal reflexes
spinal tap (lumbar puncture)
subarachnoid space
sulci
suprachiasmatic nucleus
temporal
thalamus
tracts
tricyclic antidepressants
vagus nerve
ventral horns
ventral root
ventricles
vertebrae
vertebral column
villi
visceral
visual cortex
Wernicke's area
white matter
working memory

PRACTICE MAKES PERFECT

1. You are walking to class, pondering the intricacies of physiology, when you suddenly trip over an uneven place in the sidewalk. Unhurt but embarrassed and angry, you jump up and glance around to see if anyone is watching. From your knowledge of neuroanatomy and function, *briefly* explain how the following areas of the brain might be involved in this scenario.

   a.   cerebrum

   b.   cerebellum

2. What does an electroencephalogram tell us?

3.   Give the function of the following parts of the brain:

   a.   cerebrum

   b.   hypothalamus

   c.   brain stem

   d.   cerebellum

_____4.   Meninges are
   a.   bacterial infections of the brain or spinal cord.
   b.   connective tissue coverings around the central nervous system.
   c.   synapses between the meningeal nerve fibers and the post-synaptic membranes of other neurons.
   d.   non-neuronal cells in the brain and spinal cord that help regulate the ionic concentrations of the extracellular space.

_____5.   The cerebellum
   a.   if destroyed would result in the loss of all voluntary skeletal muscle activity.
   b.   initiates voluntary muscle movement.
   c.   is essential for the performance of smoothly coordinated muscular activity.
   d.   contains the center responsible for the regulation of body temperature.
   e.   is of no importance in the control of posture and balance.

_____6.   Which of the following is TRUE of REM sleep?
   a.   The average level of activity of brain cells decreases.
   b.   Students sleeping peacefully through a physiology lecture are awakened and report that they had no dreams.
   c.   This type of sleep occurs periodically through the night and accounts for 25% of total sleeping time.
   d.   The blood flow to the brain decreases markedly.
   e.   REM sleep develops immediately after a person has fallen asleep.

MAPS

Create a map of the central nervous system using the following terms:

amygdala
arachnoid membrane
association areas
auditory cortex
axons
basal ganglion
blood-brain barrier
brain
brain stem
cell bodies
central nervous system
cerebellum
cerebral cortex
cerebrospinal fluid
choroid plexus
columns
cranium
diencephalon
dorsal horns

dorsal root ganglia
dorsal roots
dura mater
efferent signals
ependyma
ganglia
gray matter
hippocampus
hypothalamus
limbic system
medulla oblongata
meninges
mesencephalon
midbrain
nuclei
nuclei
olfactory cortex
pia mater
pons

primary motor cortex
primary somatic sensory
    cortex
propriospinal tracts
pyramids
reticular activation system
sensory information
skull
spinal reflex
suprachiasmatic nucleus
thalamus
tract
ventral horn
ventral root
ventricle
vertebral column
visual cortex
white matter

10 Sensory Physiology

LEARNING OBJECTIVES

When you complete this chapter, you should be able to:

- Differentiate between chemoreceptors, mechanoreceptors, thermoreceptors, and photoreceptors and identify the adequate stimuli for each.

- Explain how sensory receptors transduce stimulus energy into electrical signals that can be used by the nervous system.

- Compare receptor potentials and thresholds in sensory receptors with graded potentials and thresholds in neurons.

- Explain how receptive fields help the CNS to integrate sensory information, and identify the role that convergence might play in this process.

- Explain how the CNS distinguishes these properties of a stimulus: sensory modality, stimulus location, stimulus intensity, and stimulus duration. Include the following where appropriate: labeled line coding, topographical organization, lateral inhibition, population coding, tonic receptors, and phasic receptors.

- Diagram the following information for each of the somatic and special senses: stimulus energy, energy transduction at the sensor, pathway(s) from sensor to CNS, CNS integration of information into perception. Include all relevant accessory structures, cell membrane receptors, signal transduction pathways, second messenger systems, afferent nerves, and CNS anatomy. You should make a diagram for each of the following:

  - Somatic Senses: Touch, Temperature, Nociception (Proprioception is discussed in Ch. 13, so you can save that diagram for then.)

  - Special Senses: Taste, Smell, Hearing, Equilibrium, and Vision

SUMMARY

Think of this chapter as an application of principles you've learned in the chapters leading up to it. Remember the basic reflex pattern presented in Ch. 6: Stimulus → Receptor → Afferent Pathways → Integration/Perception → Efferent Pathways → Response. To that pathway, you will start to integrate specific applications of membrane physiology (Ch. 5), signal transduction and second messenger systems (Ch. 6), neuron anatomy and physiology (Ch. 8), and CNS anatomy and physiology (Ch. 9). This is a BIG chapter, so don't wait until the last minute. Begin by getting comfortable with the basics presented in "General Properties of Sensory Physiology."

The somatic senses are touch, temperature, nociception, and proprioception. The five special senses are vision, hearing, taste, smell, and equilibrium. Sensory stimuli are converted into electrical potentials by specialized receptors. Receptors have an adequate stimulus, a particular form of energy to which they are most responsive. There are four types of sensory receptors: chemoreceptors, mechanoreceptors, thermoreceptors, and photoreceptors. If the stimulus depolarizes the receptor membrane potential past threshold, then an action potential results and sensory information is sent to the CNS. (If this is unfamiliar, go back to Ch. 8 and review neuron function.) The CNS integrates the information and creates our perception of the stimulus.

Receptive fields are created when primary neurons converge on secondary neurons. Lateral inhibition enhances contrast within the receptive field so that sensation is more easily located (except in hearing and smell). Auditory localization is accomplished by interpreting the timing of stimuli. Stimulus intensity is coded by the number of activated receptors and by frequency of the action potentials transmitted. Tonic receptors send information for the duration of a stimulus; phasic receptors are attuned specifically to changes in a parameter and stop firing if stimulus strength remains constant.

The facts above summarize the themes in sensory physiology. The rest of the chapter provides details about the special senses and somatic senses. Study the chapter one sense at a time and become comfortable with how each sensory organ/receptor transduces energy, how that information is sent to the CNS, where the information is integrated in the CNS, and how the information becomes perception. Then, think about what responses the body might manifest upon perceiving specific sensory information.

TEACH YOURSELF THE BASICS
GENERAL PROPERTIES OF SENSORY SYSTEMS

1. Distinguish between the special senses and the somatic senses.

Receptors Are Sensitive to Particular Forms of Energy

2. Name and describe the simplest sensory receptors. (Fig. 10-1a)

3. Sensory receptors are (circle all that are correct) specialized neurons/specialized non-neural cells.

4. Name the four major groups of receptors and describe the stimuli that activate each receptor type. (Table 10-2)

Sensory Transduction Converts Stimuli into Graded Potentials

5. Describe the general idea of sensory transduction.

6. Explain the concept of adequate stimulus for a receptor.

7. Define threshold. What happens in a sensory neuron if a stimulus is above threshold?

8. How do stimuli create electrical signals in sensory receptors?

   ▶ Review electrical signals in neurons and the role of ions in Ch. 5 and Ch. 8).

9.   Explain receptor potentials. (Review graded potentials in Ch. 8.)

A Sensory Neuron Has a Receptive Field

10.   What is a receptive field? (Fig. 10-2)

11.   Distinguish between primary and secondary sensory neurons.

12.   Explain how different parts of the body can show different sensitivities to stimuli. (Fig. 10-3a)

The Central Nervous System Integrates Sensory Information

13.   Name the parts of the brain that process the following types of sensory information. (Fig. 10-4)

visual information = _____   sound = _____

somatic senses = _____   smell = _____

equilibrium = _____   taste = _____

14.   Which part of the brain is involved in the routing of all sensory information except that of smell?

15.   What happens to the perceptual threshold for a stimulus when we "tune it out"? Explain the mechanism that allows us to do this.

Coding and Processing Distinguish Stimulus Modality, Location, Intensity, and Duration

16.   Name four attributes of stimuli that must be preserved during nervous system processing.

Sensory Modality

17.   What indicates the modality of a signal?

18.   Explain what is meant by labeled line coding.

Location of the Stimulus

19. How can the brain tell which part of the body is sending sensory information?

20. Give an example of the topographical organization of the cerebral cortex.

21. How does the brain determine where sound stimuli originate? (Fig. 10-5)

22. Explain how lateral inhibition enhances contrast, allowing better localization of stimuli. (Fig. 10-6)

Intensity and Duration of the Stimulus

23. Name two ways that stimulus intensity is coded. (Fig. 10-7)

24. Describe how population coding allows for transmission of stimulus intensity.

25. How does frequency of action potentials in the primary sensory neuron code for stimulus intensity?

26. Compare the response of tonic receptors and phasic receptors to a constant stimulus. (Fig. 10-8)

27. What is the advantage of phasic receptors? Give some examples.

28. Describe adaptation in general terms, and then give examples of mechanisms by which adaptation might be accomplished.

SOMATIC SENSES

29. Name the four somatosensory modalities.

Pathways for Somatic Perception Project to the Somatosensory Cortex and Cerebellum

30. Distinguish between primary, secondary, and tertiary sensory neurons, and identify the location of synapses between them for different receptor types.

31. Sensations from the left side of the body are processed in the (left/right?) hemisphere of the brain because: (Fig. 10-9)

32. How does the somatosensory cortex recognize where ascending sensory tracts originate? (Fig. 10-10)

Touch Receptors Respond to Many Different Stimuli (Fig. 10-11; Table 10-4)
33. Where are touch-pressure receptors found?

34. Diagram the structure and describe the functionality of the Pacinian corpuscle. (Figs. 10-11, 10-1b)

35. Are Pacinian corpuscles tonic or phasic?

Temperature Receptors Are Free Nerve Endings
36. Describe the anatomy of temperature receptors.

37. Compare cold receptors and warm receptors.

38. Do temperature receptors adapt? Explain.

Nociceptors Initiate Protective Responses
39. What are nociceptors? How is nociceptive pain mediated?

40. Why is it inaccurate to call nociceptors "pain receptors"?

41. What are some of the stimuli that activate nociceptors? Give specific examples.

42. What are the two pathways that nociceptors might activate?

43. Draw the complete reflex pathway for the withdrawal reflex when a frog's leg is placed in hot water. (Stimulus, receptor, afferent path, etc.) You might not have all the specific information you need, but give it your best shot.

Pain and Itching Are Mediated by Nociceptors

44. Distinguish between fast pain, slow pain, and itch.

45. What are the three types of primary sensory fibers involved with nociception? (Table 10-5)

46. Where do ascending pain pathways cross the midline? Where do they terminate in the brain? (Fig. 10-9)

47. Explain how pain perception can be modulated by the CNS. (Fig. 10-12)

48. Diagram the gating theory of pain modulation. (Fig. 10-12c)

49. What is referred pain and why does it occur? (Fig. 10-13)

50. What is pathological pain? How is it different from other types of pain?

51. Briefly describe some analgesics and their mechanisms of action.

CHEMORECEPTION: SMELL AND TASTE

52. List the five special senses.

Olfaction Is One of the Oldest Senses

53. Where is the olfactory bulb located? (Fig. 10-14)

54. Where do primary olfactory neurons terminate? (Fig. 10-15)

55. Why is there such a strong link between smells and emotions?

56. What is the VNO, and what it its function?

57. Describe the olfactory epithelium and olfactory receptor cells. (Fig. 10-14a, c)

58. By what mechanism do odorant receptors transduce sensory information? (See Ch. 6 for review of signal transduction pathways.)

Taste Is a Combination of Five Basic Sensations

59. List and briefly describe the five taste sensations.

60. What is the adaptive significance of our taste sensations?

61. Where are taste cells located? Diagram and describe taste buds, taste cells, and taste pores. (Fig. 10-16)

62. What must first happen to a substance in order for it to be tasted?

63. Distinguish between type II (receptor) cells and type III (presynaptic) cells.

Taste Transduction Uses Receptors and Channels

64. Create a map that represents the current understanding of taste transduction, afferent pathways, and CNS processing for the five different taste sensations. Be sure to include ligands, receptors, second messenger systems, and specific nerves wherever appropriate. For those areas where the research is still unsettled, mark that spot in your map with a question mark. (Use a separate sheet of paper.) (Figs. 10-17, 10-4)

65. What is specific hunger? Give an example.

THE EAR: HEARING

66. What are the two distinct functions the ear is specialized to sense?

67. Diagram the anatomy of the ear. (Fig. 10-18)

Hearing Is Our Perception of Sound

68. Define hearing.

69. What is sound? What three attributes of sound waves are being sensed?

70. What is pitch, and how is it measured? What is loudness, and how is it measured? (Fig. 10-19)

Sound Transduction Is a Multistep Process

71. List the transductions required for hearing sounds.

72. Trace the anatomical path followed by sound wave energy as it moves from air through the inner ear. (Fig. 10-20)

73. Diagram the steps by which the energy of sound waves in air is converted into action potentials in the sensory neuron.

The Cochlea Is Filled with Fluid

74. Diagram the structure of the cochlea, naming all fluids, windows, and ducts. (Fig. 10-21)

75. Compare the composition of perilymph and endolymph.

76. Describe the location and structure of the organ of Corti. (Fig. 10-21)

77. Explain how hair cells convert fluid waves into action potentials. Include a description of stereo-cilia, ion channels, protein bridges, and neurotransmitter. (Fig. 10-22)

Sounds Are Processed First in the Cochlea

78. List the three properties of sound waves used for sound discrimination. Where is each property processed?

79. What role does the basilar membrane play in sound processing? (Fig. 10-23)

80. How is loudness coded by the auditory system?

Auditory Pathways Project to the Auditory Cortex

81. Trace the anatomical path that action potentials follow from the auditory sensory neurons to their final destination in the brain. (Use a separate sheet of paper, if necessary.) (Fig. 10-24)

82. How does the brain localize sound?

Hearing Loss May Result from Mechanical or Neural Damage

83. List and explain the three different forms of hearing loss.

THE EAR: EQUILIBRIUM

84. Define equilibrium. What are the two components of this special sense?

85. What receptors in the body provide sensory information about body position?

86. Compare hair cells of the inner ear to hair cells of the cochlea. How do cochlear hair cells transduce movement into electrical signals?

The Vestibular Apparatus Provides Information About Movement and Position in Space

87. Diagram the anatomy of the vestibular apparatus. (Fig. 10-25)

88. Compare the sensory functions of the semicircular canals and otolith organs.

89. Describe secretion of endolymph in the vestibular apparatus.

The Semicircular Canals Sense Rotational Acceleration

90. Why is the presence of endolymph in the inner ear key to the body's ability to sense rotation? Explain how we sense rotation. (Figs. 10-25b, c; 10-26)

The Otolith Organs Sense Linear Acceleration and Head Position

91. Diagram and describe how the otolith organs sense linear forces and head position. (Fig. 10-27)

Equilibrium Pathways Project Primarily to the Cerebellum

92. Trace the anatomical pathway that action potentials follow from the vestibular hair cells to their final destination in the brain. (Fig. 10-28)

93. Describe the role of descending (efferent) pathways from the brain to motor neurons in an equilibrium reflex.

THE EYE AND VISION

94. Define vision. List the three steps.

The Skull Protects the Eye

95. Diagram the anatomy of the eye and its associated structures. (Figs. 10-29, 10-30, 10-31)

Light Enters the Eye Through the Pupil

96. How does the eye control the amount of light hitting the retina?

97. How does the eye focus light onto the retina?

98. Explain the pupillary reflex. How is the consensual reflex related? (Fig. 10-31c)

99. Explain how the eye creates depth of field.

The Lens Focuses Light on the Retina

100. What is refraction? When/where does it occur in the eye?

101. What two factors influence the angle of refraction?

102. Compare concave and convex lenses. (Fig. 10-32)

103. Compare the focal point and the focal length (distance). (Fig. 10-32b)

104. How do you change the focal distance for a lens? (Fig. 10-33)

105. Define accommodation. How is accommodation achieved in the human eye?

106. Ciliary muscles are attached to the lens by _____. (Fig. 10-34)

107. To make the lens more round, the ciliary muscles (contract/relax?) which (increases/decreases?) tension on the zonules.

108. When the lens is more rounded, the focal distance becomes (shorter/longer?). (Fig. 10-33)

109. Explain the following vision problems and tell what shape lens would correct for each. (Fig. 10-35)

    presbyopia

    myopia

    hyperopia

110. What is astigmatism?

Phototransduction Occurs at the Retina

111. What is the frequency range of the EM spectrum that contains visible light? What is the wavelength range?

112. Diagram the layers of the retinal neurons. What embryonic tissue relationship explains the layered organization of the retina? (Fig. 10-37d)

113. What is phototransduction?

114. What is the function of melanin in the pigment epithelium?

115. What are photoreceptors? Compare the organization of photoreceptors in most of the retina with the organization of photoreceptors in the fovea. Why is there a difference? (Figs. 10-37c, 10-38)

116. The image projected onto the retina is upside down. Why then do we see things in the correct orientation? Explain how this happens. (Fig. 10-38)

117. What is the optic disk and why is it also called the blind spot?

Photoreceptors Transduce Light into Electrical Signals

118. Distinguish between rods and cones. Describe their sensitivities.

119. Diagram the three-segment structure of rods and cones and tell what process(es) occurs in each segment. (Fig. 10-39)

120. What is the role of the visual pigments? Where are they located?

121. Why does each cone have a particular wavelength to which it is most sensitive? (Fig. 10-40)

122. The perceived color of an object depends on the color(s) of light that the object (reflects/absorbs?).

123. What is color-blindness?

Phototransduction

124. Diagram the rhodopsin molecule and explain how it changes when activated by light. (Figs. 10-39, 10-41)

125. Diagram the mechanism of phototransduction in rods. Begin with darkness, show the changes that occur upon exposure to light, and then what happens during the recovery phase. Include in your diagram: transducin, rhodopsin, retinal, opsin, CNG channels, K^+ channels, voltage-gated Ca^{2+} channels, cGMP, Na^+, and K^+. (Use a separate sheet of paper if necessary.)

126. Why do our eyes require some time to adjust to changes in light intensity?

Signal Processing Begins in the Retina

127. Explain why signal processing in the retina is an excellent example of convergence. (Fig. 10-37e; also see the discussion of convergence in Ch. 8)

Bipolar Cells

128. What is the neurotransmitter released from photoreceptors onto bipolar neurons?

129. This neurotransmitter excites some bipolar neurons but inhibits others. Explain how one signal molecule can have opposing effects.

Ganglion Cells

130. Explain the relationship between a ganglion cell and its visual field. What is the organization of a visual field? What are the two types of ganglion visual fields? (Fig. 10-42)

131. Visual acuity is greatest when a ganglion cell has (many/few?) photoreceptors in its visual field.

132. The retina uses (light intensity/contrast?) to identify objects in the environment. Explain.

133. Describe M cells and P cells.

Processing Beyond the Retina

134. The optic nerves enter the brain at the optic _____. At this point, (all/some?) fibers from the right field of vision cross to the left side of the brain. (Fig. 10-43)

135. Explain how binocular vision allows us to see things in three dimensions.

136. Fibers projecting from the optic chiasm to the midbrain serve what purpose? (Fig. 10-31)

137. Fibers from the optic chiasm to the lateral geniculate body of the thalamus serve what purpose?

138. Describe the topographical organization of the lateral geniculate body—how does it relate to the visual field, and how does this organization relate to the visual cortex?

139. Describe the processing of visual information in the visual cortex.

TALK THE TALK

A-beta (Aβ) fibers
accessory structures
accommodation
adapt
A-delta (AΔ) fibers
adequate stimulus
amacrine cells
amplitude
ampulla
analgesic drugs
aqueous humor
astigmatism
basilar membrane
beta-endorphin
binocular zone
bipolar neurons

bleaching
blind spot
C fibers
capsaicin
central hearing loss
chemoreception
chemoreceptors
ciliary muscle
CNS opioid receptors
cochlea
cochlear duct
cochlear nerve
cold receptors
color-blindness
complex neural receptors
concave lens

conductive hearing loss
cones
consensual reflex
contralateral
convergence
convex lens
cornea
crista
cupula
cyclic nucleotide-gated channel (CNG channel)
decibels (dB)
depth of field
diabetic neuropathy
dynorphin
ear canal

endolymph
endolymphatic hydrops
enkephalins
equilibrium
eustachian tube
extrinsic eye muscles
fast pain
focal length (distance)
focal point
fovea
frequency
ganglion cells
gate control theory
gustation
gustatory neuron
gustducin
helicotrema
hertz (Hz)
horizontal cells
hyperopia
incus (anvil)
inflammatory pain
inhibitory modulation
ipsilateral
ischemia
itch
kinocilium
labeled line coding
labyrinth
lacrimal apparatus
lateral geniculate body
lateral inhibition
lens
loudness
macula
magnocellular ganglion cells
   (M cells)
malleus
mechanoreceptors
melanin
melanopsin
membranous labyrinth
modality
monocular zone
myocardial infarction
myopia
near point of accommodation
neuropathic pain
nociceptive pain
nociceptors
odorant
odorant receptor
off-center/on-surround fields

olfaction
olfactory
olfactory bulb
olfactory cortex
olfactory epithelium
on-center/off-surround field
opsin
optic chiasm
optic disk
optic nerve
organ of Corti
otitis media
otolith membrane
otolith organs
otoliths
oval window
pacinian corpuscles
pain
parvocellular ganglion (P cells)
pathological pain
perceptual threshold
perilymph
phasic receptors
phantom limb pain
photoreceptors
pigment epithelium
pinna
pitch
population coding
presbycusis
presbyopia
primary sensory neuron
proprioception
pupillary reflexes
receptive field
receptor potential
referred pain
refract
retina
retinal
rhodopsin
rods
rotational acceleration
round window
saccule
salt appetite
scala media
scala tympani
scala vestibuli
sclera
secondary receptive field
secondary sensory neuron
semicircular canals

sense organs
sensorineural hearing loss
simple receptors
slow pain
somatic senses
somatosensory cortex
somatosensory receptors
special senses
special senses receptors
specific hunger
stapes (stirrup)
stereocilia
substance P
surround
tastant
taste buds
taste cells
taste pore
tectorial membrane
tertiary sensory neurons
thermoreceptors
threshold
tip links
TIR receptor proteins
tonic receptors
topographical organization
transducin
transduction
two-point discrimination test
tympanic duct
tympanic membrane
type II or receptor cells
type III or presynaptic cells
umami
utricle
vanilloid receptors
vestibular apparatus
vestibular duct
vestibular nerve
vestibulocochlear nerve
vestibulum
visible light
vision
visual cortex
visual fields
visual pigments
vitreous body
vitreous chamber
vomeronasal organ
warm receptors
withdrawal reflex
zonules

PRACTICE MAKES PERFECT

1. Why aren't you constantly aware of your clothing touching your body?

2. If a salty solution is placed at the rear of the tongue, it cannot be tasted. Why not?

3. TRUE/FALSE? Defend your answer.

    Light travels through nerves and blood vessels before striking the rods and cones.

_____ 4.   Chemoreceptors

     a.   are involved in the perception of taste.
     b.   are found in the olfactory mucosa.
     c.   cover the retina except at the optic discs.
     d.   are found in the cochlea.
     e.   can be described by a and b.

5. How does astigmatism differ from nearsightedness?

_____ 6.   A sensory nerve is termed

     a.   efferent.
     b.   afferent.
     c.   monoefferent.
     d.   none of the above

7. If a person with hearing impairment has good bone conduction and no air conduction, where would you predict the problem to be?

8. Stare hard at a bright light, then shift your gaze to a white sheet of paper. What do you see on the paper? Can you explain this in terms of what you know about rhodopsin and bleaching?

9. You are sitting up straight on a stool and are spinning to your left. Suddenly, you are stopped and you remain sitting up straight. Which one of the three semicircular canals is involved in this sensation?

10. You would expect to find more of which type of visual receptor cell if you dissected the retina of a nocturnal (i.e., active at night) animal? Why?

_____11. The organ of Corti sends electrical information about sound to the brain. The pitch of sound is determined by

    a.   the location of the activated hair cells on the basilar membrane.
    b.   the frequency of the action potentials received by the brain.
    c.   the amplitude of the action potentials received by the brain.
    d.   a and b

MAPS

1. Create a map of the inner ear using the following terms and any functions/actions that you wish to add.

ampulla	inner ear	saccule
crista	macula	semicircular canals
cupula	otolith membrane	utricle
endolymph	otolith organs	vestibular apparatus
hair cells	otoliths	vestibular nerve

2. Using the vocabulary list from Talk the Talk as a starting point, select related terms and create similar maps for vision, hearing, taste, and smell.

BEYOND THE PAGES

Here's a relevant book you might enjoy:

> Levitin, D. J. *This Is Your Brain on Music: The Science of a Human Obsession.* Penguin Group (USA): 2006.

Levitin's book explores the intersection of pop culture, music theory, and neuroscience, and does so in a way that both entertains and informs. The author runs the Laboratory for Musical Perception, Cognition, and Expertise at McGill University.

TRY IT

Here are some fun demonstrations of sensory physiology.

Taste and Smell

The relationship between taste and smell can be shown with the following experiment. Cut a small slice of potato and a small slice of apple. Remove the skin or peel. Close your eyes and pinch your nose shut. Have a friend put either a piece of apple or potato in your mouth. Chew it with your nose pinched shut. Can you tell whether it is apple or potato? Repeat the experiment with your nose open so that you can smell what you are eating.

Mapping Taste Receptors

Can you map the location of taste receptors on the tongue? Old textbooks said you could...but do the areas for different tastes overlap? Try this experiment.

Assemble the following materials: cotton swabs, 1 tsp. sugar dissolved in 1 T. water, 1 tsp. salt dissolved in 1 T. water, vinegar diluted 1:1 with water, monosodium glutamate (Accent®) dissolved in a little water. Dry the tongue with a tissue. Dip a cotton swab in one of the solutions. Touch the swab to the tip, sides, and back of the tongue, and notice where you taste the dissolved solute. On the drawing of the tongue below, record the area where you noticed the taste. Rinse your mouth with water, dry the tongue with a tissue, and repeat using each of the different solutions.

The Blind Spot

In each field of vision we have a *blind spot* at the *optic disc*, the point on the retina where the fibers from the rods and cones converge to form the *optic nerve*. Normally we are not aware of this blind spot, but the following test will allow you to find it. Hold the X in the figure below directly in front of your right eye and about 20 inches from your face. Close your left eye and focus your right eye on the X. You should see both the X and the black dot. Keeping your left eye closed and focusing on the X, bring the page closer to your face until the dot disappears from your field of vision. With a ruler, have a friend measure the distance from the page to your eye at which the dot disappears.

X   ●

Focal distance for the blind spot: _____ cm = _____ mm

You can now calculate the distance from the blind spot to the fovea in the eye, using the following formula:

A/B = a/b

where   A = distance between the center of the X and the center of the dot above (in mm)
        B = the focal distance for the blind spot (in mm)
        a = distance between the optic disk and the fovea (unknown)
        b = distance from the lens of the eye to the retina (assume 20 mm)

What was the distance between the optic disk and the fovea?

Blindsight

An interesting phenomenon that was first identified in people with partial vision loss is blindsight, the ability to "see" something moving in the visual field without being consciously aware that you see it. Blindsight is believed to result from neural pathways that project to part of the visual cortex without passing through the lateral geniculate body. For a more detailed description and an online test for whether you have blindsight, go to serendip.brynmawr.edu/bb/blindsight.html.

Negative After-Images

Stare at a bright light like a penlight for about 30 seconds. Rapidly shift your gaze to a plain white surface. You should see a dark reverse image of the light. This after-image occurs because the cones have become fatigued (adapted) to the bright light.

Complementary After-Images

You can also get color after-images. Place a red or green cardboard square under a bright light and stare at it intently for about 30 seconds. Then transfer your gaze to a piece of white paper. The colors will change and you will see the *complement* of the color at which you were staring. The complement of a color is the color that appears opposite the color on a traditional color wheel. The three main pairs of complimentary colors are red-green, purple-yellow, and blue-orange. When you stare at one color, certain cones become fatigued, leaving the other cones to react to the light. If you looked at a red square, you'll see green on the paper, and vice versa.

Eye Dominance

With both eyes open, hold a cardboard cone or tube with both hands at arm's length directly in front of your nose. Center an object in the opening. Without moving the tube, close your right eye. Does the object move out of the opening? Now repeat with your left eye. If the object stayed in the opening when your right eye was open, then you are right-eye dominant. If it stayed in the opening with your left eye open, then you are left-eye dominant. Eye dominance usually matches hand dominance, that is, if you are right-handed, you will be right-eyed.

Depth Perception

Because the images seen with each eye are not identical, the brain can interpret the two views into a three-dimensional representation of the image. With one eye closed, the field of vision becomes two-dimensional, and the relative distances between near and far objects become much harder to gauge. Try the following to demonstrate depth perception.

1. Extend your arms to the side, palms facing front and index fingers extended. Close one eye. Now try to touch the tips of the index fingers to each other about one foot in front of your face. Without input from both eyes, this can be more difficult than it would seem.

2. Close one eye and pick up a very large needle and thread. Keeping one eye closed, try to thread the needle. Repeat with both eyes open. Which was easier?

Binaural Localization of Sound

Our ability to pinpoint the location of the source of sounds depends on the reception of sound waves by both ears. Our brain uses two parameters in processing the information received: the difference in loudness between the two ears and the difference in the time the sound reaches each ear.

Have the subject close his or her eyes. The tester should move around the subject with a "cricket clicker," metronome, or other noisemaker, turning it on periodically and asking the subject to identify the location of the sound source. You should experiment with the distance the noisemaker should be placed from the subject's head, as this will vary with the loudness of other masking noises in the room.

Repeat at the same distances, but this time have the subject block one ear with a finger. With one ear blocked, was the subject as accurate in pinpointing the location of the sound source?

Two-Point Discrimination

Two pins touched to the skin close together will sometimes be perceived as a single point if only a single secondary receptive field is stimulated (see Fig. 10-3). In order for two points to be perceived, two different receptive fields must be stimulated. The distance by which the points must be separated in order for them to be felt as distinct points is known as the *two-point threshold*.

You need a small ruler and a compass with two points or two *dull* pins.

1. Have the subject close his or her eyes. Use a single point of the compass or set the points 5, 15, 25, and 50 mm apart. In random order, touch the compass to the skin. Try to keep the pressure equal each time. *Do not push so hard that you break the skin!*

2. Note the distance between pins when the subject first feels two distinct points. Vary the distance between the points randomly. Repeat several times for accuracy. Record the results below.

Site	Minimum distance for two-point discrimination (mm)
fingertip	
back of neck	
cheek	
back of calf	
lateral surface of forearm	

How do these distances relate to the functions of these body parts?

THERMORECEPTORS

Our perception of temperature is related to stimulation of cold receptors, warmth receptors, and pain receptors. Most parts of the body have many more cold receptors than warmth receptors. All of the thermoreceptors *adapt* very rapidly, i.e., the intensity of the signal decreases with time so that the perception of the stimulus decreases. This means that our temperature receptors are much more sensitive to *changes in temperature* than to constant temperatures. (Are thermoreceptors tonic or phasic?) The following experiment shows adaptation of temperature receptors.

Adaptation of Receptors

1. Place one hand in 40° C. water and the other hand simultaneously into the large tub of ice water. Leave them there for one minute.

2. Now place both hands into a tub of room temperature water. Do they feel the same? Can you explain what you feel based on the past thermal history for each hand?

Referred Pain

In certain cases when a pain receptor fires, it will stimulate other neurons that run in the same nerve. This can lead to sensations of pain far from the actual site of the stimulus. One example of referred pain is irritation of the abdominal side of the diaphragm that is sensed as a pain in the shoulder. Another example can be demonstrated by placing your bent elbow into ice water. Leave it there for a minute or until you cannot stand the feeling any more. Was the sensation in your elbow or elsewhere?

VISUAL REFLEXES

Pupil Dilation Reflex

This reflex helps regulate the amount of light that enters the eye and strikes the retina.

With a penlight, shine light into one eye of the subject. What happens to the pupil? Repeat, this time watching both pupils. Do they respond simultaneously?

Have the subject sit facing the window or a lighted area of the room. Note the pupil size. Keep watching while the subject places a hand over one eye. What happens to the pupil of the uncovered eye? This is known as the *consensual reflex*.

Accommodation Reflex

The ability of the eye to focus at different distances depends on the accommodation reflex. Smooth muscle fibers (*ciliary muscle*) attach to ligaments around the edge of the lens of the eye. In normal distance vision, the ciliary muscle is relaxed and objects that are about 20' away are in focus. Because light rays from objects that are closer than 20' are not parallel, these rays are refracted to a principal focus point that is behind the retina. As a result, objects closer than 20' will be blurred.

In the accommodation reflex, as the ciliary muscles contact, the lens becomes more spherical. The *focal distance* shortens so that it strikes the retina and the close objects come into focus. As we age, the lens loses its elasticity and we lose our ability to accommodate for near and far vision. This is the reason people starting around age 40 may need to use reading glasses or bifocals.

1. Have subject look at a distant object. Note the size of the pupils.

2. As you continue to watch, ask the subject to shift focus to a nearer object on the same line of sight. What happens to the pupils? The nervous pathways that cause the lens to accommodate also cause a change in pupillary size at the same time.

Near Point of Accommodation

1. Lay a meter or yardstick flat on the table, perpendicular to the edge. In one hand take a 3×5 card with a letter "e" from the newspaper pasted on it. Close your right eye. Place your nose at the end of the meter stick.

2. Starting at arm's length, move the card in toward your face. Note the distance at which you can no longer see the "e" clearly. This distance is known as the *near point of accommodation*.

As discussed previously, the near point of accommodation increases with age as the lens loses elasticity. Average distances for near point of accommodation are 10 cm at age 20 and 13 cm at age 30. By age 70, the near point averages 100 cm.

Find the near point of accommodation for each eye. Left _____ cm   Right _____ cm

Ciliospinal Reflex

While watching the subject's pupils, suddenly pinch the skin at the nape of his or her neck. Watch the pupil of the eye on the side you pinched. (This is called the ipsilateral side). What happens? Sudden pain stimulates the sympathetic nerves and causes a pupillary response.

Efferent Division: Autonomic and Somatic Motor Control

11

LEARNING OBJECTIVES

When you complete this chapter, you should be able to:

- Explain how the autonomic nervous system accomplishes antagonistic control of physiological parameters, and contrast antagonistic control with tonic control.

- Compare the anatomy of the sympathetic and parasympathetic branches; diagram their organization and highlight the differences between the two.

- Compare chemical signaling in the two branches, including ganglionic and postganglionic synapses, all neurotransmitters, receptors and their affinity for different neurotransmitters, target tissues and their responses, and second messenger mechanisms associated with each receptor type.

- Explain direct and indirect agonists and antagonists and use specific examples to demonstrate the clinical significance of therapeutic agents based on these principles.

- Diagram the anatomy of a somatic motor pathway and the events at the neuromuscular junction. Highlight differences between the somatic motor branch and the autonomic branch..

SUMMARY

This chapter starts filling in the details of the peripheral nervous system—mainly the autonomic nervous system. Learn the details so that you can apply them later as you're exposed to specific examples of efferent function. Break out the colored pens and paper! Design color-coded pictures and diagrams to help organize the material in this chapter. Re-create figures in the book and, if necessary, change them to fit your learning style.

Remember from Ch. 8 that the peripheral nervous system is divided into the somatic motor division and the autonomic division. The autonomic division is further divided into the sympathetic and parasympathetic divisions. Sympathetic and parasympathetic divisions are often antagonistic, allowing tight control over homeostasis. Review Cannon's postulates on homeostasis (see Ch. 6) and see how the two autonomic divisions fulfill all four postulates. Somatic motor neurons always innervate skeletal muscle; autonomic neurons innervate smooth muscle, cardiac muscle, glands, and some adipose tissue.

An autonomic pathway consists of a preganglionic neuron from the CNS to an autonomic ganglion, and a postganglionic neuron from the ganglion to a target. Preganglionic divergence onto multiple synapses allows rapid control over many targets. Sympathetic pathways have ganglia near the spinal cord, while parasympathetic pathways have ganglia near their target. The synapse between a postganglionic neuron and its target is called a neuroeffector junction. Autonomic axons have varicosities from which neurotransmitter is released. CNS control over the autonomic division is linked to centers in the hypothalamus, pons, and medulla. Autonomic responses can be spinal reflexes, and the cerebral cortex and limbic system can also influence autonomic output.

Here are some points to remember: All preganglionic neurons secrete acetylcholine (ACh). Sympathetic postganglionic neurons secrete norepinephrine; parasympathetic postganglionic neurons secrete ACh. The type and concentration of neurotransmitter plus the receptor type at the target determine autonomic response. Cholinergic receptors respond to ACh and come in two isoforms: nicotinic at the ganglia and muscarinic at parasympathetic neuroeffector junctions. Adrenergic receptors respond to epinephrine and norepinephrine and come in three isoforms: α on most sympathetic tissue, β_1 on heart muscle and kidney, β_2 on tissues not innervated by sympathetic neurons.

Somatic motor pathways have only one neuron that originates in the CNS and projects to a skeletal muscle. Somatic motor neurons always excite muscles to contract, and muscle contraction can only be inhibited by inhibiting the somatic motor neuron. One motor neuron may branch to control many muscle fibers. The synapse between the neuron and the skeletal muscle is called the neuromuscular junction. ACh is the neurotransmitter at the neuromuscular junction, and the motor end plates of the muscle cell have a high concentration of nicotinic ACh receptors.

TEACH YOURSELF THE BASICS

THE AUTONOMIC DIVISION

1. What are the two efferent divisions of the peripheral nervous system?

2. Why is the autonomic division also called the visceral nervous system?

3. Characterize and compare the parasympathetic and sympathetic divisions. (Fig. 11-1)

Autonomic Reflexes Are Important for Homeostasis

4. Identify some brain regions involved with autonomically regulated functions. (Figs. 11-2, 11-3)

5. List some autonomic functions that do not require any input from the brain. Collectively, what are these reflexes called?

Antagonistic Control Is a Hallmark of the Autonomic Division

6. Using the example of heart rate, describe how the autonomic nervous system achieves antagonistic control. Contrast this with tonic control and give an example.

7. Explain how epinephrine (a catecholamine) will cause some blood vessels to constrict but other vessels to dilate.

Autonomic Pathways Have Two Efferent Neurons in Series

8. Draw and label an autonomic pathway. Include the following terms: autonomic ganglion, CNS, preganglionic neuron, postganglionic neuron. (Use a separate sheet of paper.) (Fig. 11-4)

9. What composes an autonomic ganglion? What happens within a ganglion?

10. What is the significance of divergence in autonomic regulation?

Sympathetic and Parasympathetic Branches Exit the Spinal Cord in Different Regions

11. Describe the anatomical differences between sympathetic and parasympathetic pathways. (Use a separate sheet of paper if necessary.) (Fig. 11-5)

12. Which nerve makes up the primary parasympathetic tract? (Fig. 11-6)

The Autonomic Nervous System Uses a Variety of Chemical Signals

13. List the rules for distinguishing the sympathetic and parasympathetic branches of the autonomic nervous system. (Fig. 11-7)

14. What are the exceptions to the rules?

15. What are nonadrenergic, noncholinergic neurons, and what are some of the communication chemicals they employ?

Autonomic Pathways Control Smooth and Cardiac Muscle and Glands

16. The synapse between a postganglionic autonomic neuron and its target cell is called the

     _____ junction. (See discussion of synapse in Ch. 8.)

17. List the targets of autonomic neurons:

18. Describe and diagram the components of an autonomic synapse. (Use a separate sheet of paper.) (Fig. 11-8)

19. Cite examples of modulation in autonomic neurotransmitter release.

Autonomic Neurotransmitters Are Synthesized in the Axon

20. Where does neurotransmitter synthesis take place in neurons of the autonomic nervous system? What are the primary autonomic neurotransmitters?

21. Outline the steps of neurotransmitter release, beginning with the arrival of an action potential. (Fig. 11-9)

22. Give examples of ways in which the various neurotransmitters are removed from the synapse.

23. Why is neurotransmitter concentration an important factor in autonomic function?

24. What is the name of the main enzyme responsible for degradation of catecholamines?

Most Sympathetic Pathways Secrete Norepinephrine onto Adrenergic Receptors

25. Give location and catecholamine-binding affinities of the various α and β adrenergic receptors. (Table 11-2)

26. Create a chart that compares the second messenger mechanisms for the various adrenergic receptor types.

The Adrenal Medulla Secretes Catecholamines

27. Where are the adrenal glands? Describe their anatomy and locate the adrenal medulla. (Fig. 11-10a, b)

28. Why is the adrenal medulla often described as a modified sympathetic ganglion? (Fig. 11-10c)

Parasympathetic Pathways Secrete Acetylcholine onto Muscarinic Receptors

29. As a rule, parasympathetic neurons release _____ onto muscarinic _____ receptors at their targets.

30. What are the signal tranduction pathways used by all muscarinic receptors?

Autonomic Agonists and Antagonists Are Important Tools in Research and Medicine

31. Why has the discovery and synthesis of autonomic agonists and antagonists led to better understanding of the function of the autonomic nervous system? (Table 11-3)

32. Give an example of a pharmaceutical that is an autonomic agonist/antagonist. Describe in general terms how it works.

Primary Disorders of the Autonomic Nervous System Are Relatively Uncommon

33. List some examples of autonomic nervous system disorders.

34. What is the result of continued diminished sympathetic output?

Summary of Sympathetic and Parasympathetic Divisions

35. Refer to Fig.11-11 and Table 11-4. Create a map, an outline, a diagram, or table to compare the two branches of the autonomic nervous system. Use specific details.

THE SOMATIC MOTOR DIVISION

36. Anatomically and functionally, how do somatic motor pathways differ from autonomic pathways? (Table 11-5)

A Somatic Motor Pathway Consists of One Neuron

37. Describe the anatomy of the somatic motor pathways. (Fig. 11-11)

38. Describe and diagram the components of the neuromuscular junction (Fig. 11-12)

The Neuromuscular Junction Contains Nicotinic Receptors

39. Outline the events that result in neurotransmitter release, beginning with the arrival of an action potential at the axon terminal.

40. Which neurotransmitter is used, and how does it get to its receptors?

41. Give an overview of the cellular mechanism employed by nicotinic receptors on the skeletal muscle membrane. (Fig. 11-13)

42. ACh acting on the motor end plate is always (excitatory/inhibitory?) and creates muscle contraction.

43. How is skeletal muscle relaxation achieved?

TALK THE TALK

acetylcholinesterase (AChE)
adrenal cortex
adrenal glands
adrenal medulla
adrenergic receptor(s)
α_1-receptors
α_2-receptors
antagonistic control
anticholinesterases
autonomic ganglion
autonomic motor neurons
autonomic nervous system
β_1-receptors
β_2-receptors
β_3-receptors
chromaffin cells
denervation hypersensitivity

divergence
fight-or-flight
impotence
incontinence
mixed nerves
modified sympathetic
  ganglion
monoamine oxidase (MAO)
motor end plate
multiple system atrophy
muscarinic cholinergic
neuroeffector junction
neuromuscular junction
nicotinic cholinergic receptors
non-adrenergic neurons
non-cholinergic neurons
organophosphate insecticides

parasympathetic branches
postganglionic neuron
preganglionic neuron
somatic motor neurons
spinal reflexes
sympathetic branches
sympathetic cholinergic neurons
sympathetic ganglia
terminal boutons
vagotomy
vagus nerve
varicosity
vegetative nervous system
visceral nervous system

PRACTICE MAKES PERFECT

_____1.   The parasympathetic nervous system is characterized by

    a.   long preganglionic and short postganglionic nerve fibers.

    b.   short preganglionic and long postganglionic nerve fibers.

    c.   direct connections (no synapses) between the central nervous system and the inner-
vated organs.

    d.   mediation by adrenaline (epinephrine).

    e.   the axons coming into immediate contact with the effector organs lying inside the
central nervous system.

2.   Review agonists and antagonists. In the following scenarios, would the experimental drug be
an agonist or antagonist?

    a.   You have just discovered a new vertebrate species while doing field work in the rain
forest. After obtaining proper permission, you begin studying the effects of different
known neurotransmitters on the heart rate of this organism. You inject ACh alone and
observe that it decreases heart rate. After letting the heart rate return to resting values,
you inject GABA. You see that GABA increases the heart rate. After letting the creature rest
again, you inject ACh + GABA and notice that the heart rate does not change. Without
further experiments you can't be certain, but at this point, do you think that GABA is an
agonist or antagonist to ACh? Explain your reasoning.

b.   You're having a very successful expedition, and the next day you find another new vertebrate species that has some very interesting properties. The creature has bright red skin
early in the morning, but by midday, the creature has dull brown skin. From other obser-
vations, you suspect that this process is under the control of the autonomic nervous system. Therefore, you decide to test this idea by injecting the animal with various
chemicals. The results of your experiments are in the table.

	Initial skin color	Inject	Resulting skin color
1	fading from red to brown	Norepinephrine	color returns to bright red
2	dull brown, beginning to change to red	ACh	skin returns to the dull brown
3	beginning to turn from red to brown	chemical that binds ACh and prevents it from acting for several hours	skin stays red
4	red	chemical named Compactine	skin changes to dull brown

From these experiments, do you think Compactine is acting as an agonist or antagonist to
ACh?

3.   In what way(s) is autonomic neurotransmitter action different from that of somatic motor neuron neurotransmitter action?

4.   Certain drugs inhibit the activity of monoamine oxidase. In general terms, predict what these
drugs would do to autonomic activity.

5.   TRUE/FALSE? Defend your answer.

The adrenocorticosteroid hormones, such as cortisol, are useful in fight-or-flight situations.

MAPS

1. Create a map that shows all efferent divisions of the nervous system. Show all neurons, all receptor types on their respective tissues, and all neurotransmitters.

12 Muscles

LEARNING OBJECTIVES

When you complete this chapter, you should be able to:

- Define the following terms as they relate to skeletal muscle: origin, insertion, joint, flexor, extensor, antagonistic muscle groups.

- Draw a series of diagrams to show the different levels of organization of skeletal muscle. Start with a muscle, and then show the muscle fascicle, then the muscle fiber. Then show a myofibril, a sarcomere, thick and thin filaments, actin, myosin, troponin, tropomyosin, nebulin, and titin.

- Diagram the sliding filament theory of contraction.

- Diagram the molecular events of the contractile cycle. Be sure to highlight the roles of the calcium and ATP/ADP.

- Diagram the molecular events of excitation-contraction coupling. Be sure to include receptors and channels where appropriate.

- Describe how muscles meet their constant energy requirements.

- Discuss the differences between central fatigue and peripheral fatigue and provide examples that highlight current thinking about the causes of fatigue.

- Discuss the differences between slow-twitch fibers, fast-twitch oxidative-glycolytic fibers, and fast-twitch glycolytic fibers. Then, describe how skeletal muscle plasticity can blur these distinctions.

- Diagram and describe why each sarcomere contracts with optimal force if it is at optimal length.

- Explain how summation increases force of contraction.

- Explain tetanus and differentiate between incomplete (unfused) tetanus and complete (fused) tetanus.

- Describe what a motor unit is and how skeletal muscles can create graded contractions of varying force and duration.

- Compare and contrast isometric and isotonic contractions.

- Describe how bones and muscles form fulcrums and levers and discuss how this affects the mechanics of body movements.

- Diagram smooth muscle anatomy and explain why smooth muscle is more difficult to study than skeletal muscle.

- Differentiate between single-unit smooth muscle and multi-unit smooth muscle.

- Compare and contrast the contractile, regulatory, and structural components of smooth muscle and skeletal muscle.

- Diagram the molecular events of smooth muscle contraction and relaxation and contrast these events with those of skeletal muscle contraction. Be sure to highlight the roles of calcium (including sources) and ATP/ADP.

- Describe myogenic contraction/relaxation, slow wave potentials, pacemaker potentials, and pharmacomechanical coupling.

■   Describe the regulation of smooth muscle activity by chemical signals.

■   Compare and contrast cardiac muscle with skeletal and smooth muscle.

SUMMARY

Lots of good details here. Draw pictures and make maps to help you really grasp this material.

There are three types of muscles: skeletal, smooth, and cardiac. Remember from the last chapter that skeletal muscle contracts in response to signals from somatic motor neurons. Smooth and cardiac muscles respond to neural and chemical signals.

Take some time to learn the basic anatomy of skeletal muscle—both macro and micro. Be able to recount information about all the components of a sarcomere. According to the sliding filament theory, contraction boils down to stationary myosin fibers (thick filaments) pushing along mobile actin fibers (thin filaments), driven by the power of ATP consumption. Here are the molecular events in a nutshell (you should be able to elaborate on these): an action potential created by opening an acetylcholine-gated ion channel depolarizes the muscle fiber. Depolarization spreads across the muscle membrane and t-tubule system, opening Ca^{2+} channels in the sarcoplasmic reticulum. Ca^{2+} flows into the cytosol, where it binds to troponin. Ca^{2+}-binding moves tropomyosin, which uncovers actin-binding sites on myosin. Myosin can now bind fully to actin and complete the power stroke. For relaxation, Ca^{2+} is pumped back into the sarcoplasmic reticulum, decreasing intracellular Ca^{2+} concentrations. Ca^{2+} unbinds from troponin and is removed from the cytosol, actin-myosin binding is partially blocked again, and the muscle fiber relaxes.

Skeletal muscle fibers are classified by their speed of contraction and their resistance to fatigue. Learn the difference between fast-twitch glycolytic fibers, fast-twitch oxidative fibers, and slow-twitch oxidative fibers. Where would you expect to find each? Fatigue is the condition of not being able to generate or sustain muscle power. It is affected by a variety of factors.

The tension possible in a single muscle fiber is determined primarily by the length of its sarcomeres before contraction. Force in a muscle can be increased by increasing the stimulus, up to a point of maximal contraction, or tetanus. Isotonic contraction creates force and moves loads, while isometric contraction creates force without movement. A motor unit is a group of muscle fibers and the somatic motor neuron that controls them. Recruitment is a means of creating more force within a skeletal muscle. When done asynchronously, recruitment helps prevent fatigue.

Smooth muscle is more complex than other muscle types. In general, it is much more fatigue-resistant. Many of the same intracellular fibers create contraction in both skeletal and smooth muscle, but the organization of the contractile unit and the molecular events are different for the two muscle types. In contrast to skeletal muscle, smooth muscle contraction is influenced by extracellular Ca^{2+} influx in addition to SR Ca^{2+} release, and contraction may be graded according to this influx. Smooth muscle molecular events in a nutshell: Ca^{2+} binds to calmodulin, and the Ca^{2+}-calmodulin complex activates myosin light chain kinase (MLCK). Active MLCK phosphorylates myosin light chain proteins near the head, which in turn enhances myosin ATPase activity. This results in actin binding and crossbridge cycling, which increases tension in the muscle. For relaxation, Ca^{2+} is pumped out of the cytoplasm, and phosphatase dephosphorylates myosin. Smooth muscle can have unstable membrane potentials that can cycle as slow wave potentials or, in some cells, can become pacemaker potentials. Smooth muscle contraction can also be influenced by a number of other factors, like chemical modulation and muscle stretch, making it a difficult tissue to study.

Cardiac muscle has qualities of both skeletal and smooth muscle. It is striated like skeletal muscle, but is under autonomic and endocrine control like smooth muscle. Cardiac muscle has intercalated disks, which contain gap junctions that allow action potentials to spread throughout the tissue.

TEACH YOURSELF THE BASICS

1.   List the three types of muscles. (Fig. 12-1)

SKELETAL MUSCLE

IP *Muscular Physiology*

2. How do skeletal muscles attach to bones?

3. Define the following terms:

    origin

    insertion

    joint

    flexor

    extensor

    antagonistic muscle groups (Fig. 12-2)

Skeletal Muscles Are Composed of Muscle Fibers

4. What is the difference between a muscle and a muscle fiber? (Fig. 12-3a)

5. What is a fascicle?

6. What do you find in skeletal muscle besides muscle fibers?

Muscle Fiber Anatomy

7. Define the following terms: (Table 12-1)

    sarcolemma

    sarcoplasm

    myofibril

    sarcoplasmic reticulum (SR) (Figs. 12-3b, 12-4)

    transverse tubules (t-tubules) (Fig. 12-4)

8. Why does muscle tissue have many glycogen granules in the cytoplasm?

Myofibrils Are the Contractile Structures of a Muscle Fiber

9. Name the two contractile proteins of the myofibril.

10. Name the two regulatory proteins.

11. Name two giant accessory proteins.

12. Describe and diagram the structure of myosin. (Fig. 12-3e)

13. Diagram thin filaments as you would expect to find them in skeletal muscle. Distinguish between G-actin and F-actin. (Fig. 12-3f)

14. Describe and diagram the relationship between actin and myosin in a myofibril. (Fig. 12-3d)

15. What is a crossbridge?

16. What is a sarcomere? Sketch a diagram of one. (Figs. 12-1a, 12-3c, 12-5)

17. Explain what creates the following elements in a sarcomere. How many of each are in one sarcomere?

    A band

    H zone

    I band

    M line

    Z disk

18. What is the function of titin? (Fig. 12-6)

19. What is the function of nebulin? (Fig. 12-6)

Muscle Contraction Creates Force

20. What is muscle tension?

21. What is meant by the term *load*?

22. What are the major steps leading to a muscle contraction? (Fig. 12-7)

Actin and Myosin Slide Past Each Other During Contraction

23. Give a brief overview description of the sliding filament theory of contraction.

24. According to this theory, tension generated in a muscle fiber is directly proportional to:

25. What happens to the Z disks as thick and thin filaments slide past each other? (Fig. 12-8)

26. Explain why the A bands of the sarcomere do not shorten during contraction.

27. Explain why the I band and H zone almost disappear during contraction.

Myosin Crossbridges Move Actin Filaments

28. What is a motor protein?

29. Describe and diagram the roles of actin and myosin in the power stroke.

Myosin ATPase

30. Diagram the ATPase activity of myosin.

Calcium Signals Initiate Contraction

31. Diagram and describe the regulatory role of troponin, tropomyosin, and calcium in muscle contraction and relaxation. (Use a separate sheet of paper if necessary.) (Fig. 12-10)

Myosin Heads Step Along Actin Filaments

32. Diagram and describe the molecular events of a contractile cycle, beginning with the rigor state. (Use a separate sheet of paper if necessary.) (Fig. 12-9)

The Rigor State

33. Why do muscles freeze in the state known as rigor mortis after death?

Acetylcholine Initiates Excitation-Contraction Coupling

34. What is excitation-contraction coupling? (Fig. 12-11)

35. Diagram the detailed events of excitation-contraction coupling. (Fig. 12-11)

36. What is the immediate signal for contraction: ACh, an action potential, Na^+, or Ca^{2+}?

37. A single contraction-relaxation cycle is called a(n) _____.

38. What is the latent period and what creates it? (Fig. 12-12)

Skeletal Muscle Contraction Requires a Steady Supply of ATP

39. What is the role of phosphocreatine in muscle? (Fig. 12-13)

40. Compare the ATP yield for aerobic and anaerobic metabolism of one glucose molecule. (See Fig. 4-18.)

41. TRUE/FALSE? Defend your answer.

    Muscles can only use glucose for energy.

Muscle Fatigue Has Multiple Causes

42. What is muscle fatigue?

43. What factors are believed to contribute to muscle fatigue? (Fig. 12-14)

44. What is central fatigue and why is it considered to be a protective mechanism?

45. Contrast central fatigue with peripheral fatigue.

46. What are some possible causes of fatigue in cases of short-duration maximal exertion?

47. During maximal exercise, K^+ (enters/leaves?) the muscle fiber with each action potential. What effect will this have on the fiber's membrane potential?

Skeletal Muscle Is Classified by Speed and Resistance to Fatigue

48. List the three classifications of muscle fibers and describe their speed of contraction and resistance to fatigue. (Fig. 12-15; Table 12-2)

49. What factor determines speed of tension development?

50. What factor determines duration of contraction?

51. Which fiber type has the longer duration of contraction?

52. Compare the types of movements for which the three fiber types are best suited. (Table 12-2)

53. What contributes to fatigue in fast-twitch fibers?

54. What is the role of myoglobin in skeletal muscle? Which fiber type has the most myoglobin?

55. What are the differences in the two types of fast-twitch muscle? (Fig. 12-15)

Tension Developed by Individual Muscle Fibers Is a Function of Fiber Length

56. At the molecular level, what determines the tension a muscle fiber can generate? (Fig. 12-16)

57. For a single muscle fiber, explain why long and short sarcomeres develop less tension than sarcomeres at optimal length. (Fig. 12-16)

Force of Contraction Increases with Summation of Muscle Twitches

58. Explain the process of summation in a muscle fiber. (Fig. 12-17; also see neuronal summation in Ch. 8)

59. Would you consider summation in a muscle to be an example of spatial or temporal summation? Why?

60. Describe the difference between unfused (incomplete) tetanus and fused (complete) tetanus. (Fig. 12-17c, d)

A Motor Unit Is One Somatic Motor Neuron and the Muscle Fibers It Innervates

61. What is a motor unit? (Fig. 12-18)

62. A muscle responsible for fine muscle movements will have (more /fewer?) muscle fibers in its motor units than a muscle responsible for gross motor actions.

63. Fibers in a motor unit are of the (same/different?) fiber type.

64. During embryological development, what determines the type of muscle fibers that will be in each motor unit?

65. Do muscle fibers exhibit plasticity, the ability to alter metabolic characteristics with use?

Contraction Force Depends on the Types and Numbers of Motor Units

66. Explain the two ways the body can vary force of contraction in a muscle. (Fig. 12-18)

67. What is recruitment and how does it take place?

68. What is asynchronous recruitment? Why is it helpful for avoiding fatigue during a sustained contraction?

MECHANICS OF BODY MOVEMENT

69. What does the term *mechanics* mean when applied to muscle physiology?

Isotonic Contractions Move Loads; Isometric Contractions Create Force Without Movement

70. Compare and contrast isotonic and isometric contractions. (Fig. 12-19)

71. What are the series elastic elements (Fig. 12-20), and what role do they play in isometric contractions?

Bones and Muscles Around Joints Form Levers and Fulcrums

72. The body uses lever and fulcrum systems to move loads. What are the levers of the body and what are the fulcrums? (Fig. 12-21)

73. What advantage does the body gain by using lever and fulcrum systems?

74. What is the disadvantage of a lever system in which the fulcrum is positioned near one end of the lever? What is the advantage of this type of lever system? (Fig. 12-22)

75. For a given muscle, what is the relationship between the speed of contraction and the load being moved by the muscle? (Fig. 12-23)

Muscle Disorders Have Multiple Causes

76. List examples of skeletal muscle disorders and identify the cause(s) for the disorder.

SMOOTH MUSCLE

77. Where in the body are smooth muscles found?

78. Compare the speed and duration of contraction in smooth and skeletal muscle. (Fig. 12-24)

79. Why is smooth muscle more difficult to study than skeletal muscle?

Smooth Muscles Are Much Smaller Than Skeletal Muscle Fibers

80. Compare the neuromuscular junction in neurally controlled smooth muscle to that of skeletal muscle. (See Fig. 11-9.)

81. Contrast single-unit smooth muscle and multi-unit smooth muscle. (Fig. 12-25) Give examples of where each can be found.

Smooth Muscle Has Longer Actin and Myosin Filaments

82. The isoform of myosin found in smooth muscle has (faster/slower?) ATPase activity. This (increases/decreases?) the rate of crossbridge cycling and (lengthens/shortens?) the contraction phase.

83. Myosin light chains are:

84. Compare the amounts of actin, tropomyosin, and troponin in smooth muscle with the amounts of these elements in skeletal muscle.

85. Smooth muscle has (less/more?) sarcoplasmic reticulum than skeletal muscle.

86. What is the primary Ca^{2+} release channel in smooth muscle sarcoplasmic reticulum?

87. What role might caveolae play in smooth muscle? (Fig. 12-26; also see caveolae in Ch. 5)

Smooth Muscle Is Not Arranged in Sarcomeres

88. Describe and diagram the arrangement of contractile fibers in smooth muscle. (Fig. 12-27) How does this compare to the organization of contractile fibers in skeletal muscle?

Protein Phosphorylation Plays a Key Role in Smooth Muscle Contraction

89. Where does the Ca^{2+} for contraction in smooth muscle come from? (Fig. 12-28; Table 12-3)

90. In skeletal muscle, Ca^{2+} entering the cytoplasm binds to _____. In

    smooth muscle, Ca^{2+} entering the cytoplasm binds to _____.

91. In skeletal muscle, Ca^{2+} binding to troponin does what?

92. In smooth muscle, Ca^{2+} binding to calmodulin does what?

93. Diagram and explain the molecular events of smooth muscle contraction. (Fig. 12-28)

Relaxation in Smooth Muscle Has Several Steps

94. What events are necessary for relaxation in smooth muscle? (Fig. 12-29)

95. What is a latch state and what are its advantages?

Calcium Entry Is the Signal for Smooth Muscle Contraction

96. From which two sources does smooth muscle calcium originate?

97. How does this create graded contractions? How is the contraction force related to the calcium signal(s)?

98. Compare the release of calcium from smooth muscle sarcoplasmic reticulum to release of calcium from skeletal muscle sarcoplasmic reticulum.

99. How are internal calcium stores replenished in smooth muscle cells?

100. A smooth muscle is stretched to the point that a myogenic contraction is initiated. Yet over time, the muscle begins to relax even though it remains stretched. In this scenario, explain how the myogenic contraction is initiated and how it is terminated.

Some Smooth Muscles Have Unstable Membrane Potentials

101. What are the differences between slow wave potentials and pacemaker potentials? (Fig. 12-31)

102. What is pharmacomechanical coupling? (Fig. 12-31c)

Smooth Muscle Activity Is Regulated by Chemical Signals

Autonomic Neurotransmitters

103. Describe autonomic innervation differences in smooth muscle. What is the difference between those under antagonistic autonomic control and those under tonic control? Why are different receptor types important?

Hormones and Paracrines

103. Describe some of the effects hormones and paracrines can have on smooth muscle contraction.

CARDIAC MUSCLE (Table 12-3)

105. How is a cardiac muscle fiber similar to a skeletal muscle fiber?

106. How is a cardiac muscle fiber different from a skeletal muscle fiber?

107. How is a cardiac muscle fiber similar to a smooth muscle fiber?

108. How is a cardiac muscle fiber different from a smooth muscle fiber?

109. What are intercalated disks?

TALK THE TALK

A band (A is for anisotropic)
actin
anaerobic glycolysis
antagonistic muscle groups
asynchronous recruitment
attachment plaques
beta-oxidation
biceps brachii
caldesmon
calmodulin
cardiac muscle
caveolae
central fatigue
complete (fused) tetanus
contraction
creatine kinase
creatine phosphokinase
crossbridge tilting
crossbridges
dense bodies
dihydropyridine (DHP)
Duchenne's muscular dystrophy
dystrophin
endothelium-derived relaxing
     factor
end-plate potential (EPP)
extension
extensor
F-actin
fascicles
fast-twitch muscle fiber (type 2)
fatigue
flexion
flexor
fulcrum
gastrointestinal
H zone (H is for helles)
I band (I is for isotropic)

incomplete (unfused) tetanus
insertion
intercalated disks
IP_3-receptor channel
isometric contractions
isotonic contraction
L-type Ca^{2+} channels
latch state
latent period
lever
load
M line (M is for mittel)
McArdle's disease
motor unit
multi-unit smooth muscle
muscle cramp
muscle fibers
muscle tension
myofibrils
myogenic contraction
myoglobin
myophosphorylase deficiency
myosin
myosin light chain
myosin light chain kinase
myosin phosphatase
nebulin
nitric oxide
ocular
origin
pacemaker potentials
peripheral fatigue
pharmacomechanical coupling
phosphocreatine
power stroke
recruitment
red muscle
relaxation

reproductive
respiratory
rigor mortis
ryanodine receptors (RyR)
sarcolemma
sarcomere
sarcoplasm
sarcoplasmic reticulum
series elastic elements
single-unit smooth muscle
skeletal muscle
sliding filament theory of con-
     traction
slow wave potentials
slow-twitch fibers (type 1)
smooth muscle
sphincters
store-operated Ca^{2+} channels
striated muscle
summation
tendons
terminal cisternae
tetanus
thick filament
thin filaments
titin
tonic control
transverse tubules (t-tubules)
triad
triceps brachii
tropomyosin
troponin
twitch
urinary
vascular
visceral smooth muscle
white muscle
Z disks (Z is for zwischen)

BEYOND THE PAGES

▶   Larach, M. G. "Making Anesthesia Safer: Unraveling the Malignant Hyperthermia Puzzle." *FASEB Break-throughs in Bioscience.* (Free access to FASEB at opa.faseb.org/pages/Publications/breakthroughs.html.)

Excerpt from this publication: "Why should halothane—in 1968 the most widely used anesthetic in the world—kill these pigs? The multidisciplinary research team of anesthesiologists, surgeons, internists, and laboratory technologists interrupt their liver transplantation research in order to find out."

PRACTICE MAKES PERFECT

1. What is the primary ion required for muscle contraction?

2. Which neurotransmitter is secreted by somatic motor neurons?

3. TRUE /FALSE? Defend your answer.

   A single motor neuron can have synapses on more than one muscle fiber.

4. Why does Na^+ entry exceed K^+ efflux when ACh-gated channels open at the motor end plate?

5. Which of the following characteristics are typical of vertebrate skeletal muscle fibers? (circle all that apply)
   a. When stimulated, they produce Ca^{2+}-dependent action potentials.
   b. Depolarization of the muscle fiber results in an influx of Ca^{2+} from ECF, which causes contraction.
   c. They twitch spontaneously.
   d. Each muscle fiber is a multinucleated single cell.
   e. None of the above.

6. Why is muscle contraction faster than relaxation?

7. Describe and explain the difference between temporal summation in muscle and temporal summation in nerves.

8. Why does a fatigued muscle take longer to relax than a fresh muscle?

9. What would be the disadvantage of having gap junctions between skeletal muscle cells like there are between cardiac muscle cells?

10. TRUE/FALSE? Defend your answer.

    Muscle length changes during an isotonic muscle contraction.

11. Which of the following statements applies to isometric muscle contraction? (circle all that apply)
    a. Thick and thin filaments slide past each other.
    b. An isometric twitch lasts longer than an isotonic twitch.
    c. Thick and thin filaments do not slide past each other.
    d. Maximum tension is always generated, irrespective of initial fiber length.

12. In order to move a load, isotonic muscle tension should be:
    a. greater than the load
    b. equal to the load
    c. less than the load
    d. equal to the square root of the load

13. Can you give some examples of motor proteins that you studied in the nervous system? In organelles?

14. In the muscular disorder myasthenia gravis, an autoimmune response causes a reduction in the number of ACh receptor sites at the motor end plate. Predict the symptoms of this disease and explain them based on your knowledge of the neuromuscular junction.

15. You have isolated the leg muscle of a frog and the nerve that innervates the muscle. To obtain a muscle contraction, you can either electrically stimulate the muscle itself (direct stimulation) or you can stimulate the nerve (indirect stimulation), creating an action potential in the nerve. In both instances, your recording shows a latency period, a delay between the time of stimulation and the peak of the contraction. Do you expect the latency with direct stimulation to be greater or less than the latency with indirect stimulation? Explain briefly.

MAPS

1. Map the anatomical organization of skeletal muscle (see map in Anatomy Summary), then add functions where appropriate.

2. Use a map to relate the following words into a cohesive, orderly description of muscle contraction. You must use all the words. You may add other words to make the relationships clear.

acetylcholine	end plate potential	t-tubule
actin	ion channel	thick filament
action potential	myosin	thin filament
binding site	myosin ATPase	tropomyosin
calcium	power stroke	troponin
crossbridge	sarcoplasmic reticulum	
dihydropyridine receptor	swing	

13 Integrative Physiology I: Control of Body Movement

LEARNING OBJECTIVES

When you complete this chapter, you should be able to:

▨ Classify neural reflexes according to the efferent division that controls the response, the CNS location where the reflex is integrated, whether the reflex is innate or learned, and the number of neurons in the reflex pathway.

▨ Discuss the role of CNS integration and modulation in autonomic reflexes.

▨ Discuss the conversion of emotional stimuli into visceral responses.

▨ Match the specific components of a skeletal muscle reflex to the general steps of a reflex.

▨ Draw a muscle spindle, diagram the stretch reflex, and explain the purpose of this reflex.

▨ Explain alpha-gamma coactivation.

▨ Draw a Golgi tendon organ, diagram the Golgi tendon reflex, and explain the purpose of this reflex.

▨ Define a myotatic unit.

▨ Demonstrate a monosynaptic stretch reflex with a diagram of the knee jerk reflex and its reciprocal inhibition.

▨ Demonstrate a polysynaptic reflex pathway with a diagram of the flexion reflex and its associated crossed extensor reflex.

▨ Explain the differences between reflex movement, voluntary movement, and rhythmic movement. You should also be able to describe the roles of central pattern generators and feedforward responses in skeletal muscle movement.

▨ Outline the roles played by different regions of the CNS in integrating body movements.

▨ Describe ways in which control of movement in visceral muscles differs from control of movement in skeletal muscles.

SUMMARY

This chapter integrates all the information you've learned so far and then adds some. If the old material doesn't immediately click, now's the time to go back and review.

Remember the components of a neural reflex? If not, look in this chapter and back to Ch. 6 to see where it all began. Neural reflexes are classified in different ways. There are somatic reflexes and visceral reflexes, named for the neurons involved. There are spinal reflexes and cranial reflexes, named for the location where information is integrated. There are monosynaptic and polysynaptic reflexes, named for the number of neurons involved. Some neural reflexes are innate, others are learned. (Go to Ch. 9 and review the process of learning.) Neural activity can be altered by neuromodulators. Presynaptic modulation is more specific than postsynaptic modulation. Think about how modulation relates to divergence and convergence.

Some autonomic reflexes are spinal reflexes integrated in the spinal cord and modulated by the brain. Other reflexes are integrated in the brain itself, in the hypothalamus, thalamus, and brain stem. (Go to Ch. 9 and review CNS anatomy.) Many homeostatic reflexes are controlled by the brain. All autonomic reflexes are polysynaptic, and many exhibit tonic activity. (Go to Ch. 11 and review autonomic nervous system anatomy and physiology.)

Now for the new material. Remember that skeletal muscle contraction is controlled by somatic motor neurons. (See Ch. 12, skeletal muscle contraction.) Contractile fibers are called extrafusal fibers, and they are controlled by alpha motor neurons. There are receptors within skeletal muscles that respond to stretch and length stimuli to protect the muscle from damage. Muscle spindles, made of intrafusal fibers, report information about length to the CNS. Sensory neurons wrap around the noncontractile centers of spindles, sending information to the CNS. Gamma motor neurons return commands to the contractile ends of the intrafusal fibers. Stretching the muscle fiber initiates a stretch reflex that creates contraction. Alpha-gamma coactivation ensures that the spindle stays active during contraction. Golgi tendon organs are found at the junction of muscle and tendon. They respond to both stretch and contraction, yielding a reflexive relaxation.

A myotatic unit is a set of synergistic and antagonistic muscles that control a single joint. Reciprocal inhibition is necessary to allow free movement by one member of the antagonistic pair. Flexion reflexes move a limb away from a harmful stimulus. Crossed extensor reflexes are postural reflexes that help maintain balance while one foot is off the ground. Central pattern generators, like those for breathing, are neural networks that spontaneously generate rhythmic muscle movements.

There are three categories of movement: reflex movements, voluntary movements, and rhythmic movements. Reflex movements are the least complex, and rhythmic movements are the most complex. The CNS coordinates and plans movements.

TEACH YOURSELF THE BASICS
NEURAL REFLEXES

1. List the steps in a neural reflex. (See Ch. 6 if you need a refresher.)

2. Compare the roles of negative feedback and feedforward responses in the control of body movement.

Neural Reflex Pathways Can Be Classified Different Ways (Table 13-1)
3. What is the difference between somatic reflexes and autonomic (visceral) reflexes?

4. What is the difference between a spinal reflex and a cranial reflex?

5. What is the difference between an innate reflex and a learned reflex?

6. What is the difference between a monosynaptic reflex (Fig. 13-1a) and a polysynaptic reflex? (Fig. 13-1b)

7. Compare convergence and divergence.

AUTONOMIC REFLEXES

8. Where are autonomic reflexes integrated? (Fig. 11-3)

9. Describe the influences that emotions and higher brain centers can have on autonomic reflexes.

10. Autonomic reflexes are (always monosynaptic/always polysynaptic/may be either?). (Fig. 13-2)

SKELETAL MUSCLE REFLEXES

11. What is the only way to inhibit skeletal muscle contraction?

12. What are the components of skeletal muscle reflexes?

13. What are proprioceptors? What are the three types found in the body?

14. What are alpha motor neurons? (Fig. 13-3a)

15. What are extrafusal muscle fibers?

Muscle Spindles Respond to Muscle Stretch

16. What is the function of muscle spindles? (Fig. 13-3a)

17. What is the difference between intrafusal and extrafusal muscle fibers? (Fig. 13-3b)

18. What is the difference between alpha and gamma motor neurons?

19. What is muscle tone? (Fig. 13-4a)

20. Describe the stretch reflex, using the standard steps of a reflex. (Fig. 13-4b)

    stimulus _____   receptor _____

    afferent path _____   integrating center _____

    efferent path _____   effector _____

    tissue response _____   systemic response _____

21. Explain alpha-gamma coactivation. (Fig. 13-5)

22. How do muscle spindles work during a stretch reflex? (Fig. 13-6a, c)

Golgi Tendon Organs Respond to Muscle Tension

23. Describe the structure of the Golgi tendon organs. (Fig. 13-3c)

24. Describe the Golgi tendon reflex, using the standard steps of a reflex. (Fig. 13-6d, e)

    stimulus _____   receptor _____

    afferent path _____   integrating center _____

    efferent path _____   effector _____

    tissue response _____   systemic response _____

25. What is the function of this reflex? (Fig. 13-6d, e)

Stretch Reflexes and Reciprocal Inhibition Control Movement Around a Joint

26. What is a myotatic unit?

27. Describe a monosynaptic stretch reflex, the knee jerk reflex, using the standard steps of a reflex. (Fig. 13-7)

    stimulus _____   receptor _____

    afferent path _____   integrating center _____

    efferent path _____   effector _____

    tissue response _____   systemic response _____

28. Describe reciprocal inhibition in the knee jerk reflex, using the standard steps of a reflex. (Fig. 13-7)

stimulus _____   receptor _____

afferent path _____   integrating center _____

efferent path _____   effector _____

tissue response _____   systemic response _____

29. How can a single stimulus, transmitted through a single sensory neuron, create two opposing responses?

Flexion Reflexes Pull Limbs Away from Painful Stimuli

30. Describe the flexion reflex (withdrawal reflex), using the standard steps of a reflex. There are two efferent pathways because this reflex involves reciprocal inhibition. (Fig. 13-8)

stimulus _____   receptor _____

afferent path _____   integrating center _____

efferent path 1 _____   effector 1 _____

efferent path 2 _____   effector 2 _____

tissue response 1 _____ tissue response 2 _____

systemic response _____

31. Why does this reflex take longer than the knee jerk reflex?

32. Now describe the crossed extensor reflex that often accompanies the flexion reflex. (Fig. 13-8)

stimulus _____   receptor _____

afferent path _____   integrating center _____

efferent path 1 _____   effector 1 _____

efferent path 2 _____   effector 2 _____

tissue response 1 _____ tissue response 2 _____

systemic response _____

THE INTEGRATED CONTROL OF BODY MOVEMENT

Movement Can Be Classified as Reflex, Voluntary, or Rhythmic (Table 13-2)

33. Name and describe the three basic types of movement. (Use a separate sheet of paper.)

34. What are central pattern generators, and what role do they play in movement?

35. What role do feedforward reflexes play in voluntary movement?

The CNS Integrates Movement

36. What three levels of the nervous system control movement? (Table 13-3)

37. Describe the role of the following in planning and executing movement: (Figs. 13-9, 13-10, 13-11; Table 13-3)

     thalamus

     cerebral cortex

     basal ganglia

     cerebellum

38. Describe the role of the corticospinal tract in control of voluntary movement. (Fig. 13-12)

39. Explain and give an example of a feedforward postural reflex. (Fig. 13-13)

Symptoms of Parkinson's Disease Reflect the Functions of the Basal Ganglia

40. Describe the pathophysiology behind Parkinson's disease.

41. What functions of the basal ganglia were dicovered through research into Parkinson's?

42. What are some current treatments?

CONTROL OF MOVEMENT IN VISCERAL MUSCLES

43.   How does reflex control of visceral muscles differ from reflex control of skeletal muscles?

TALK THE TALK

alpha motor neurons	gamma motor neurons	postural reflexes
alpha-gamma coactivation	Golgi tendon organ	proprioceptors
autonomic reflexes	inhibit	pyramidal tract
capsule	inhibitory	pyramids
central nervous system	innate	reciprocal inhibition
central pattern generators	intrafusal fibers	reflex
conditioned reflex	joint receptors	reflex movements
convergence	learned reflex	rhythmic movements
corticospinal tract	monosynaptic reflex	sensory neurons
cranial reflexes	monosynaptic stretch reflex	sensory receptors
crossed extensor reflex	muscle tone	somatic motor neurons
divergence	myotatic unit	somatic reflexes
effectors	negative feedback	spinal reflexes
extrapyramidal system	Parkinson's disease	stretch reflex
extrapyramidal tract	(Parkinsonism)	visceral reflexes
feedforward postural reflexes	piloerection	voluntary movements
feedforward reflexes	polysynaptic reflexes	

PRACTICE MAKES PERFECT

1.   In the left-hand column, arrange the words below in the proper order:

tissue response, efferent pathway, afferent pathway, receptor, integrating center, systemic response

In the right-hand column, give the parts of any *biological* reflex discussed or demonstrated in the textbook except the knee jerk reflex.

Steps of a reflex	Example
Stimulus	

2. Proprioceptors monitor: (circle the best one)

    a.   limb and muscle state.
    b.   taste perception.
    c.   blood pressure in the carotid artery.
    d.   a and c

3. Look at the pathways diagramed in Figs. 13-9 and 13-11. Can you find examples of convergence and divergence in these figures?

BEYOND THE PAGES

TRY IT

Proprioreception

Proprioreceptors are special receptors in the muscles and joints that send messages to the central nervous system about the position of the body parts relative to each other. Proprioreception normally works with visual cues. These exercises will show you the effectiveness of proprioreception alone.

Get a partner and try the following exercises with your eyes closed.

1. Have your partner time how long you can stand on one leg with your eyes closed and arms extended. Note the extent of swaying. Repeat with the eyes open.

2. Extend your arms to the side at shoulder level, palms down. With your eyes closed and without bending your elbows, try to bring your arms together in front of you, palms down, so that the index fingers meet side by side. Repeat with your eyes open.

3. Extend your arms to the side again, palms facing front. With eyes closed, bend your elbows and try to touch the tips of your index fingers together in front of you. Repeat with your eyes open.

    In all three tests, which was easier? Why? What can you conclude from these exercises?

14 Cardiovascular Physiology

LEARNING OBJECTIVES

When you complete this chapter, you should be able to:

- Define all terms in the following mathematical relationships and describe why these relationships are valid:

  - Flow $\propto \Delta P/R$

  - $R = 1/r^4$

- Apply these mathematical relationships to predict changes in blood flow and velocity when a pressure gradient changes and/or when blood vessel diameter changes.

- Diagram the anatomy of the cardiovascular system and trace a drop of blood as it moves from the left ventricle, to a destination (e.g., a muscle in the foot), and eventually back to the left ventricle again.

- List the key functions of the cardiovascular system.

- Distinguish between myocardial contractile cells and myocardial autorhythmic cells and diagram the mechanisms by which action potentials are generated in each cell type (including ions, channels, and direction of ion movement).

- Diagram excitation-contraction coupling in cardiac muscle and compare it to E-C coupling in skeletal and smooth muscle.

- Diagram the heart and explain the structure and function of the four chambers, the different heart valves, and the electrical conduction system.

- Diagram the electrical and mechanical events of the cardiac cycle and apply this information to interpret the pressure-volume graph (Fig. 14-25) and the Wiggers diagram (Fig. 14-26).

- Map factors that affect cardiac output (CO = heart rate × stroke volume), including all relevant neurotransmitters, membrane receptors, and cellular mechanisms.

- Describe the Frank-Starling law of the heart and apply it to length-tension relationships in contracting muscle.

SUMMARY

This chapter discusses how the heart creates blood flow and examines the variables that influence blood flow through the circulatory system. Blood flow is a result of pressure gradients, resistance, volume, vessel length, and the viscosity of the fluid. The heart—the pump in this system—consists of striated contractile muscle cells and specialized myocardial autorhythmic cells that act as pacemakers. The most important pacemaker is the sinoatrial (SA) node, which sets the intrinsic heart rate by its firing rate. Depolarization of the SA node begins a transmission of action potentials along the heart's conduction system (AV node, bundle of His, Purkinje fibers) to generate cardiac muscle contraction. Pacemaker activity is controlled by specialized ion channels and is influenced by the autonomic nervous system. Calcium, Na^+, and K^+ ions are involved in contraction of the myocardial contractile cells—compare the role of each ion in both skeletal muscle contraction and cardiac muscle contraction. The catecholamines, epinephrine and norepinephrine, affect the strength and duration of contraction by altering Ca^{2+} influx/efflux.

The pumping action of the heart can be examined from both electrical and mechanical standpoints. An ECG records the electrical events and provides information about the mechanical events that follow. The mechanical pumping action of the heart is a result of the electrical signals produced by the autorhythmic cells. Blood flow through the heart during a cardiac cycle can be described in terms of pressure and volume changes created as the chambers contract and relax. See the pressure-volume graph (Fig. 14-25) and the Wiggers diagram (Fig. 14-26) to review the electrical and mechanical events of a cardiac cycle.

Cardiac output, the volume of blood pumped per ventricle per minute, is determined by heart rate and stroke volume. Heart rate is intrinsically controlled by the autorhythmic cells and is varied by autonomic influence. Stroke volume is influenced by neural and endocrine modulation of heart contractility as well as by the end-diastolic volume, which is a function of venous return (review the Frank-Starling law of the heart).

TEACH YOURSELF THE BASICS

OVERVIEW OF THE CARDIOVASCULAR SYSTEM

IP *Cardiovascular—Anatomy Review: The Heart*

1. In simplest terms, describe the basic structure of a cardiovascular (CV) system.

The Cardiovascular System Transports Materials Throughout the Body

2. List at least five substances transported by the blood. (Table 14-1)

3. What are the key functions of the CV system?

The Cardiovascular System Consists of the Heart, Blood Vessels, and Blood

4. How do arteries differ from veins?

5. What ensures one-way flow of blood through the system?

6. Diagram the structure of the heart.

7. Trace a drop of blood from the left ventricle to the stomach and back to the left ventricle. (Fig. 14-1)

8. Compare the pulmonary circulation with the systemic circulation.

9. What is a portal system?

10. Name the three portal systems of the body. (See Ch. 7.)

PRESSURE, VOLUME, FLOW, AND RESISTANCE

11. Liquids and gases flow from areas of _____ pressure to areas of _____ pressure.

12. How does the cardiovascular system create a region of higher pressure?

13. As blood moves away from the heart, what happens to the pressure? Why?

14. The highest pressure in the blood vessels is found in the _____ and the lowest

    pressures are found in the _____. (Fig. 14-2)

The Pressure of Fluid in Motion Decreases over Distance
15. What is the difference between hydrostatic pressure and hydraulic pressure? (Fig. 14-3)

16. What units are used to measure pressure in the cardiovascular system?

Pressure Changes in Liquids Without a Change in Volume
17. What happens to the pressure inside a water-filled balloon when you squeeze on it? Why?

18. What is driving pressure?

19. What happens to pressure when the heart relaxes or the blood vessels dilate?

Blood Flows from an Area of Higher Pressure to One of Lower Pressure

20. What is a pressure gradient? What is the relationship between a pressure gradient and flow? (Fig. 14-4)

21. Fluid is flowing through two identical tubes. In tube A, the pressure at one end is 150 mm Hg and the pressure at the other end is 100 mm Hg. In tube B, the pressure at one end is 75 mm Hg and the pressure at the other end is 10 mm Hg. Which tube will have the greatest flow? (Answer at the end of the workbook.)

Resistance Opposes Flow

22. What two factors contribute to friction in the CV system?

23. Define resistance.

24. When resistance (R) increases, flow (increases/decreases?).

25. Express this relationship mathematically:

26. Name the three parameters that influence resistance for fluid flowing through a tube.

27. The relationship between these factors and resistance was expressed mathematically by Jean Poiseuille. Write the equation known as Poiseuille's law:

28. In humans, which of these factors are relatively constant and which play a significant role in determining resistance to blood flow?

29. When the radius of a tube decreases, what happens to the resistance of that tube?

    What happens to flow through that tube?

30. Write the mathematical expression for the relationship between resistance and radius.

31. If the radius of a tube doubles, what happens to the resistance? (Fig. 14-5)

32. Define vasoconstriction and vasodilation in terms of diameter and resistance.

33. Write the mathematical expression for the relationship between flow, resistance, and the pressure gradient.

Velocity Depends on the Flow Rate and the Cross-Sectional Area

34. What is the difference between flow (flow rate) and the velocity of flow? (Give units.) (Fig. 14-6)

35. What factors have the biggest influence on the flow rate? On the velocity of flow?

36. Fluid is flowing through a tube at a constant rate of flow. What happens to the velocity of flow if the tube suddenly narrows?

37. Write the equation that expresses the relationship between the flow rate (Q), the velocity of flow (v), and the cross-sectional area (A) of a tube.

38. What is mean arterial pressure (MAP), and what are the two main parameters influencing MAP?

CARDIAC MUSCLE AND THE HEART

IP *Cardiovascular—Cardiac Action Potential*

The Heart Has Four Chambers

39. The heart is a muscle that lies in the center of the _____ cavity, surrounded by

    a membrane called the _____. (Fig. 14-7)

40. TRUE/FALSE? Defend your answer.

    The base of the heart is the pointed end that angles downward.

41. The medical term for cardiac muscle is _____.

42. The heart has (how many?) _____ chambers, separated by a wall known as the _____.

43. The _____ are the lower chambers and the _____ are the upper

    chambers. Which chambers have the thickest walls?

44. Diagram the blood vessels that connect to each chamber and tell where they are bringing blood
    from or taking blood to.

45. Name the vessels that supply blood to the heart muscle itself.

46. Name two functions of the fibrous connective tissue rings that surround the four heart valves.
    (Fig. 14-9a)

Heart Valves Ensure One-Way Flow in the Heart

47. Diagram the location of the heart valves. (Figs. 14-7g, 14-9)

48. What are the chordae tendineae and what is their function? How are chordae tendineae related
    to papillary muscles? (Fig. 14-9)

49. Compare and contrast the two AV valves and two semilunar valves. (Fig. 14-9)

Cardiac Muscle Cells Contract Without Nervous Stimulation

50. How are myocardial autorhythmic cells different from myocardial contractile cells?

51. What is a pacemaker potential? (See Ch. 12.)

52. What are intercalated disks? (Figs. 14-7g, 14-10b)

53. What role do desmosomes and gap junctions play in myocardial contractile cells? (See Ch. 3.)

54. Compare the myocardial contractile cells to skeletal muscle cells. (See Table 12-3 for a comparison of the three muscle types.)

55. Why do myocardial cells have a high rate of oxygen consumption? Because of this, what organelle occupies about one-third of the cell volume?

Cardiac E-C Coupling Combines Features of Skeletal and Smooth Muscle

56. In what ways does cardiac muscle E-C coupling blend aspects of skeletal muscle and smooth muscle?

57. Diagram the mechanism for E-C coupling in cardiac muscle. Why is this also called Ca^{2+}-induced Ca^{2+} release?

58. How is Ca^{2+} removed from the cytoplasm of a cardiac muscle cell?

Cardiac Muscle Contraction Can Be Graded

59. How does a single cardiac muscle fiber create graded contractions?

When Cardiac Muscle Is Stretched, It Contracts More Forcefully

60. TRUE/FALSE? Defend your answer.

    The force of cardiac muscle contraction depends on the length of the muscle fiber when contraction begins. (Fig. 14-12; also see Ch. 12)

61. In the intact heart, the length of the sarcomeres at the beginning of a contraction is a reflection of:

Action Potentials in Myocardial Cells Vary According to Cell Type

62. What ion is important in cardiac muscle action potentials but plays no significant role in skeletal muscle or neuronal action potentials?

Myocardial Contractile Cells (Fig. 14-13)

63. Do myocardial contractile cells have a stable or unstable membrane potential?

64. In the myocardial contractile cell, the rapid depolarization phase is due to the entry of _____.

65. The rapid repolarization phase is due to _____ (influx/efflux?).

66 In Fig. 14-13, what causes the small fall in the membrane potential between points 1 and 2 (initial repolarization)?

67. How does a longer action potential in cardiac muscle (compared to skeletal muscle) prevent tetanus? (Fig. 14-4)

Myocardial Autorhythmic Cells

68. Diagram the mechanisms (ions, ion channels) involved in the constantly fluctuating membrane potential of myocardial autorhythmic cells. Be sure to distinguish between the pacemaker potential and the action potential. (Fig. 14-15)

69. The rapid depolarization phase is due to the entry of _____. How does this compare with a neuron?

70. The repolarization phase in autorhythmic cells is due to _____ (influx/efflux?). How does this compare to a neuron?

Autonomic Neurotransmitters Modulate Heart Rate

71. How do the catecholamines affect the rate of depolarization in pacemaker cells? Describe in terms of ion movement. (Fig. 14-16a)

72. To which receptor are the catecholamines binding?

73. How does ACh affect the rate of depolarization in pacemaker cells? Describe in terms of ion movement. (Fig. 14-16b)

74. To which receptor does ACh bind?

▶ Table 14-3 compares AP generation in the two types of myocardial muscle with skeletal muscle.

THE HEART AS A PUMP

IP *Cardiovascular—Intrinsic Conduction System*

IP *Cardiovascular—The Cardiac Cycle*

IP *Cardiovascular—Cardiac Output*

Electrical Conduction in the Heart Coordinates Contraction

75. Where do electrical signals in the heart originate? (Figs. 14-17, 14-18)

76. If you cut all nerves leading to the heart, will it continue to beat? Explain.

77. What cell structures allow electrical signals to spread quickly to adjacent cells? (Fig. 14-17)

78. Starting at the sinoatrial (SA) node, describe, outline, or map the spread of electrical activity through the heart. Be sure to include all the following terms. (Fig. 14-18)

atrial conducting system   bundle of His   right atrium
atrioventricular (AV) node   internodal pathway   septum
AV node delay   left atrium   ventricle
bundle branches   Purkinje fibers

79. Why is it necessary to direct the electrical signals through the AV node?

80. What is the purpose of AV node delay?

Pacemakers Set the Heart Rate

81. If the SA node is damaged, will the heart continue to beat? At the same rate? Explain.

The Electrocardiogram Reflects the Electrical Activity of the Heart

82. What is an electrocardiogram (ECG)?

83. What information does an ECG show?

84. Does the ECG directly show ventricular contraction? (Fig. 14-21)

85. Define cardiac cycle.

86. What is Einthoven's triangle? (Fig. 14-19)

87. Name the waves of the ECG, tell what electrical event they represent, and name the mechanical event with which each wave is associated. (Use a separate sheet of paper if necessary.) (Figs. 14-20, 14-21)

88. Why is an ECG not the same as a single action potential? (Fig. 14-22)

89. Can you tell if an ECG is showing depolarization or repolarization simply by looking at the shape of a wave relative to the baseline? Why or why not?

90. What are the four questions you should ask as you begin to interpret an ECG?

91. What are the medical terms for rapid and slow heart rate? What is considered the normal range for resting heart rate?

92. When looking at an ECG, how can you determine the heart rate?

93. Why is it important to see that the P-R segment on an ECG has a constant duration? (Fig. 14-23b)

The Heart Contracts and Relaxes Once During a Cardiac Cycle (Fig. 14-24)

94. Define systole and diastole.

95. Is the heart in atrial and ventricular systole at the same time? Explain.

Step 1: The heart at rest is: atrial and ventricular diastole (Fig. 14-24)
96. As the ventricles relax, which valves open?

97. Is blood flowing into the heart? Which chambers?

Step 2: Completion of ventricular filling: atrial systole (Fig. 14-24)
98. In a person at rest, how much of ventricular filling depends on atrial contraction?

99. Is there a valve between the atria and the veins emptying into them?

Step 3: Early ventricular contraction and the first heart sound (Fig. 14-24)
100. What electrical event precedes ventricular systole?

101. As the ventricles contract, why do the AV valves close?

102. What creates the heart sounds?

103. Explain what is happening during isovolumic ventricular contraction.

104. What happens to pressure within the ventricles during isovolumic ventricular contraction?

105. What is happening to the atria during this phase of the cycle?

Step 4: The heart pumps: ventricular ejection (Fig. 14-24)
106. Why do the semilunar valves open, allowing blood to be ejected into the arteries?

107. Blood flows out of the ventricles; therefore you know that ventricular pressure must be (lower/higher?) than arterial pressure.

Step 5: Ventricular relaxation and the second heart sound (Fig. 14-24)
108. As the ventricles relax, the ventricular pressure (increases/decreases?).

109. Why do the semilunar valves close?

110. What creates the second heart sound?

111. During this phase (name it), are the AV valves open or closed? Explain.

112. When do the AV valves open?

Pressure-Volume Curves Represent One Cardiac Cycle
Answer the following questions using the pressure-volume loop shown in Fig. 14-25.

113. At point A, is the atrium relaxed or contracting? Is the ventricle relaxed or contracting?

114. From point A to point B, the ventricular volume is increasing. Why doesn't ventricular pressure also increase substantially?

115. Define end-diastolic volume (EDV). What is the EDV for our 70 kg man at rest?

116. What event begins at point B and continues until point D?

117. From point B to point C, why does pressure increase without a change in ventricular volume?

118. What factors are affecting pressure and volume from point C to point D?

119. At point D, why does the aortic valve close?

120. Define end-systolic volume (ESV). What is the ESV for our 70 kg man at rest?

121. At what point(s) on the graph are both the aortic and mitral valves closed? What happens to ventricular pressure? volume?

▶ The Wiggers diagram (Fig. 14-26) summarizes electrical and mechanical cardiac events.

Stroke Volume Is the Volume of Blood Pumped per Contraction

122. Define stroke volume. (Give units.)

123. How do you calculate stroke volume?

124. If the end-diastolic volume increases and the end-systolic volume decreases, has the heart pumped more or less blood?

Cardiac Output Is a Measure of Cardiac Performance

125. Define cardiac output (CO). (Give units.)

126. What information does CO tell us? What does it not tell us?

127. What are average values for stroke volume and cardiac output in the 70 kg man at rest?

128. TRUE/FALSE? Defend your answer.

     The right side of the heart pumps blood only to the lungs, so its cardiac output is less than that of the left side of the heart, which must pump blood to many more tissues.

Heart Rate Is Modulated by Autonomic Neurons and Catecholamines

129. Explain the antagonistic control of heart rate by sympathetic and parasympathetic neurons. Name the neurotransmitters and receptors involved. (Fig. 14-27 and Fig. 6-21)

130. If you were to block all autonomic input to the heart, what would happen to heart rate?

131. Describe the effects of autonomic modulation of conduction through the AV node.

Multiple Factors Influence Stroke Volume

132. Stroke volume is directly related to the _____ generated by cardiac muscle during contraction.

133. Force is affected by these two factors:

134. Define contractility.

Length-Tension Relationships and the Frank-Starling Law of the Heart

135. As sarcomere length increases, what happens to force of contraction?

136. In the intact heart we cannot measure sarcomere length directly. What parameter do we use as an indirect indicator of sarcomere length?

137. State the Frank-Starling law of the heart. (Fig. 14-28)

Stroke Volume and Venous Return

138. Define venous return. What cardiac volume does it determine?

139. What are three factors that affect venous return? Name them and describe how they influence venous return.

Contractility Is Controlled By the Nervous and Endocrine Systems

140. What is an inotropic agent?

141. What effect does a positive inotropic agent have on the heart? Name two agents that create a positive inotropic effect.

142. At the cellular level, an increase in contractility occurs when _____ increases.

143. Explain how Fig. 14-29 shows that contractility is distinct from length-tension relationships in cardiac muscle.

144. Diagram the mechanisms by which catecholamines enhance the force of cardiac muscle contraction. (Fig. 14-30; also see Ch. 11)

145. When voltage-gated Ca^{2+} channels are phosphorylated, their probability of opening is (increased/decreased?) and (more/less?) Ca^{2+} enters the cell.

146. What is phospholamban and what direct effect does it accomplish upon phosphorylation?

147. Diagram the mechanisms by which catecholamines affect the duration of cardiac contraction.

148. How do cardiac glycosides enhance contractility? Name an example of a cardiac glycoside. What are the dangers associated with these compounds?

EDV and Arterial Blood Pressure Determine Afterload

149. Define and describe afterload.

150. In which pathological condition is increased afterload a resulting factor?

151. What is the ejection fraction? What are normal ejection fraction values for our 70 kg man?

▶ Fig. 14-31 summarizes the factors that determine cardiac output.

TALK THE TALK

afterload
ankyrin-B
aorta
aortic valve
apex
arrhythmia
arteries
atrioventricular (AV) bundle
atrioventricular (AV) node
atrioventricular valves
atrium
autorhythmic cells
AV node delay
base
bicuspid valve
bradycardia
bundle branches
calcium-induced calcium
   release
capillaries
cardiac cycle
cardiac output
cardiovascular system
chordae tendineae
complete heart block
contractility
coronary arteries
coronary sinus
coronary veins
cyanosis
diastole
ejection fraction
electrical events
electrocardiograms
end-diastolic volume (EDV)
end-systolic volume (ESV)
first heart sound
flow
Frank-Starling law of the heart
friction rub

funny current
gap junctions
graded contractions
heart failure
hepatic portal vein
hydrostatic pressure
hyperpolarization-activated
   cyclic nucleotide-gated
   (HCN) channels
hypoxia
iatrogenic
I_f channels
inferior vena cava
inotropic agent
inotropic effect
intercalated disks
internodal pathway
intervals
isovolumic relaxation
isovolumic ventricular
   contraction
isovolumic ventricular
   relaxation
long QT syndrome
L-type Ca^{2+} channel
mechanical events
mitral valve
myocardium
Na^+-Ca^{2+} exchanger (NCX)
ouabain
P wave
pacemaker potential
pacemakers
papillary muscles
pericardium
phospholamban
Poiseuille's law
portal system
preload

premature ventricular contractions
pressure
pressure gradients
prolapse
pulmonary arteries
pulmonary circulation
pulmonary trunk
pulmonary valve
pulmonary veins
Purkinje fibers
QRS complex
respiratory pump
ryanodine receptor-channels
   (RyR)
second heart sound
segments
semilunar valves
septum
sinoatrial node (SA node)
skeletal muscle pump
Starling curve
sternum
stroke volume
superior vena cava
systemic circulation
systole
T wave
tachycardia
thoracic cavity
tricuspid valve
vasoconstriction
vasodilation
veins
velocity of flow
venous return
ventricle
ventricular ejection
viscosity
waves

QUANTITATIVE THINKING

Flow $\propto \Delta P$ where $\Delta P = P_1 - P_2$

Flow $\propto 1/R$ where $R = 8L\eta/\pi r^4$ and in most cases, $R = 1/r^4$

Flow $\propto \Delta P/R$ or Flow $\propto \Delta P r^4$

stroke volume = EDV − ESV

cardiac output (CO) = heart rate × stroke volume

1. If the small intestines need more blood, what should happen to the resistance and radius of the blood vessels supplying their blood?

2. Compare the flow rates between tubes A and B below. L = length, r = radius, P1 = pressure at one end of the tube and P2 = the pressure at the other end.

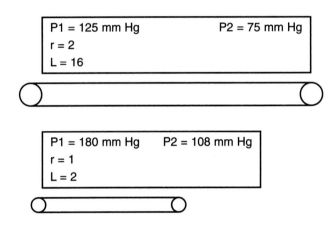

P1 = 125 mm Hg   P2 = 75 mm Hg
r = 2
L = 16

P1 = 180 mm Hg   P2 = 108 mm Hg
r = 1
L = 2

3. At rest, Juliet's heart rate is 72 beats/min and her cardiac output is 5.0 L/min. However, when she sees Romeo, her heart rate increases to 120 beats/min and her cardiac output reaches 15 L/min. What is Juliet's stroke volume before and after seeing Romeo?

PRACTICE MAKES PERFECT

1. Write out the full words for the following abbreviations:

    ACh

    AV (node or valves)

    CO

    ECG

    EDV

    ESV

    SA (node)

    SV

2. Describe the electrical and mechanical events associated with the electrocardiogram.

	Electrical event	Mechanical event
P wave		
QRS complex		
T wave		
PQ segment		
ST segment		
TP segment		

3. In a person at rest, only 20% of ventricular filling depends on atrial contraction. If heart rate speeds up, would this percentage become greater or less? Explain.

4. Compare the action potentials of cardiac and skeletal muscles.

	Skeletal muscle	**Contractile**	**Autorhythmic**
Membrane			
Events leading to threshold potential			
Rising phase of action potential			
Repolarization phase			
Hyperpolarization			
Duration of action potential			
Refractory period			

5. What happens to the kinetic energy of blood flow as it leaves the heart and travels through the blood vessels of the circulatory system?

6. You need to design two different drugs, one that increases and another that decreases the force of a cardiac muscle contraction. What molecular and cellular mechanisms of the cardiac muscle cell might you try to alter with these drugs? (There are several possible answers; be creative. You are not limited to what has already been developed.)

7. Explain how the Frank-Starling law of the heart, venous return, and input from the autonomic nervous system affect cardiac output.

8. Below is a diagram of the heart. Draw and/or label the atrioventricular node, atrial conducting system, bundle of His, Purkinje fibers, sinoatrial node. Show the direction of current flow.

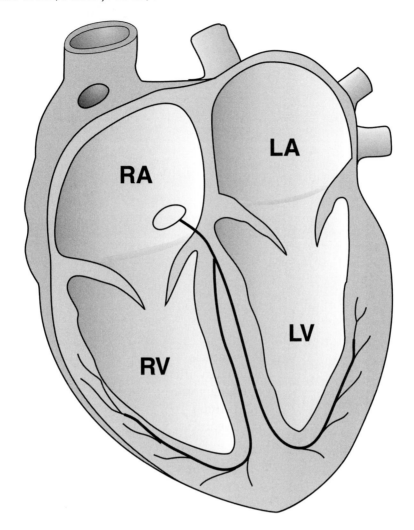

9. The total amount of blood in the circulatory system is about 5 liters. If the cardiac output of the left ventricle is 4.5 L/min at rest, what would be the cardiac output of the right ventricle? Explain your answer.

10. Answer the following questions using the pressure-volume loop shown in Fig. 14-25.

    At point A, what can you say about atrial pressure relative to ventricular pressure?

    At what point on the curve does ventricular pressure match aortic pressure?

    Estimate the highest pressure in the aorta.

11. Draw a graph that has end-diastolic volume (EDV) on the *x*-axis and stroke volume on the *y*-axis. Draw a point that represents a stroke volume at rest of 70 mL with an EDV of 135 mL. Draw a curve that would represent an increase in stroke volume with the same EDV.

MAPS

1. Construct a flow chart using the following terms to trace a drop of blood through the cardiovascular system:

aorta	left ventricle	vein
arteriole	pulmonary artery	vena cava
artery	pulmonary vein	venule
capillary	right atrium	
left atrium	right ventricle	

2. The cardiovascular control center located in the medulla oblongata signals that an increase in cardiac output is required. Use the terms below to trace the sequence of events that would produce an increase in cardiac output.

acetylcholine	end-diastolic volume	parasympathetic activity
β_1 receptors	epinephrine	stroke volume
cardiac output	heart rate	sympathetic activity

3. Construct a diagram or map that provides the sequence of events that occurs during a cardiac cycle. Include in your answer the following: all electrical events, opening and closing of the valves, atrial and ventricular filling, atrial and ventricular systole and diastole, isovolumetric ventricular contraction, ventricular pressure increase, ventricular ejection, and isovolumetric relaxation.

4. Construct a flow chart or map that outlines the sequence of events leading to and occurring during excitation-contraction coupling in cardiac cells. Include membrane events, second messenger systems, the role of ions, and the contractile proteins in your answer. How is the contractile process modulated?

BEYOND THE PAGES
TRY IT

William Harvey's Experiments

Duplicate William Harvey's experiment that led him to believe that blood circulated in a closed loop:
1. Take your resting pulse. 2. Assume that your heart at rest pumps 70 mL per beat and that 1 mL of blood weighs 1 gram. Calculate how long it would take your heart to pump your weight in blood. (1 kilogram = 2.2 pounds)

15 Blood Flow and the Control of Blood Pressure

LEARNING OBJECTIVES

When you complete this chapter, you should be able to:

- Define all terms in the following mathematical relationships and describe why these relationships are valid:

  - Pulse pressure = systolic pressure – diastolic pressure

  - MAP = diastolic P + 1/3(systolic P – diastolic P)

  - $MAP \propto CO \times R_{arteriole}$

  - $Flow_{arteriole} \propto 1/R_{arteriole}$

- Apply these mathematical relationships to show what happens to blood pressure and distribution of blood when arteriolar resistance, blood volume, heart rate, or stroke volume change.

- Explain why velocity of blood flow is fastest in the arteries and slowest in the capillaries.

- Compare the anatomy and roles of arteries, arterioles, capillaries, venules, and veins in the cardiovascular system.

- Explain the role of vascular smooth muscle in regulating peripheral resistance and diagram the myogenic autoregulation, paracrine, and autonomic reflexes that influence resistance. (Hormonal control will be discussed in later chapters.)

- Distinguish between the paracellular pathway and transepithelial transport.

- Explain how colloid osmotic pressure (π) and hydrostatic pressure lead to net filtration at the arterial end of a capillary and net absorption at the venous end.

- Describe the lymphatic system and how it interacts with the cardiovascular system.

- Diagram the baroreceptor reflex for increased blood pressure and decreased blood pressure.

- Describe how atherosclerosis and hypertension develop, and how they contribute to cardiovascular disease.

- Map the factors that affect blood flow in the cardiovascular system and show the local control mechanisms and autonomic reflexes that modulate key factors. You might need to review Ch. 14 in order to supply some of the additional details. (A good place to start: Fill in the gaps in Fig. 15-13 and integrate even more details about cardiac cycle events, sensory receptors, integrating centers, cytokines, neurotransmitters, cellular receptors, ions, ion channels.)

SUMMARY

Blood vessels are composed of an inner layer of endothelium, surrounded by elastic and fibrous connective tissues and smooth muscle. Not all blood vessels have the same characteristics. For example, arterioles contain more smooth muscle than veins, and capillaries of the kidney are more porous than those of the brain. Form and function are directly related, as usual. More smooth muscle indicates a greater role in resistance and blood pressure regulation; a more porous endothelium indicates participation in filtration and fluid balance, and so on. Movement across the endothelium of capillaries is accomplished by diffusion, transcytosis, and bulk flow. Filtration and absorption take place at the capillaries. Net filtration (bulk flow) is a result of the interaction of the hydrostatic and osmotic forces that exist in the capillaries and the interstitial space. The lymphatic system plays an important role in restoring fluid lost through capillary filtration in addition to its immune system function.

Blood pressure is a result of the pressure and volume changes associated with cardiac function as well as the physical properties of the blood vessels (vessel diameter, total cross-sectional area, composition of vessel structure, and so on). The main factors influencing mean arterial pressure (MAP) are cardiac output (CO) and peripheral resistance. CO is determined by stroke volume and heart rate (review Ch. 14 for the factors affecting CO). Peripheral resistance is variable—affected by vasoconstriction and vasodilation of the arterioles (primarily). Various chemicals such as norepinephrine, epinephrine, and angiotensin II control vasodilation and vasoconstriction. The medullary cardiovascular control center (CVCC) integrates sensory information from carotid and aortic baroreceptors and regulates blood pressure through autonomic activity at the heart, vascular smooth muscle, and kidneys. (Here's a chance to apply your knowledge of tonic and antagonistic control of homeostasis!) The baroreceptor reflex is the primary homeostatic control for blood pressure. However, paracrines can also have significant local control over blood pressure.

TEACH YOURSELF THE BASICS

1. What is variable resistance?

2. If blood flow through the aorta is 5 L/min, what is blood flow through the pulmonary artery?

THE BLOOD VESSELS

IP *Cardiovascular—Anatomy Review: Blood Vessel Structure & Function*

3. Diagram the layered composition of blood vessel walls. (Fig. 15-2)

4. What is the *tunica intima*?

Blood Vessels Contain Vascular Smooth Muscle

5. What is vascular smooth muscle?

6. What is the term for a decrease in the diameter of a blood vessel? What is the term for an increase in blood vessel diameter?

7. What is muscle tone? What is it dependent upon? What can influence it?

Arteries and Arterioles Carry Blood Away from the Heart

8. Describe the physical characteristics of the aorta and major arteries. (Fig. 15-2)

9. Blood flow from arteries to arterioles is best described as (divergent/convergent?).

10. Describe the key characteristic of arterioles.

11. How do metarterioles differ from arterioles? (Fig. 15-3)

12. What is the function of metarterioles? How do they interact with precapillary sphincters?

13. What vessels make up the microcirculation?

Exchange Takes Place in the Capillaries

14. Describe the capillary wall and explain how this structure allows the capillaries to carry out their function. (See Ch. 3.)

15. What are pericytes?

16. How do the capillaries of the blood-brain barrier differ from those in the rest of the systemic circulation?

Blood Flow Converges in the Venules and Veins

17. Blood flow from capillaries to venules is best described as (divergent/convergent?).

18. Compare the walls of veins with those of arteries. (Fig. 15-2)

19. How much of the blood in the circulatory system is found in the veins?

20. Are the bluish blood vessels you see under the skin arteries or veins?

Angiogenesis Creates New Blood Vessels

21. Define angiogenesis.

22. What is the reason for angiogenesis in children? In adults?

23. If we can find a way to stop angiogenesis, why might this become useful in treating cancer?

24. In coronary artery disease, what happens to the arteries? Why would a drug that stimulates angiogenesis be useful for treating this condition?

25. List some of the key substances involved in angiogenesis and antiangiogenesis.

BLOOD PRESSURE

IP *Cardiovascular—Measuring Blood Pressure*

26. What property of artery walls plays a key role in the ability of arteries to sustain the driving pressure created by the heart? (Fig. 15-4b)

   ▶ Review Table 15-1 for the rules for fluid flow that were introduced in Ch. 14.

Systemic Blood Pressure Is Highest in the Arteries and Lowest in the Veins

27. Why does blood pressure decrease as blood flows through the circulatory system? (Fig. 15-5)

28. Define systolic pressure and diastolic pressure. Provide average aortic systolic and diastolic values (with units).

29. TRUE/FALSE? Defend your answer.

   The pulse is created by a wave of blood flowing through the arteries.

30. The pulse amplitude (increases/decreases?) over distance from the heart due to what factor(s)? (Fig. 15-5)

31. What is pulse pressure and how do you calculate it?

32. How can low-pressure venous blood in the feet flow uphill against gravity to get back to the heart? (Fig. 15-6)

Arterial Blood Pressure Reflects the Driving Pressure for Blood Flow

33. What does "blood pressure" reflect?

34. Why is blood pressure such an important parameter to know?

35. Explain mean arterial pressure (MAP).

36. Write the formula for estimating MAP:

37. What kinds of problems might result when blood pressure is too low? Too high?

Blood Pressure Is Estimated by Sphygmomanometry

38. Explain how a sphygmomanometer is used to estimate arterial pressure of the radial artery. (Fig. 15-7)

39 What makes Korotkoff sounds? How are they used to determine systolic and diastolic pressure?

40. Explain a blood pressure of 100/70.

41. What is an average value for blood pressure?

42. Blood pressure is considered too high if systolic pressure is chronically in the range of _____

    mm Hg or diastolic pressure chronically in the range of _____ mm Hg. The medical term for

    high blood pressure is _____.

Cardiac Output and Peripheral Resistance Determine Mean Arterial Pressure

43. What two main factors determine mean arterial pressure (MAP)? Write this as a mathematical
    expression. (Fig. 15-8)

44. If blood flow into the arteries increases but there is no change in blood flow out of the arteries,
    MAP will (increase/decrease?).

45. What happens to MAP if peripheral resistance increases?

Changes in Blood Volume Affect Blood Pressure

46. If the volume of blood circulating through the system decreases, blood pressure
    (increases/decreases?).

    ▶ As an analogy for blood volume exerting pressure, think of the tension on the wall of a water-filled
      balloon. One way to adjust pressure is to add or remove water from the balloon.

47. Which organ is responsible for decreasing blood volume?

48. Which two systems of the body are responsible for homeostatic regulation of blood pressure?
    (Fig. 15-9)

49. TRUE/FALSE? Defend your answer.

    If blood volume decreases, the kidneys can increase blood volume by reabsorbing water.

50. Name two ways the cardiovascular system tries to compensate for a decrease in blood volume.
    (See Fig. 14-31.)

51. If arterial pressure falls, venous constriction mediated through increased (sympathetic/parasym-
    pathetic?) activity will have what effect on blood distribution and blood pressure?

    ▶ See Fig 15-10 for a summary map of factors affecting MAP.

RESISTANCE IN THE ARTERIOLES

IP *Cardiovascular—Factors That Affect Blood Pressure*

52. Which vessels are the main site of variable resistance in the systemic circulation?

53. What property of these vessels permits them to change resistance?

54. Write the mathematical expression for the relationship between radius (r) and resistance (R).

55. What are the goals of local control of arteriolar resistance?

56. What are the goals of reflex control of arteriolar resistance?

57. What are the goals of hormonal control of arteriolar resistance?

▶   See Table 15-2 for a list of chemicals that mediate arteriolar resistance by producing vasoconstriction or vasodilation.

Myogenic Autoregulation Automatically Adjusts Blood Flow

58. What is myogenic autoregulation?

59. Diagram the mechanism of myogenic autoregulation.

60. When blood pressure in an arteriole increases, myogenic regulation causes the arteriole to

_____.

Paracrines Alter Vascular Smooth Muscle Contraction

61. Tissue and endothelial paracrines locally control arteriole resistance. List some of the important vasoactive paracrines and indicate their physiological role. (Table 15-2)

62. Diagram the reflex pathway for active hyperemia. (Fig. 15-11a)

63. Diagram the reflex pathway for reactive hyperemia. (Fig. 15-11b)

The Sympathetic Branch Controls Most Vascular Smooth Muscle

694. Most systemic arterioles are innervated by sympathetic neurons. A notable exception is the erection reflex of the _____ and _____. This reflex is controlled by _____.

65. Tonic norepinephrine + ____ receptors = myogenic tone (Fig. 15-12)

66. Norepinephrine or epinephrine + ____ receptors = reinforcement of vasoconstriction

67. Which catecholamine is favored by α receptors? (See Ch. 11.)

68. Fight-or-flight: a) epinephrine + ____ receptors (smooth muscle of heart, liver, skeletal muscle arterioles)

   b) epinephrine + ____ receptors (other arterioles)

▶ Fig. 15-13 summarizes the many factors that influece blood flow in the body.

DISTRIBUTION OF BLOOD TO THE TISSUES

69. Why don't all tissues get equal blood flow at all times?

70. At rest, which four organ systems receive most blood flow? (Fig. 15-13)

71. Are arterioles arranged in series or parallel? (Fig. 15-1)

72. At any given moment, the total blood flow through all arterioles = _____.

73. Flow through individual arterioles depends on _____. (Fig. 15-15)

74. When resistance of an arteriole increases, its blood flow (increases/decreases?).

75. When blood flow decreases through one set of arterioles, where does that blood go?

76. Capillary blood flow can be regulated by precapillary _____.

EXCHANGE AT THE CAPILLARIES

IP Cardiovascular—Autoregulation and Capillary Dynamics

77. What determines capillary density in a tissue? Which tissues have the highest capillary density?

78. Describe the physical characteristics of capillaries. (Fig. 15-2)

79. Compare the structure and function of continuous and fenestrated capillaries. (Fig. 15-17)

80. Which three tissues don't have traditional capillaries? What do they have? Why are these modified vessels necessary?

Velocity of Blood Flow Is Lowest in the Capillaries

81. Explain the relationship between total cross-sectional area and velocity of flow in the circulatory system. Specifically, how does the total cross-sectional area of capillaries compare to that of larger-diameter blood-vessels, and what effect does this have on velocity in the different vessels? (Fig. 15-18)

Most Capillary Exchange Takes Place by Diffusion and Transcytosis

82. What are the options for exchange at the capillary?

83. For substances that diffuse freely across capillary walls, what factor is most important for determining the rate of diffusion?

84. The pores of capillaries are too small to allow proteins to pass through them. How then do protein hormones and other essential proteins move out of the blood and into the interstitial fluid? (Fig. 15-17b; also see Ch. 5)

Capillary Filtration and Reabsorption Take Place by Bulk Flow

85. Define bulk flow.

86. Distinguish between filtration and absorption in capillaries.

87. What forces regulate capillary bulk flow?

88. What creates the osmotic pressure gradient between the plasma and the interstitial fluid?

89. What is colloid osmotic pressure (π)?

90. What happens to colloid osmotic pressure along the length of the capillary? (Fig. 15-19a)

91. Hydrostatic pressure pushes water (into/out of?) capillaries. This pressure decreases along the

     length of the capillary as energy is lost to _____. (Fig. 15-19a)

92. How is net fluid flow determined? Compare net fluid flow at the arterial end of a capillary with
     net fluid flow at the venous end. (Fig. 15-19a)

93. Is filtration in capillaries exactly equal to absorption? Explain. (Fig. 15-19a)

THE LYMPHATIC SYSTEM

IP *Fluids & Electrolytes—Electrolyte Homeostasis, Edema*

94. Name the three systems with which the lymphatics interact and explain the role of the lym-
     phatics in each system.

95. Compare the anatomy of the lymphatic system to that of the circulatory system. (Figs. 15-19b,
     15-20)

96. Bulk flow of fluid, proteins, and bacteria is (into/out of?) lymph capillaries.

97. What is lymph? What are lymph nodes?

98. Where does lymph rejoin the blood?

99. Name the factors that influence fluid flow through the lymphatics. (Does the lymph system have a pump like the heart?)

100. What is edema?

101. Explain why disruption of the osmotic gradient between the plasma and the interstitial fluid causes edema.

Edema Results from Alterations in Capillary Exchange

102. Outline the mechanisms behind two different causes of edema.

REGULATION OF BLOOD PRESSURE

IP Cardiovascular—Blood Pressure Regulation

103. Where in the brain is the main integrating center for regulation of blood pressure homeostasis?

The Baroreceptor Reflex Is the Primary Homeostatic Control for Blood Pressure

104. What type of sensory receptor responds to changes in blood pressure? (Fig. 15-22)

105. Where are the two main receptors for blood pressure located? What is significant about these locations?

106. If you are monitoring the electrical activity of the sensory neurons linking these baroreceptors to the cardiovascular control center, would you observe any electrical activity when a person's blood pressure is in the normal range? Are these receptors tonic or phasic?

107. Identify the components of the baroreceptor reflex in response to increased blood pressure. (Figs. 15-22, 15-23)

     stimulus

     receptor

afferent path

integrating center

all efferent pathways

all effectors; match the effectors to their efferent pathways

responses of the effectors

systemic response

108. A decrease in blood pressure results in (increased/decreased?) sympathetic activity and (increased/decreased?) parasympathetic activity.

109. An increase in sympathetic activity will have what effect on heart rate, force of contraction, and arteriolar diameter?

110. An increase in parasympathetic activity will have what effect on heart rate, force of contraction, and arteriolar diameter?

111. Vasoconstriction will (increase/decrease?) peripheral resistance and (increase/decrease?) blood pressure.

112. List, and briefly describe, other mechanisms, in addition to the baroreceptor reflex, that modulate cardiovascular function.

113. Explain the integration of breathing rate and cardiac output.

Orthostatic Hypotension Triggers the Baroreceptor Reflex

114. What is orthostatic hypotension? Why does blood pressure initially fall when standing up after lying flat? (Fig. 15-24)

115. Map the reflex response to orthostatic hypotension. (Fig. 15-24) Be sure to include all the steps of the reflex pathway (stimulus, receptor(s), afferent path, integrating center, efferent pathways, all effectors matched to their efferent pathways, responses of the effectors, systemic response).

CARDIOVASCULAR DISEASE

116. Why are millions of dollars yearly spent trying to determine causes and optimal treatment of cardiovascular disease?

Risk Factors Include Smoking and Obesity

117. List the uncontrollable risk factors for cardiovascular disease.

118. List the controllable risk factors for cardiovascular disease.

119. Explain how blood elevated lipids and diabetes mellitus have both an uncontrollable genetic component and a modifiable lifestyle component.

Atherosclerosis Is an Inflammatory Process

120. What is atherosclerosis? Describe how it arises. What disease is a result of atherosclerosis? (Fig. 15-25)

121. What is LDL-cholesterol and what is its normal function? Why is elevated LDL-cholesterol in the blood undesirable?

122. What is HDL-cholesterol and why is it treated differently than LDL-cholesterol?

123. Compare stable plaques and vulnerable plaques and describe their role in cardiovascular disease.

124. Why is atherosclerosis considered an inflammatory process?

125. Describe how a blood clot in a coronary artery can lead to the development of an arrhythmia or a myocardial infarction.

Hypertension Represents a Failure of Homeostasis

126. Hypertension means chronically elevated blood pressure, with systolic pressures greater than

     _____ mm Hg or diastolic pressures greater than _____ mm Hg. For every 20/10 mm Hg

     increase in blood pressure over a baseline of 115/75, the risk for CVD _____. (Fig. 15-26)

127. Differentiate between essential (primary) hypertension and secondary hypertension.

128. Explain why we say that hypertension represents failure of homeostasis.

129. How does hypertension contribute to atherosclerosis?

130. Why does high arterial blood pressure put additional strain on the heart?

131. How do you explain the fact that stroke volume remains constant in hypertensive patients?

132. What is congestive heart failure? How does it arise, and what are its effects on the body?

133. List some of the common treatments for hypertension.

TALK THE TALK

absorption
active hyperemia
angiogenisis
angiotensin II
apoB
atherosclerosis
atrial natriuretic peptide (ANP)
baroreceptor reflex
baroreceptors
basal lamina
blood-brain barrier
bulk flow
calcium channel blockers
cerebral hemorrhage
colloid osmotic pressure
congestive heart failure
continuous capillaries
coronary heart disease
diastolic pressure
diastole
edema
elastic recoil
endothelium
essential (primary)
    hypertension
fatty streak
fenestrated capillaries
fibroblast growth factor (FGF)

filtration
foam cells
high-density lipoprotein-
    cholesterol (HDL-C)
histamine
hypertension
hypertrophies
hypotension
intima
kinins
Korotkoff sounds
lesion
low-density lipoprotein-
    cholesterol (LDL-C)
lymph
lymph nodes
mean arterial pressure (MAP)
medullary cardiovascular con-
    trol centers
metarterioles
microcirculation
muscle tone
oncotic pressure
orthostatic hypotension
perfusion
peripheral resistance
plaques
precapillary sphincters

pre-eclampsia
pressure reservoir
pulmonary edema
pulse
pulse pressure
reactive hyperemia
respiratory pump
retinopathy
serotonin
sinusoids
skeletal muscle pump
sphygmomanometer
stable plaques
stroke
systolic pressure
systole
tunica intima
vascular endothelial growth
    factor (VEGF)
vascular smooth muscle
vasoconstriction
vasodilation
vasovagal response
vasovagal syncope
volume reservoir
vulnerable plaques

QUANTITATIVE THINKING

MAP = Diastolic P + 1/3 (Systolic P – Diastolic P)

Pulse pressure = Systolic pressure – Diastolic pressure

MAP ∝ Cardiac output × Resistance$_{arterioles}$

Flow$_{arteriole}$ ∝ 1/R$_{arteriole}$

1. At age 20, Missy had a blood pressure of 110/70. At 60 years old, she has a blood pressure of 125/82.

    a. In what units are these blood pressures measured?

    b. What are Missy's pulse pressures and mean arterial pressures at 20 and 60 years of age?

    c. Missy does not smoke and does not have other controllable risk factors for cardiovascular disease. Why did her pulse and mean arterial pressures change with age?

2. If the radius of an arteriole increases from 2 μm to 3 μm, how does this affect resistance and blood flow? Explain your answer using qualitative and quantitative terminology.

3. If total peripheral resistance increases and cardiac output does not change, how is MAP affected?

4. Chris has been in training for a triathlon. The kinesiology department decided to study his endurance and put him through some tests. His end-systolic volume was 50 mL, his end-diastolic volume was 160 mL, his heart rate was 140 beats/min, and his arterial blood pressure was 135/78. What was his cardiac output?

PRACTICE MAKES PERFECT

1. Match the blood vessel with its main characteristics.

        _____ arteries   a.   lots of smooth muscle

        _____ arterioles   b.   low compliance and high recoil

        _____ capillaries   c.   high compliance and high recoil

        _____ veins   d.   high compliance and low recoil

                                                e.   contains endothelium only

2. Complete the table below. Your answer should include the relative amounts of the various types of tissues that each vessel contains.

Blood vessel	Physical characteristics	Function(s)
arteries		
arterioles		
capillaries		
veins		

3. A 45-year-old woman has a ventricular systolic pressure of 130. How high must you inflate the cuff of a sphygmomanometer on her left arm in order to stop blood flow through the brachial artery? Explain your reasoning.

4. Compare and contrast the response of a healthy artery with a diseased artery (atherosclerosis) during systole and diastole. How would pulse pressure be affected by atherosclerosis?

5. Even though the radius of a single capillary is smaller than that of an arteriole, the peripheral resistance to blood flow through the capillaries is less than that of blood flow through the arterioles. Explain why this is TRUE.

6. During exercise, blood flow to skeletal muscles is increased, but flow to the digestive system is decreased. How is this achieved?

7. Match the neurotransmitter/neurohormone and receptor with its target(s). Answers may be used more than once and more than one answer may apply to any target.

   a. norepi on α receptors
   b. ACh on nicotinic receptors
   c. epi on β_2 receptors

   d. ACh on muscarinic receptors
   e. norepi on β_1
   f. none of the above

   SA node _____   ventricular myocardium _____   skeletal muscle capillary _____

   cardiac vasculature _____   renal arterioles _____   brain arterioles _____

8. The arterioles of the kidneys constrict as a result of local control mechanisms. Assuming that no compensatory homeostatic mechanisms are triggered, what happens (increases/decreases/no change) to each of the following? Be able to defend your answer.

   blood flow through the kidneys

   mean arterial pressure

   blood flow through skeletal muscle arterioles

   cardiac output

   total peripheral resistance

   blood flow through the venae cavae

   through the lungs

9. TRUE/FALSE? Defend your answer.

   In a fight-or-flight reaction, epinephrine from the adrenal cortex will combine with β_1 receptors in the heart and cause vasodilation.

10. You are a doctor and have just prescribed a calcium-channel blocking drug for a patient with high blood pressure. The patient asks how the drug works at a molecular and cellular level. How would you answer?

11. The figure below shows the cardiovascular control center (CVCC) in the medulla oblongata, the heart, aortic arch, and carotid artery and carotid sinus, and an arteriole with a capillary bed. Cardiac output can be influenced by reflexes that alter heart rate, force of contraction, and peripheral resistance. Draw the anatomical components of the reflex pathways, including sensory receptors, sensory neurons, integrating centers, and efferent neurons that control cardiac function and peripheral blood pressure. Use different colors to represent different parts of the system. Where neurons terminate on targets, write in the appropriate neurotransmitters and receptors.

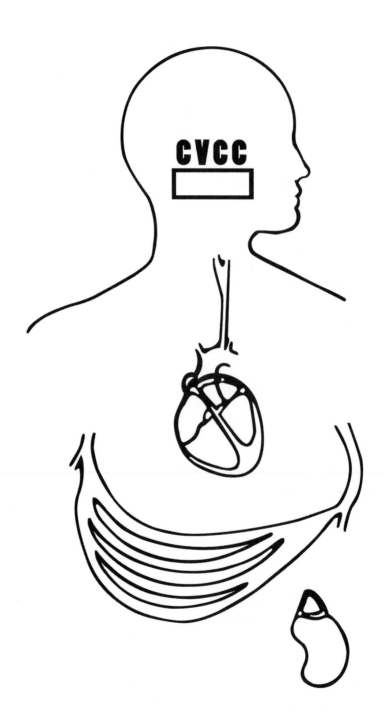

12. Fill in the following reflex pathways.

stimulus <u>increased blood pressure</u>

receptor _____   afferent path _____

integrating center _____   efferent path 1_____

effector 1 _____   efferent path 2 _____

effector 2_____   tissue/organ response 1 _____

tissue/organ response 2 _____

systemic response <u>decreased blood pressure</u>

13. A man has developed thromboangiitis obliterans, a condition in which the arteries in his legs (only) have become calcified and partially obstructed. He comes to his physician complaining of pain when walking; the pain subsides when he stops walking and rests. The man's blood pressure, taken in his left arm, is normal.

a.   What is the blood flow in his legs compared to normal? On the basis of this answer, explain why he has pain when walking but not at rest.

b.   What would happen to blood flow in this man's legs if you: (answer and explain your reasoning)

administer a peripheral vasodilator?

cut the sympathetic nerves innervating the blood vessels in his legs?

MAPS

Reckless Ronnie was brought to the emergency room following a motorcycle accident. His blood pressure was 80/25 and pulse was 135/min. Tests showed the presence of intra-abdominal bleeding. Draw a *complete* reflex map that explains all Ronnie's physical findings, beginning with bleeding as the stimulus. Use the following terms and add any additional terms you wish to add.

α receptor

β_1 receptor

β_2 receptor

acetylcholine

aorta

baroreceptor

cardiac output

cardiovascular control center

carotid artery

decreased blood pressure

decreased blood volume

end-diastolic volume

heart rate

hemorrhage

increased blood pressure

norepinephrine

parasympathetic activity

peripheral resistance

sensory neuron

stroke volume

sympathetic activity

vasoconstriction

venous return

BEYOND THE BASICS

TRY IT

Reactive Hyperemia

A temporary increase in blood flow into a tissue that has been deprived of flow can be easily demonstrated without any equipment. Wrap the fingers of your right hand around the base of your left index finger. Squeeze tightly for at least one minute to shut off blood flow into the finger. It is important to shut off as much arterial flow as you can. After one minute, release the finger and watch for color changes. It should flush red for a few seconds, then gradually fade to normal color as the vasodilators are washed away by the restored blood flow.

The Baroreceptor Reflex

You can demonstrate the baroreceptor reflex easily with a friend. Find the subject's pulse at the radial artery of the wrist. While monitoring the pulse, have the subject find the pulse point in the carotid artery (just to the side of the Adam's apple) and press *gently* on it for a few seconds. You should notice a decrease in the subject's pulse as the increased pressure created by pressing on the carotid artery is sensed by the cardiovascular control center. The baroreceptor reflex is one reason that taking a carotid pulse in an exercise class is not the most accurate indicator of heart rate.

Additional Reading

The following documents can be found online (free) at this URL: opa.faseb.org/pages/Publications/breakthroughs.htm

- ▶ Dustan H. P., Roccella, E. J., and Garrison, H. H. "Controlling hypertension: A research success story." *Archives of Internal Medicine 156*(17) (Sept. 23, 1996): 1926–1935.

- ▶ Porro, J. "Cardiovascular disease and the endothelium." opa.faseb.org/pages/Publications/breakthroughs.htm.

- ▶ Patlak, M. "From viper's venom to drug design: Treating hypertension." *The FASEB Journal 18* (2004): 421E.

LEARNING OBJECTIVES

When you complete this chapter, you should be able to:

▪ Describe the composition of plasma and list the major functions of plasma proteins.

▪ Map the differentiation of blood's cellular elements, starting from a pluripotent hematopoietic stem cell. Include key cytokines involved in development as well as the function(s) and distinguishing characteristics of each cellular element (cover all five types of leukocytes).

▪ Define hematocrit, describe how a person's hematocrit is determined, and identify the clinical relevance of this value.

▪ Describe the molecular structure of hemoglobin and create a map of iron metabolism and hemoglobin synthesis.

▪ Distinguish between the different types of anemias.

▪ Diagram the key steps of hemostasis, the coagulation cascade, and fibrinolysis.

SUMMARY

This chapter gives some detail about our blood, its components and their functions. To help you organize these details, consider making a large map where the main headings are the components of blood (e.g., plasma, RBCs, WBCs, platelets).

Plasma resembles interstitial fluid except that it contains plasma proteins. Plasma proteins include albumins, globulins, fibrinogen, and immunoglobulins. Plasma proteins raise the osmotic pressure of plasma, thus pulling water from the interstitial fluid into the capillaries (see Ch. 15).

Red blood cells (RBCs, erythrocytes) transport O_2 and CO_2 between the lungs and tissues with the help of hemoglobin (Hb). Hb production requires that iron be ingested in the diet. Mature RBCs are biconcave disks that lack a nucleus and membranous organelles. Thus, they are essentially membranous bags filled with enzymes and Hb. Because RBCs lack a nucleus, they cannot make new proteins or membrane molecules and therefore die after about 120 days. Their flexible membrane and flattened shape allow RBCs to change shape in response to osmotic fluctuations.

White blood cells (WBCs, leukocytes) consist of five mature cell types: lymphocytes, monocytes, neutrophils, eosinophils, and basophils. Neutrophils and monocytes (macrophage precursors) are collectively called phagocytes because of their ability to ingest foreign particles. Basophils, eosinophils, and neutrophils are called granulocytes because of cytoplasmic inclusions that give a granular appearance.

Platelets are anuclear cell fragments that have broken off from megakaryocytes in the bone marrow. Platelets are always present in the blood, but they are inactive unless there is damage to the circulatory system. Ruptured blood vessels expose collagen fibers that activate platelets, and the activated platelets initiate coagulation with the goal of hemostasis.

All blood cells are derived from pluripotent hematopoietic stem cells that develop according to chemical signals into the mature cells already discussed. These chemical signals include cytokines (see Ch. 6), growth factors, and interleukins. Table 16-2 lists some of these chemicals and how they affect blood cell development. Some examples include erythropoietin, which influences erythropoiesis; various colony-stimulating factors, which influence leukopoiesis; and thrombopoietin, which influences platelet production.

TEACH YOURSELF THE BASICS

PLASMA AND THE CELLULAR ELEMENTS OF BLOOD

1. How many liters of blood does our 70 kg man have? What percentage of his body weight does this comprise?

Plasma Is Composed of Water, Ions, Organic Molecules, and Dissolved Gases

2. What is plasma? (Fig. 16-1)

3. Describe the composition of plasma.

4. Compare plasma with interstitial fluid.

5. Where are most plasma proteins made? In what functions do plasma proteins participate?

6. Name four main groups of plasma proteins and give their functions. (Table 16-1)

7. What role do plasma proteins play in capillary filtration? (See Ch. 15.)

Cellular Elements Include Red Blood Cells, White Blood Cells, and Platelets

8. List the three main cellular elements of the blood and describe their primary function(s).

9. Why do we call them "cellular elements" rather than "cells"?

10. What are the "formal" names for red blood cells and white blood cells?

11. What are the parent cells of platelets called?

12. List the five mature WBCs found in blood and give the function(s) of each type.

13. Which of the WBCs are known as phagocytes, and why are they given that name? (See Ch. 5.)

14. Why are lymphocytes also called immunocytes?

15. Which WBCs are called granulocytes and why?

BLOOD CELL PRODUCTION

16. What are pluripotent hematopoietic stem cells? Where are they found? (Fig. 16-2)

17. Describe the differentiation of pluripotent hematopoietic stem cells into specialized blood cells.

Blood Cells Are Produced in the Bone Marrow

18. Define hematopoiesis. Where does it take place in embryos, children, and adults?

19. Describe red bone marrow.

20. Describe yellow bone marrow.

21. Of all blood cells produced, _____% will become RBCs and _____% will become WBCs.

22. Compare the lifespans of RBCs and WBCs.

Hematopoiesis Is Controlled by Cytokines

23. What are cytokines? (See Ch. 6.)

24. What are some cytokines involved in hematopoiesis? Describe their roles.

▷ Table 16-2 lists some cytokines involved in hematopoiesis.

Colony-Stimulating Factors Regulate Leukopoiesis

25. Where are the CSFs that regulate reproduction and development of WBCs made?

26. How do existing WBCs participate in the regulation of leukopoiesis? Why is this clever?

27. What is a differential white cell count? What is its clinical utility? (Fig. 16-3)

28. What is leukemia?

29. What is neutropenia?

Thrombopoietin Regulates Platelet Production

30. What is thrombopoietin (TPO) and where is it made?

Erythropoietin Regulates Red Blood Cell Production

31. What is RBC production called?

32. What is erythropoietin (EPO) and where is it made?

33. What is the stimulus for EPO synthesis and release?

▷ Share your thoughts on the potential therapeutic value of various recombinant hematopoietic cytokines (like EPO, filgrastim, or sargramostim).

RED BLOOD CELLS

34. What is the red blood cell count of a μL of whole blood? How does this compare to white blood cell and platelet count?

35. Define hematocrit. How is it determined? (Fig. 16-3)

Mature Red Blood Cells Lack a Nucleus

36. Name two immature forms of an erythrocyte. How are they different from the mature RBC? (Fig. 16-4c)

37. Describe the structure and contents of the mature RBC. (Figs. 16-5, 16-6b)

38. Without mitochondria, RBCs cannot carry out (anaerobic/aerobic?) metabolism.

39. What is the primary source of ATP for mature RBCs?

40. How is the lack of a nucleus related to the limited life span of the RBC?

41. What holds the RBC in its unique shape? (Fig. 16-5c)

42. Why does an erythrocyte need to be flexible?

43. When placed in a hypotonic solution, an RBC will do what? What will it do in a hypertonic solution? (Fig. 16-6; also see Ch. 5)

44. What clues can RBC morphology give us about disease states?

Hemoglobin Synthesis Requires Iron

45. Describe the structure of hemoglobin (Hb).

46. Describe hemoglobin synthesis and metabolism. (Fig. 16-7)

Red Blood Cells Live About Four Months

47. What is the average life span of an RBC?

48. How are old RBCs destroyed and what happens to the RBC components?

49. What is the relationship between heme, bilirubin, and bile? (Fig. 16-7)

50. How are bilirubin and its metabolites excreted?

51. What is jaundice?

Red Blood Cell Disorders Decrease Oxygen Transport

52. List four common causes of anemia. (Table 16-3)

53. Why are people with anemia often weak and fatigued?

54. Anemias in which the RBCs are destroyed at a high rate are called _____ anemias.

55. Lack of adequate iron in the diet results in _____ anemia.

56. Smaller-than-normal RBCs are said to be _____, while paler-than-normal

     RBCs are said to be _____.

57. What makes the hemoglobin of sickle cell disease abnormal? (Fig. 16-8)

58. In polycythemia vera, patients have higher than normal RBC production. Why is this harmful? (See Ch. 14.)

59 What is relative polycythemia?

PLATELETS AND COAGULATION

60. What is the challenge our bodies must overcome when repairing a damaged blood vessel?

61. What temporarily seals the hole in a broken blood vessel until it can be repaired?

Platelets Are Small Fragments of Cells

62. Describe how platelets are formed from megakaryocytes. (Figs. 16-4c, 16-9a)

63. What intracellular components do platelets contain?

64. What is the typical lifespan of a platelet?

65. Why are megakaryocytes polyploid? (Fig. 16-9a)

Hemostasis Prevents Blood Loss from Damaged Blood Vessels

66. Define hemostasis. What are the three major steps in this process? (Fig. 16-10)

67. What is the difference between platelet adhesion and platelet aggregation?

68   Give an overview of the coagulation cascade. What initiates the cascade? What pieces are involved?

Platelet Activation Begins the Clotting Process

69   Describe how platelets are activated, both directly and indirectly. (Figs. 16-10, 16-11)

70.   What vasoconstrictive compounds are released/created as platelets are activated and begin adhering to exposed collagen fibers? (Table 16-4)

71.   Activated platelets release chemicals that activate additional platelets. This is an example of what kind of pathway?

72.   How is formation of the platelet plug restricted to the damaged region? (Fig. 16-11)

Coagulation Converts a Platelet Plug into a Clot

73.   Outline the process of coagulation. (Use a separate sheet of paper.) (Fig. 16-12)

74   How does initiation of the intrinsic pathway differ from initation of the extrinsic pathway? (Fig. 16-12)

75.   Where do the two pathways unite, and what is significant about the enzyme activated at this step? (Fig. 16-13)

76.   Why is it necessary to cross-link fibrin?

77.   What is fibrinolysis? How is this process accomplished? (Fig. 16-13)

Anticoagulants Prevent Coagulation

78.   What are two mechanisms that limit the extent of blood clotting within a vessel? (Table 16-6)

79. What are anticoagulants? How do they work?

80. List some known anticoagulants, both those produced by our body and others—indicate whether each is endogenous or exogenous.

81. What is hemophilia? What are some causes of the disorder? What are some of the treatments?

BEYOND THE PAGES

The following documents can be found online (free) at this URL: opa.faseb.org/pages/Publications/breakthroughs.htm.

▶ Roberts, S. S. "Blood safety in the age of AIDS." *FASEB Journal 10*(4) (Mar. 1996): 391–402.

▶ Delude, C. "Clot busters! Discovery of thrombolytic therapy for heart attack & stroke." *FASEB Journal 19* (2005): 671.

▶ Patlak, M. "Targeting leukemia: From bench to bedside." *FASEB Journal 16* (2002): 273.

TALK THE TALK

albumins
anemia
antibodies
anticoagulants
antithrombin III
basophils
bile
bilirubin
bone marrow
clot
coagulation cascade
colony-stimulating factors
   (CSFs)
common pathway
coumarin
differential white cell count
eosinophils
erythroblasts
erythrocytes
erythropoiesis
erythropoietin (EPO)
extrinsic pathway
ferritin

fibrin
fibrinogen
fibrinolysis
globulins
granulocytes
hematocrit
hematological disorders
hematopoiesis
hemoglobin
hemolytic anemias
hemophilia
hemorrhage
hemostasis
heparin
hereditary spherocytosis
hypochromic
hypoxia
hypoxia-inducible factor 1
   (HIF-1)
immunocytes
immunoglobulins
infarct
integrins

interleukins
intrinsic pathway
iron-deficiency anemia
jaundice
leucopoiesis
leukemias
leukocytes
macrophages
mast cells
mean red cell volume (MCV)
megakaryocytes
monocytes
morphology
myocardial infarctions
neutropenias
neutrophils
packed red cells
phagocytes
plasma proteins
plasmin
platelet adhesion
platelet aggregation
platelet plug

platelet-activating factor (PAF)
platelets
pluripotent hematopoietic
   stem cell
polycythemia vera
polyploidy
progenitor cells
prostacyclin
protein C

red blood cells
relative polycythemia
reticulocyte
serotonin
sickle cell disease
thrombin
thrombocytes
thrombopoietin (TPO)
thromboxane A2

thrombus
tissue factor
tissue plasminogen activator
   (tPA)
tissue thromboplastin
transferrin
uncommitted stem cells
warfarin
white blood cells

PRACTICE MAKES PERFECT

1. Describe two major functions of blood.

2. Give at least two characteristics that are used to identify different types of leukocytes.

3. A blood sample from a patient shows normal white cell count, low red cell count, and more than normal reticulocytes present in the blood. Would you suspect that this person's problem is a result of a defect in the bone marrow or a problem with the circulating red blood cells? Defend your choice.

4. A person has a total blood volume of 4.8 L and a hematocrit of 40%. What is her plasma volume?

5. Which one of the following is FALSE?

   Neutrophils, eosinophils, and basophils

   a. are leukocytes.
   b. are lymphocytes.
   c. are granulocytes.
   d. are polymorphonuclear (have bi-lobed or tri-lobed nucleus).

6. Erythropoietin (EPO) is traditionally considered to be a hormone. Based on what you have learned about EPO, is it most accurately classified as a hormone or as a cytokine?

MAPS

1. Create a map showing hemostasis and coagulation. Include the terms below and any others you wish to add.

blood	fibrin	prostacyclin
clot	fibrinogen	thrombin
coagulation	fibrinolysis	thrombocyte
coagulation cascade	hemostasis	thrombus
collagen	intrinsic path	tissue factor
cytokine	plasmin	tissue plasminogen activator
endothelium	platelet	(tPA)
extracellular matrix	platelet adhesion	vasoconstriction
extrinsic path	platelet aggregation	
factor	platelet plug	

2. Create a map of the blood cells, their synthesis, and their basic functions, using the terms below and any others you wish to add.

basophil	granulocyte	neutrophil
blood	hematopoiesis	phagocyte
bone marrow	immunocyte	phagocytosis
colony-stimulating factor (CSF)	immunoglobulin	platelet
cytokine	interleukin	reticulocyte
eosinophil	leukocyte	stem cell
erythroblast	lymphocyte	stroma
erythrocyte	macrophage	thrombocyte
erythropoiesis	megakaryocyte	thrombopoietin
erythropoietin (EPO)	monocyte	umbilical cord blood
	multipotent progenitor	

3. Create a map for red blood cell synthesis and destruction, using the terms below and any others you wish to add.

amino acids	erythropoietin (EPO)	iron
bile	ferritin	jaundice
bilirubin	fetal hemoglobin	protein
erythroblast	hematopoiesis	transferrin
erythrocyte	heme group	
erythropoiesis	hemoglobin	

Mechanics of Breathing

LEARNING OBJECTIVES

When you complete this chapter, you should be able to:

- List the key functions of the respiratory system.

- Define all terms in the following mathematical relationships and explain how each is relevant to respiratory physiology:

  - Flow $\propto \Delta P/R$

  - $R \propto L\eta/r4$

  - Partial pressure of an atmospheric gas = $P_{atm} \times$ % of gas in atmosphere

  - $P_1V_1 = P_2V_2$

  - $P = 2T/r$

- Apply these mathematical relationships to predict the behaviors of gases and potential changes in air flow if variables were altered.

- Diagram the anatomy of the respiratory system and identify the role each structure plays in external respiration (Creates a pressure change? Site of gas exchange? Etc.).

- Define and describe the lung volumes and lung capacities.

- Diagram the alveolar and intrapleural pressure changes that occur during inspiration and expiration.

- Explain the significance of intrapleural pressure and use the example of pneumothorax to contrast normal intrapleural conditions.

- Compare and contrast compliance and elastance in respiratory physiology. Give examples of disease states that arise from changes in compliance and/or elastance.

- Explain the role of surfactants in respiratory physiology.

- Diagram the factors affecting airway resistance and highlight the local and reflex control mechanisms involved with variable resistance (bronchodilation and bronchoconstriction).

- Define anatomic dead space and diagram how it affects ventilation.

- Compare and contrast total pulmonary ventilation and alveolar ventilation.

- Explain why gas composition in the alveoli remains relatively constant during normal breathing, and demonstrate how it might change during other breathing patterns.

- Contrast pulmonary capillaries with other capillaries in the body.

- Diagram the mechanisms by which ventilation and alveolar blood flow are matched.

- Identify key diseases and conditions that can affect respiratory function.

SUMMARY

Cellular respiration, which you studied in Ch. 4, refers to the metabolic processes that consume oxygen and nutrients and produce energy and CO_2. External respiration is the exchange of gases between the atmosphere and the cells. Ventilation (breathing) is the process by which air is moved into and out of the lungs. O_2 is transported via the blood to cells, and CO_2 is removed by the blood and taken to the lungs.

Air movement in the respiratory system highlights its anatomy: air goes from nasopharynx to trachea to bronchi, bronchioles, and finally alveoli. The lungs are contained within the thoracic cage. Each lung is surrounded by a double-walled pleural sac, and the pleural fluid that exists between the pleural membranes holds the lungs against the thoracic wall. This pleural fluid also helps the membranes slip past each other as the lungs move during respiration. The diaphragm, a sheet of skeletal muscle, forms the bottom of the thoracic cage, and its movement creates volume changes that in turn create air movement. The other muscles involved in respiration include the intercostal muscles, the scalenes, the sternocleido-mastoids, and the abdominals.

Air moves according to a set of physical laws collectively known as the gas laws. Boyle's law describes how, in a closed system, volume increases as pressure decreases (and vice versa). Air moves from areas of higher pressure to areas of lower pressure. Therefore, when the thoracic volume is increased by respiratory muscle movement, the pressure in the thoracic cavity drops and air moves in down its pressure gradient (inspiration). Likewise, as thoracic volume decreases, the alveolar pressure increases and air moves out of the body (expiration).

Movement of individual gases depends on their partial pressure gradients. Just as water and solutes move down their concentration gradients, so do gases. Dalton's law describes how the total pressure of a gaseous mixture is the sum of the pressures of the individual gases. The pressure of an individual gas is called a partial pressure, and gases move from higher partial pressures to lower partial pressures. Blood flow and alveolar ventilation are closely matched, with partial pressures serving as primary stimuli, to ensure efficient delivery and removal of gases.

Other respiratory functions include pH regulation, vocalization, and protection from foreign substances. Be sure you understand the graphs of this chapter. They will help you create a visual explanation of respiratory function. Also, be sure you understand the ways in which the cardiovascular system and respiratory system are integrated.

TEACH YOURSELF THE BASICS

1. List four primary functions of the respiratory system.

2. What is lost from the body through the respiratory system besides carbon dioxide?

THE RESPIRATORY SYSTEM

IP *Respiratory System: Anatomy Review*

3. Distinguish between cellular respiration (see Ch. 4) and external respiration. (Fig. 17-1)

4. List the four integrated processes of external respiration (three exchanges and one transport).

5. Distinguish between inspiration and expiration.

6. Name the three major components of the respiratory system. (Fig. 17-2)

7. Name the structures of the upper and lower respiratory system.

Bones and Muscles of the Thorax Surround the Lungs

8. What bones and muscles form the thoracic cage? The floor? (Fig. 17-2a)

9. Name the two additional sets of muscles associated with the thoracic cage.

10. Name the three sacs enclosed within the thorax. What is in each sac? (Fig. 17-2d)

11. What thoracic structures are not contained within these three sacs?

Pleural Sacs Enclose the Lungs

12. The lungs are light, spongy tissue mostly occupied by _____-filled spaces. (Fig. 17-2a, c)

13. What is the relationship between the lungs, the pleura, and the pleural fluid? (Fig. 17-3)

14. What purposes does pleural fluid serve?

Airways Connect Lungs to the Environment

15. Following an oxygen molecule from the air to the exchange epithelium of the lung, name each structure the molecule passes. (Fig. 17-2)

16. As the molecule moves into the airways, the diameter of the airways gets progressively smaller and the total cross-sectional surface area of the airways (increases / decreases?). (Fig. 17-4)

17. The velocity of air flow is highest in the _____ and lowest in the _____.
    (See Chs. 14–15.)

The Airways Warm, Humidify, and Filter Inspired Air

18. Describe the reasons why and the mechanisms by which the airways warm, humidify, and filter inspired air. (Fig. 17-5)

19. What conditions/diseases can occur if these processes are not functioning properly?

Alveoli Are the Site of Gas Exchange

20. Describe the structure of the alveoli. (Fig. 17-2f, g)

21. Describe and give the functions of the two types of epithelial cells in alveoli. (Fig. 17-2g)

22. Describe the composition of the alveolar walls and surrounding connective tissue. (Fig. 17-2g)

23. Describe the association of the alveoli and the circulatory system. (Fig. 17-2h)

The Pulmonary Circulation Is a High-Flow, Low-Pressure System

24. Trace a drop of blood through the pulmonary circulation from the (left/right?) ventricle to the (left/right?) atrium.

25. Compare the following aspects of the pulmonary circulation to those of the systemic circulation:

    volume of blood in the pulmonary vessels

    (of this volume, how much is participating in gas exchange at any moment?)

    total blood flow through the lungs in liters per minute

    pulmonary arterial pressure

    What effect does lower mean pulmonary blood pressure have on capillary fluid exchange? (See Ch. 15.)

GAS LAWS

▶ Gas laws are given in Table 17-1.

IP *Respiratory System: Pulmonary Ventilation*

26. Although we can draw many comparisons between air flow and blood flow, air and blood differ significantly in what way?

27. When we use an atmospheric pressure of 0 mm Hg, what is that value equivalent to? Why do we use this convention?

Air Is a Mixture of Gases

28. State Dalton's law.

▶ Table 17-2 summarizes partial pressures of atmospheric gases.

29. How do you calculate the partial pressure of a single gas in a mixture?

30. What happens to the partial pressures of individual gases if dry air is suddenly humidified? (Table 17-2)

Gases Move Down Pressure Gradients

31. What role does muscle contraction play in the creation of air flow in the respiratory system?

32. Describe movement of individual gases with regard to their partial pressure.

Boyle's Law Describes Pressure-Volume Relationships of Gases

33. What factors contribute to gas pressure in a closed container?

34. Describe the pressure-volume relationship of Boyle's law. (Fig. 17-6)

35. For gases in a closed container, as volume (increases/decreases?), pressure (increases/decreases?).

36. How does the respiratory system create volume changes?

VENTILATION

IP *Respiratory System: Pulmonary Ventilation*

37. Define ventilation.

Lung Volumes Change During Ventilation

38. What does a spirometer do? (Fig. 17-7)

Lung Volumes

39. Name, give the abbreviations for, and describe the four lung volumes. (Fig. 17-8)

Lung Capacities

40. What are lung capacities?

41. Name, give the abbreviations for, and describe the four lung capacities. (Fig. 17-8)

During Ventilation, Air Flows Because of Pressure Gradients

42. Name the primary muscles involved in quiet breathing.

43. What is forced breathing?

44. Restate the mathematical relationship between flow, pressure, and resistance and then describe in words the interaction between the three factors. For instance, what happens to air flow when the pressure gradient increases? When resistance decreases?

Inspiration Occurs When Alveolar Pressure Decreases

45. Diagram the role of the rib cage, diaphragm, and inspiratory muscles during inspiration. (Figs. 17-9, 17-10, 17-2b)

46. **Alveolar pressure changes:** Between breaths, is there air flow?

    Therefore, alveolar pressure must be equal to _____. (Fig. 17-11, point A_1)

47. When thoracic volume increases during inspiration, what happens to alveolar pressure?

48. At what point in the respiratory cycle is alveolar pressure lowest?

49. When does alveolar pressure again equalize with the atmospheric pressure? Why?

Expiration Occurs When Alveolar Pressure Exceeds Atmospheric Pressure

50. When thoracic volume decreases during expiration, what happens to alveolar pressure?

51. At what point in the respiratory cycle is alveolar pressure highest?

52. Are the following muscles contracting in passive expiration?

    external intercostal muscles _____   internal intercostal muscles _____

    diaphragm_____   abdominal muscles _____

53. Are the following muscles contracting in active expiration?

    external intercostal muscles _____   internal intercostal muscles _____

    diaphragm_____   abdominal muscles _____

54. What property of the respiratory system is responsible for passive expiration?

55. Give examples of diseases that can affect ventilation.

Intrapleural Pressure Changes During Ventilation

56.   Define intrapleural pressure.

57.   Explain why the intrapleural pressure is normally subatmospheric. (Fig. 17-12a)

58.   Explain why puncturing the pleural membrane causes the lung to collapse and the rib cage to move out. (Fig. 17-12b)

59.   Diagram intrapleural pressure changes during the respiratory cycle. (Fig. 17-11)

Lung Compliance and Elastance May Change in Disease States

60.   Define compliance.

61.   A high-compliance lung (requires additional force to stretch it/is easily stretched?).

62.   Define elastance.

63.   TRUE/FALSE? Defend your answer.

      A high-compliance lung always has high elastance.

64.   What happens to compliance and elastance in emphysema?

65.   Patholigical conditions in which compliance is reduced are called _____ lung diseases.

Surfactant Decreases the Work of Breathing

66. What creates resistance to stretch in the lung? (See Ch.2.)

67. State the law of LaPlace and relate it to surface tension in alveoli.

68. According to the law of LaPlace, if two bubbles have equal surface tension, the (larger/smaller?) will have a higher internal pressure. (Fig. 17-13a)

69. What is the function of surfactants?

70. Which will have a higher concentration of surfactant, a large alveolus or a small one? Explain. (Fig. 17-13b)

71. What cells in the lung secrete surfactant? (Fig. 17-2g)

72. What happens in premature babies who have not produced surfactant?

Airway Diameter Is the Primary Determinant of Airway Resistance

73. Explain the relationship between resistance to air flow, length of airways (L), viscosity of air (η), and radius of airways (r) using Poiseuille's Law. (See Ch. 14.)

74. In the respiratory system, which of these factors is usually the most significant?

75. Where in the airways does air flow normally encounter the highest resistance?

76. What part of the respiratory system is the site of variable resistance?

77. For each of the following, explain whether a bronchiole would react by constricting or dilating. (Table 17-3)

    increased CO_2

    histamine

    parasympathetic innervation

    sympathetic innervation

    epinephrine in circulation

Rate and Depth of Breathing Determine the Efficiency of Breathing

78. Define total pulmonary ventilation and show how it is calculated.

79. Give normal average values for ventilation rate _____ and tidal volume _____ in an adult.

80. Using these values, what is an average value for total pulmonary ventilation? (Give units.)

81. Define anatomic dead space. (Fig. 17-14)

82. How is alveolar ventilation different from total pulmonary ventilation?

83. Why is alveolar ventilation a more accurate indicator of breathing efficiency? How is this volume calculated?

▶ Table 17-4 compares alveolar ventilation differences resulting from different breathing patterns. Table 17-5 describes various patterns of ventilation. Table 17-6 gives normal ventilation values.

Gas Composition in the Alveoli Varies Little During Normal Breathing

84. What happens to P_{O_2} and P_{CO_2} with increased alveolar ventilation (hyperventilation)? (Fig. 17-15)

85. What happens to P_{O_2} and P_{CO_2} with decreased alveolar ventilation (hypoventilation)? (Fig. 17-15)

86   During normal breathing, partial pressures in alveoli remain constant. Why?

Ventilation and Alveolar Blood Flow Are Matched

87.   Explain what is meant by the expression: "Ventilation is matched to perfusion in the lungs."

88.   How are pulmonary capillaries different from other capillaries?

89.   How does this property help the body meet a demand for additional oxygen, such as during exercise?

▶   Local homeostatic mechanisms attempt to keep ventilation and perfusion matched in each section of the lung. (Fig. 17-16; Table 17-7)

90.   When P_{CO_2} of expired air increases, bronchioles (dilate/constrict?). When P_{CO_2} of expired air decreases, bronchioles (dilate/constrict?).

91.   When the tissue P_{O_2} around pulmonary arterioles decreases, the arterioles (dilate/constrict?). When the tissue P_{O_2} around pulmonary arterioles increases, the arterioles (dilate/constrict?). Why is this adaptive?

92.   Compare this response of pulmonary arterioles to that of systemic arterioles. (See Ch. 15.)

93.   Will these local control mechanisms always be able to correct the initial disturbance? Explain.

Auscultation and Spirometry Assess Pulmonary Function

94.   What is auscultation? How is it used in diagnosis of pathologies? (See Ch. 14.)

95.   What is the collective name given to diseases in which air flow is diminished? Give some examples.

TALK THE TALK

active expiration
airways
algorithm
alveolar fluid transport
alveolar pressure
alveolar ventilation
alveoli (singular alveolus)
anatomic dead space
atmospheric pressure
bleb
Boyle's law
bronchioles
bronchoconstriction
bronchodilation
bronchus
bulk flow
cellular respiration
chronic obstructive
    pulmonary disease (COPD)
compliance
conductive system
cystic fibrosis
Dalton's law
diaphragm
dipalmitoylphosphatidyl-
    choline
elastance
expectorate
expiration
expiratory muscles
expiratory reserve volume
    (ERV)
external respiration
fibrosis

fibrotic lung disease
forced expiratory volume in 1
    second (FEV$_1$)
friction rub
functional residual capacity
goblet cells
histamine
hyperventilation
hypoventilation
idiopathic pulmonary fibrosis
immunoglobulins
inspiration
inspiratory capacity
inspiratory muscles
inspiratory reserve volume
    (IRV)
intercostal muscles
intrapleural pressure
kilopascals (kPa)
larynx
law of LaPlace
leukotrienes
lower respiratory tract
mast cells
maximum voluntary
    ventilation
minute volume
mucociliary escalator
myasthenia gravis
newborn respiratory distress
    syndrome (RDS)
obstructive lung disease
obstructive sleep apnea
partial pressure

passive expiration
perfusion
pericardial sac
pharynx
pleura
pleural fluid
pleural membrane
pleural sacs
pneumothorax
polio
positive pressure ventilation
primary bronchi
pulmonary function test
respiratory bronchioles
respiratory cycle
respiratory system
restrictive lung disease
scalenes
spirometer
sternocleidomastoids
surfactant
thoracic cage
thoracic portion
tidal volume (V$_T$)
total lung capacity (TLC)
total pulmonary ventilation
trachea
type I alveolar cells
type II alveolar cells
upper respiratory tract
ventilation
vital capacity (VC)
vocal cords

QUANTITATIVE THINKING

1. The diagram below represents a spirometer tracing.

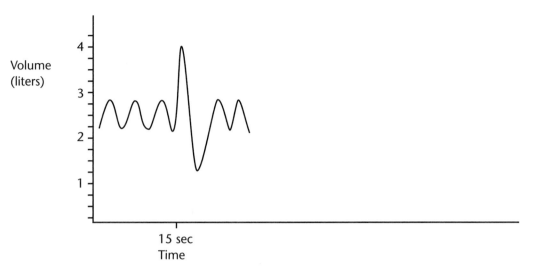

    Using the tracing above, calculate the following:

    total lung capacity _____   vital capacity _____

    expiratory reserve volume_____   residual volume _____

    total pulmonary ventilation, normal breathing (L/min)_____

2. If atmospheric pressure is 720 mm Hg and nitrogen is 78% of the atmosphere, what is the par-
    tial pressure of nitrogen?

3. A student breathes according to the following schedule (assume an anatomical dead space of
    150 mL):

        tidal volume = 300 mL/breath   breath rate = 20 breaths/min

    Calculate her pulmonary ventilation rate and her alveolar ventilation rate.

4. Patient A is breathing 12 times a minute with a tidal volume of 500 mL. Patient B is breathing
    20 times a minute with a tidal volume of 300 mL. Which patient has better alveolar ventila-
    tion? Explain.

PRACTICE MAKES PERFECT

1. Spell out the words for the following abbreviations:

   V_T

   P_{O_2}

   RV

   IRV

2. Match the neurotransmitter/neurohormone and receptor with its target(s). Answers may be used more than once and more than one answer may apply to any target.

   a. norepi on α receptors
   b. ACh on nicotinic receptors
   c. epi on β_2 receptors
   d. ACh on muscarinic receptors
   e. norepi on β_1

   diaphragm _____   external intercostals _____   bronchioles _____

3. Blood flow to a small region of lung is blocked due to a blood clot in a small pulmonary artery.

   a. What happens to the P_{O_2} and P_{CO_2} of the alveoli that are associated with that artery?

   b. What happens to the tissue P_{O_2} and P_{CO_2} around the arterioles distal to the blockage?

   c. What is the response of the bronchioles and the arterioles in this region? Will either or both of these responses be effective in compensating for the blocked artery? Explain.

4. Alveolar air has an average P_{O_2} of 100 mm Hg but expired air has an average P_{O_2} of 120 mm Hg. If the lungs are taking oxygen into the body, why is there more oxygen in the expired air?

5. Compare the following pairs of items. Put the symbols below in the space provided.

   greater than >   less than <   same as or equal =

   a _____ b   a. intrapleural pressure at the end of expiration
              b. intra-alveolar pressure at the end of inspiration

   a _____ b   a. blood flow in peripheral arterioles when surrounding tissue P_{O_2} is 70 mm Hg
              b. blood flow in pulmonary arterioles when interstitial P_{O_2} is 70 mm Hg

   a _____ b   a. resistance to air flow in the bronchioles
              b. resistance to air flow in the trachea

   a _____ b   a. compliance in alveoli with surfactant
              b. compliance in alveoli without surfactant

MAPS

Create a map using the following terms:

alveolar ventilation   depth of breathing   tidal volume
dead space volume   rate of breathing   total pulmonary ventilation

BEYOND THE PAGES

Normal values in pulmonary medicine:

Lung volumes and capacities (liters)	Men	Women
Tidal volume	0.5	0.5
Inspiratory reserve volume	3.3	1.9
Expiratory reserve volume	1.0	0.7
Residual volume	1.2	1.1
Total lung capacity	6.0	4.2

Total pulmonary ventilation: 6 L/min   Total alveolar ventilation: 4.2 L/min

Max. voluntary ventilation: 125–170 L/min   Respiration rate: 12–20 breaths/min

Blood gases

Arterial P_{O_2}: 95 mm Hg (85–100)*   Arterial P_{CO_2}: 40 mm Hg (37–43)   Arterial pH: 7.4

Venous P_{O_2}: 40 mm Hg   Venous P_{CO_2}: 46 mm Hg   Venous pH: 7.38

*Although we are considering arterial P_{O_2} to be equal to alveolar P_{O_2}, in reality the P_{O_2} drops slightly after leaving the pulmonary capillaries. This is because a small amount of deoxygenated venous blood from the nonexchange portions of the respiratory tract and from the coronary circulation combines with oxygenated blood as it returns to the left side of the heart. The actual arterial P_{O_2} value is closer to 95 mm Hg.

TRY IT

Lung Volumes and Capacities

Calculate your volumes and capacities using the table below. What will happen to your predicted vital capacity when you are 70 years old?

H = height in cm, A = age in years (Source: Medical Physiology Syllabus, University of Texas Medical Branch, Galveston)

Lung volume (L)	Subject	Formula
Vital capacity	Men	$(0.06 \times H) - (0.0214 \times A) - 4.65$
	Women	$(0.0491 \times H) - (0.0216 \times A) - 3.95$
Total lung capacity	Men	$(0.0795 \times H) + (0.0032 \times A) - 7.333$
	Women	$(0.059 \times H) - 4.537$
Functional residual capacity	Men	$(0.0472 \times H) + (0.009 \times A) - 5.29$
	Women	$(0.036 \times H) + (0.0031 \times A) - 3.182$
Residual volume	Men	$(0.0216 \times H) + (0.0207 \times A) - 2.84$
	Women	$(0.0197 \times H) + (0.0201 \times A) - 2.421$

18   Gas Exchange and Transport

LEARNING OBJECTIVES

When you complete this chapter, you should be able to:

▓ List four factors that influence the diffusion of gases.

▓ Explain the difference between the concentration of a gas in solution and the partial pressure of that gas in solution. What factors influence the movement of gas into solution?

▓ Compare and contrast the solubility of oxygen and carbon dioxide.

▓ Describe normal physiological pressures of oxygen and carbon dioxide in the following locations: alveoli, arterial blood, resting cells, and venous blood.

▓ Diagram the pressure gradients at the sites of gas exchange and show the direction of oxygen and carbon dioxide movement.

▓ Diagram the three categories of problems that might result in low arterial oxygen content and demonstrate examples of each.

▓ Diagram the factors affecting total blood oxygen content. Start with the equation: total blood oxygen content = amount dissolved in plasma + amount bound to hemoglobin. Work from there to map the factors that contribute to the amount of oxygen dissolved in plasma and the amount bound to hemoglobin.

▓ Describe the structure of hemoglobin.

▓ Draw the oxyhemoglobin dissociation curve (for normal conditions). What is the physiological significance of the shape of this curve?

▓ Draw the shifts in the oxygen-hemoglobin dissociation curve that result from changes in pH, temperature, and levels 2 and 3-DPG.

▓ Compare and contrast fetal hemoglobin with hemoglobin found in adults.

▓ Write the chemical reaction for the conversion of carbon dioxide to bicarbonate ions, including the enzyme that catalyzes the reaction.

▓ Explain how bicarbonate acts as a buffer.

▓ Map the three ways in which carbon dioxide is transported, diagramming specific mechanisms in each case. Also, be sure you can map the reverse mechanisms in order to show how these transport methods interact to remove CO_2 from the body.

▓ Diagram the current model of neural control of breathing.

▓ Map the ways in which central and peripheral chemoreceptors monitor carbon dioxide, oxygen, and pH levels for the purpose of regulating ventilation. Show specific mechanisms. What is the strategic significance of the placement of central and peripheral chemoreceptors?

▓ Describe protective reflexes that guard the lungs.

▓ Describe the influence higher brain centers can exert on breathing patterns.

SUMMARY

Recall from Ch. 17 that air moves from areas of higher pressure to areas of lower pressure. Therefore, when the thoracic volume is increased by inspiratory muscle movement, the pressure in the thoracic cavity drops and air moves in down its pressure gradient (inspiration). Likewise, as thoracic volume decreases, the alveolar pressure increases and air moves out of the body (expiration). Just as water and solutes move down their concentration gradients, so do gases. Movement of individual gases depends on their partial pressure gradients: gases move from higher partial pressures to lower partial pressures.

As the body consumes O_2 and releases CO_2, partial pressure gradients are created for each gas. Gas exchange is the result of simple diffusion down partial pressure gradients. The exchange surface in the lungs is the exchange epithelium of the alveoli. Blood flow and respiration are closely matched to ensure efficient delivery of gases. Hemoglobin (Hb) is the main facilitator of gas transport in the blood. Hb, a protein component of RBCs, consists of an iron molecule surrounded by a porphyrin ring embedded within a tetrameric (four-subunit) protein.

Hemoglobin binds O_2 and CO_2 and will carry either until a partial pressure gradient causes the gas to be released. Hemoglobin-binding ability is affected by pH, temperature, and 2,3-DPG. The relationship between P_{O_2} and Hb binding is shown by an oxyhemoglobin dissociation curve. Only 23% of CO_2 is transported bound to Hb. About 70% of CO_2 in venous blood is converted to H^+ and HCO_3^- inside the RBCs. Carbonic anhydrase is the enzyme that catalyzes this reaction:

$$CO_2 + H_2O \underset{\substack{\text{carbonic} \\ \text{anhydrase}}}{\rightleftharpoons} H_2CO_3 \rightleftharpoons H^+ + HCO_3^-$$

A network of neurons in the pons and medulla oblongata controls respiration. Researchers are still working to piece together all of the complex neural interactions that control breathing. See the textbook for the current accepted model.

Chemical factors also affect respiration. Central chemoreceptors respond to increases in H^+, due to elevated P_{CO_2}, by increasing ventilation. Peripheral chemoreceptors monitor blood pH, P_{CO_2}, and P_{O_2}. We can control our respiration consciously to a certain extent, but chemoreceptors override conscious control.

TEACH YOURSELF THE BASICS

DIFFUSION AND SOLUBILITY OF GASES

1. By what mechanism do gases move between the alveoli and the plasma?

2. List the four rules for diffusion of gases. (See Ch. 5.)

3. Why do respiratory physiologists commonly use partial pressures to express gas concentrations in solution?

Dissolved Gas Depends on Pressure, Solubility, and Temperature

4. When a gas is placed in contact with a liquid, what three factors determine how much gas will dissolve in the liquid?

5. TRUE/FALSE? Defend your answer.

   If a liquid is exposed to a P_{CO_2} of 100 mm Hg and a P_{O_2} of 100 mm Hg, equal amounts of oxygen and carbon dioxide will dissolve in the liquid. (Fig. 18-2)

6. The more soluble a gas is in a particular liquid, the (higher/lower?) the partial pressure required to force the gas into solution.

7. Gases move between liquid and gaseous phases until _____ is reached. (Fig. 18-2)

8. At equilibrium, the (concentration/partial pressure/both?) of a gas will be equal in the air and gas phases.

9. Which is more soluble in body fluids: oxygen or carbon dioxide?

GAS EXCHANGE IN THE LUNGS AND TISSUES

IP *Respiratory System: Gas Exchange*

10. Give the following partial pressures in a normal person at sea level: (Fig. 18-3; Table 18-1)

   ▶ Unless otherwise specified, the terms "arterial blood" and "venous blood" refer to blood in the systemic circulation.

   P_{O_2}: alveoli = _____ arterial blood = _____ resting cells = _____ venous blood = _____

   P_{CO_2}: alveoli = _____ arterial blood = _____ resting cells = _____ venous blood = _____

11. If the alveolar P_{O_2} is 98 mm Hg, what will the departing arterial P_{O_2} be? Why?

12. Define hypoxia and hypercapnia. (Table 18-2)

13. What are the three categories of problems that result in low arterial oxygen content?

A Decrease in Alveolar P_{O_2} Decreases Oxygen Uptake at the Lungs

14. If alveolar P_{O_2} is low, what two factors might have caused the decrease?

15. Explain the relationship between altitude and P_{O_2}.

16. What are the pathological factors that cause hypoventilation? (Fig. 18-4)

Changes in the Alveolar Membrane Alter Gas Exchange

17. What two cell layers must gases cross to go from the alveoli to the plasma? (Fig. 18-5)

18. Describe the pathological changes that adversely affect gas exchange.

19. Explain how emphysema can result in a loss of alveolar surface area. (Fig. 18-4b)

20. Explain how fibrotic lung diseases can cause decreased oxygen exchange between alveoli and blood. (Fig. 18-4c)

21. How much of the exchange epithelium must be incapacitated before arterial P_{O_2} drops?

22. What is pulmonary edema and how does it alter gas exchange? (Fig. 18-4d)

23. Explain why some patients with pulmonary edema have low arterial P_{O_2} but normal arterial P_{CO_2}.

24. Explain alveolar flooding and adult respiratory distress syndrome (ARDS).

GAS TRANSPORT IN THE BLOOD

IP *Respiratory System: Gas Transport*
(See law of mass action in Ch. 5.)

Hemoglobin Transports Most Oxygen to the Tissues

25. List two ways that gases are transported in the blood.

26. Total blood oxygen content = _____ + _____ .

27. _____% of oxygen in a given volume of blood will be carried bound to hemoglobin. (Fig. 18-6)

28. Compare the body's oxygen consumption at rest with the delivery of dissolved oxygen to the cells. Assume a cardiac output of 5 L/min. (Fig. 18-7a)

29. How much additional oxygen per minute can be delivered by hemoglobin if each liter of blood carries 197 mL O_2 bound to hemoglobin? (Fig. 18-7b)

30. What two factors determine the amount of O_2 bound to Hb?

31. List three factors that establish the arterial P_{O_2}.

32. What determines the number of binding sites for oxygen?

33. What is mean corpuscular Hb?

One Hemoglobin Molecule Binds Up to Four Oxygen Molecules

34. Hemoglobin molecules are composed of (how many?) _____ protein subunits, each with an

    O_2-binding _____ group. (See Ch. 16.) This group is based around the element

    _____ that binds weakly to oxygen. (Fig. 18-8a)

35. Describe the differences between adult and fetal hemoglobin.

Oxygen-Hemoglobin Binding Obeys the Law of Mass Action

36. Hb bound to O_2 is called _____ $(HbO_2)_{1-4}$.

37. Explain how oxygen-hemoglobin binding obeys the law of mass action.

P_{O_2} Determines Oxygen-Hemoglobin Binding

38. Diagram the steps followed by an oxygen molecule as it goes from the alveoli to its binding site on hemoglobin. (Use a separate sheet of paper if necessary.)

39. As dissolved O_2 diffuses into RBCs, what happens to the P_{O_2} of the surrounding plasma?

40. Therefore, as O_2 binds with Hb, (more/less?) O_2 can diffuse from alveoli into plasma.

41. Diagram the steps followed by a molecule of oxygen as it diffuses from the plasma (and RBCs) and into a tissue.

42. As dissolved O_2 diffuses into a tissue, what happens to the P_{O_2} of the surrounding plasma?

43. Therefore, as O_2 enters a tissue, (more/less?) O_2 is released from Hb-binding to diffuse into the plasma.

Oxygen Binding Is Expressed as a Percentage

44. Define percent saturation of hemoglobin and write the formula for how it is calculated.

45. At 100% saturation, all possible binding sites are (bound/free?).

46. In the oxyhemoglobin dissociation curve (Fig. 18-9), the (P_{O_2}/percent saturation of Hb) determines the (P_{O_2}/percent saturation of Hb).

47. Adaptively, why is it important that the slope of the curve flattens out at P_{O_2} values above 60 mm Hg?

48. Below P_{O_2} of 60 mm Hg, where the curve is steeper, small changes in P_{O_2} cause relatively (small/large?) releases of O_2 from hemoglobin.

49. With respect to the oxyhemoglobin dissociation curve, describe how O_2 is delivered to metabolically active tissues.

Temperature, pH, and Metabolites Affect Oxygen-Hemoglobin Binding

▶ Any factor that alters the hemoglobin protein may alter O_2-binding ability. (Fig. 18-10)

50. An increase in pH (increases/decreases?) hemoglobin's affinity for oxygen. What is the Bohr effect? (Fig. 18-10a)

51. An increase in temperature (increases/decreases?) hemoglobin's affinity for oxygen. (Fig. 18-10b)

52. An increase in P_{CO_2} (increases/decreases?) hemoglobin's affinity for oxygen. (Fig. 18-10c)

53. The metabolite 2,3-DPG (increases/decreases?) hemoglobin's affinity for oxygen. What triggers an increase in 2,3-DPG production? (Fig. 18-11)

54. Fetal Hb has a/an (increased/decreased?) affinity for oxygen. (Fig. 18-12)

55. A left shift in the curve indicates (increased/decreased?) binding affinity.

56. A right shift in the curve indicates (increased/decreased?) binding affinity.

57. Why is it significant that a shift in the O_2-Hb dissociation curve is more pronounced at low P_{O_2} and less pronounced at higher P_{O_2}?

▶ Fig. 18-13 summarizes all the factors that influence oxygen transport in the blood.

Carbon Dioxide Is Transported in Three Ways

58. Why is it important that CO_2 be removed from the body?

59. List the three ways that CO_2 is transported in the blood. (Fig. 18-14)

CO_2 and Bicarbonate Ions

60. What two purposes does the conversion of CO_2 to HCO_3^- serve?

61. Write the equation in which CO_2 is converted into bicarbonate ion (HCO_3^-) and H^+.

62. What enzyme catalyzes this reaction, and where in the blood is it found?

63. Explain why CO_2 forms HCO_3^- and H^+ in the systemic capillaries, but HCO_3^- and H^+ form CO_2 in pulmonary capillaries. (Fig. 18-14)

64. Why must H^+ and HCO_3^- be removed from RBC cytoplasm?

65. What is the chloride shift and what does it accomplish?

66. In what way does HCO_3^- act as a buffer? (See Ch. 2.)

Hemoglobin and H^+

67. What is respiratory acidosis?

68. How does Hb help prevent this condition?

69. Constant removal of CO_2 from plasma (increases/decreases?) P_{CO_2} and allows (more/less?) CO_2 to leave cells.

Hemoglobin and CO_2

70. What facilitates the formation of carbaminohemoglobin? How? (Fig. 18-10)

CO_2 Removal at the Lungs

71. Diagram CO_2 removal at the lungs and how this removal influences other CO_2 transport mechanisms to effectively remove CO_2 from the body. (Fig. 18-15)

▶ O_2 and CO_2 transport are summarized in Fig. 18-15.

REGULATION OF VENTILATION

IP *Respiratory System: Control of Respiration*

72. Compare the rhythmicity and control of breathing to that of the heartbeat.

73. What is a central pattern generator? (See Ch. 13.)

74. Based on the contemporary model, describe what we currently know about neural control of breathing. (Fig. 18-16)

Neurons in the Medulla Control Breathing

75. Diagram the complex interactions of medullary neurons that exert influence on respiration. Your diagram should include the following terms: nucleus tractus solitarius, dorsal respiratory group (DRG), phrenic nerves, intercostal nerves, intercostal muscles, vagus and glossopharyngeal nerves, pontine respiratory groups, ventral respiratory group (VRG), pre-Bötzinger complex, quiet breathing, forced breathing.

76. In the context of respiratory neurophysiology, what is ramping?

Carbon Dioxide, Oxygen, and pH Influence Ventilation

77. List the location and the chemical factor(s) monitored by each group of respiratory chemo-receptors.

78. What is the primary chemical stimulus for changes in ventilation?

79. Explain the strategic significance of the location of the peripheral chemoreceptors. (Fig. 18-16)

Peripheral Chemoreceptors

80. To what chemical signals do the carotid and aortic bodies respond?

81. Diagram the basic mechanism by which the carotid and aortic bodies respond to stimuli. (Fig. 18-19)

82. Where do the sensory neurons leading from these receptors send their signals?

83. Using the oxygen-hemoglobin dissociation curve in Fig. 18-9, explain the adaptive significance of the fact that the peripheral chemoreceptors do not respond to decreases in P_{O_2} until the P_{O_2} drops below 60 mm Hg.

Central Chemoreceptors

84. Diagram how the central chemoreceptors respond to elevated blood P_{CO_2}. (Fig. 18-20)

85. An increase in P_{CO_2} will trigger a/an (decrease/increase?) in ventilation. (Fig. 18-21) How do central chemoreceptors respond to decreased P_{CO_2}?

86. If P_{CO_2} is chronically elevated, the sensory receptors will _____ and the venti-lation rates will (increase/decrease?).

87. If a person has chronic hypercapnia and hypoxia, is CO_2 the primary chemical drive for venti-lation? Why?

88. What will happen to ventilation if this person with chronic hypercapnia and hypoxia is given pure O_2 to breathe? Explain.

Protective Reflexes Guard the Lungs
89. Write the reflex response to an inhaled irritant.

receptor _____   afferent path _____

integrating center _____   efferent path _____

effector _____   tissue response _____

systemic response _____

90. Describe the Hering-Breuer inflation reflex.

Higher Brain Centers Affect Patterns of Ventilation
91. Give two examples of how higher brain centers can influence ventilation.

TALK THE TALK

2,3-diphosphoglycerate	fetal hemoglobin (HbF)	oxyhemoglobin dissociation
acidosis	Fick's law of diffusion	curves
adult respiratory distress syn-	globin	partial pressure
drome (ARDS)	glomus cells	percent saturation of hemoglo-
alveolar flooding	glossopharyngeal nerves	bin
aortic bodies	heme group	peripheral chemoreceptors
Bohr effect	Hering-Breuer inflation reflex	phrenic nerve
bronchoconstriction	hypercapnia	pontine respiratory group
carbaminohemoglobin	hypoventilation	porphyrin ring
carbonic acid	hypoxia	pre-Bötzinger complex
carbonic anhydrase (CA)	intercostal nerve	pulmonary edema
carotid bodies	irritant receptors	pulse oximeter
central chemoreceptors	law of mass action	ramping
central pattern generator	membrane thickness	respiratory acidosis
chloride shift	nucleus tractus solitarius (NTS)	solubility
chronic hypoxia	obstructive sleep apnea	surface area
concentration gradient	oxygen-hemoglobin binding	transcutaneous oxygen sensor
diffusion distance	reaction	vagus nerve
dorsal respiratory group (DRG)	oxyhemoglobin	ventral respiratory group (VRG)
emphysema		

QUANTITATIVE THINKING

1. During exercise, a man consumes 1.8 L of oxygen per minute. His arterial oxygen content is 190 mL/L and the oxygen content of his venous blood is 134 mL/L. What is his cardiac output?

2. You are given the following data on a person:

    arterial plasma P_{O_2} = 95 mm Hg

    blood volume = 4.2 liters

    hematocrit = 38%

    hemoglobin concentration = 13 g/dL whole blood

    maximum oxygen-carrying capacity of hemoglobin = 1.34 ml oxygen/g hemoglobin

    At a P_{O_2} of 95 mm Hg, plasma contains 0.3 mL oxygen per deciliter (dL) and hemoglobin is 97% saturated.

    Using the data above, calculate the total amount of oxygen that could be carried in the person's blood.

HINTS

Total blood oxygen = amount dissolved in plasma + amount bound to hemoglobin

To determine amount of oxygen dissolved in plasma:

What is this person's plasma volume? (Use hematocrit to determine.)

What is the solubility of oxygen in plasma?

To determine the amount of oxygen bound to hemoglobin:

How much hemoglobin is in this person's blood?

You are given total blood volume, hematocrit, and hemoglobin content/dL whole blood. Which of these parameters will you use?

Maximum oxygen-carrying capacity represents what percent saturation?

What is the percent saturation in this person's blood?

PRACTICE MAKES PERFECT

In the following questions, mark each answer as either TRUE or FALSE.

1. Oxygen
   a. is mainly transported in the blood while bound to the hemoglobin in red blood cells. _____
   b. is as soluble as carbon dioxide in plasma. _____
   c. is the primary chemical drive for ventilation. _____

2. Carbon dioxide
   a. is primarily transported as a gas dissolved in the plasma. _____
   b. binds to hemoglobin in erythrocytes. _____
   c. is converted to carbonic acid through the action of carbonic anhydrase. _____

3. The P_{O_2} of the blood
   a. is a measure of the amount of oxygen dissolved in the plasma. _____
   b. is the most important factor determining the percent saturation of hemoglobin. _____
   c. is normal in anemia. _____
   d. is an accurate indicator of the total oxygen content of blood. _____
   e. determines P_{O_2} of the alveoli. _____

In the questions, pick the single best answer.

4. Which of the following would decrease the ability of oxygen to diffuse across the alveolar/capillary membrane? (circle all that are correct)
   a. an increase in thickness of the alveolar membrane
   b. increased hemoglobin concentration in erythrocytes
   c. an increase in the partial pressure of oxygen in the alveoli
   d. CNS depression by drugs or alcohol
   e. a decrease in the surface area of the alveoli

5. Compare the following pairs of items. Put the symbols below in the space.

    greater than >   less than <   same as or equal =

    a _____ b
        a.   oxygen released from hemoglobin at a cell whose P_{O_2} is 40 mm Hg when the plasma is at pH 7.4
        b.   oxygen released from hemoglobin at a cell whose P_{O_2} is 40 mm Hg when the plasma is at pH 7.2

    a _____ b
        a.   arterial oxygen transport in a person with a hemoglobin of 10 g Hb/dL blood and P_{O_2} of 140 mm Hg
        b.   arterial oxygen transport in a person with a hemoglobin of 11 g Hb/dL blood and P_{O_2} of 100 mm Hg

    a _____ b
        a.   arterial P_{O_2} in a person with anemia
        b.   arterial P_{O_2} in a normal person

6. You are an astronomer who has been invited with colleagues to work for a week at the observatory on the summit of Mauna Kea, in Hawaii. The summit of this extinct volcano is 13,796 feet above sea level. Describe how each of the following parameters will change by the end of your journey to the summit, and explain the stimulus and pathway for each change.

    a.   partial pressure of oxygen in the air

    b.   barometric pressure

    c.   arterial P_{O_2}

    d.   arterial P_{CO_2}

    e.   arterial pH

    f.   When you arrive at the observatory, you meet some resident astronomers who have been living there for years. How does the hemoglobin content of their blood compare with that of you and your colleagues?

    g.   Within a day of arrival, one of your colleagues complains of difficulty breathing and a severe headache. The emergency oxygen tank has run out of oxygen. What should you do?

    h.   When you first arrive at the observatory, you notice that you begin breathing more rapidly. But within a day, your breathing has returned to a more normal rate, although you know that your body has not had time to manufacture more hemoglobin. How could you explain this phenomenon?

7. A person hyperventilates. What effect will this have on the total oxygen content of her blood? Explain.

MAPS

Compile the following terms into a concept map showing the relationships between them. The major concept of your map is *total arterial O₂ content*. This should be your starting point.

2,3-DPG
adequate perfusion of alveoli
airway resistance
alveolar surface area
alveolar ventilation
amount of interstitial fluid
composition of inspired air

diffusion distance
dissolved in plasma
hemoglobin (Hb) content
lung compliance
membrane thickness
number of Hb binding sites
number of RBCs

O_2 diffusion between alveoli
   and blood
pH
P_{O_2}
rate and/or depth of breathing
temperature

BEYOND THE PAGES

SOME TYPICAL CLINICAL VALUES IN RESPIRATORY PHYSIOLOGY

total pulmonary ventilation = 6 L/min   total alveolar ventilation = 4.2 L/min
max. voluntary ventilation = 125–170 L/min   respiration rate = 12–20 breaths/min

Blood gases:
arterial P_{O_2} = 95 mm Hg (85–100)*   arterial P_{CO_2} = 40 mm Hg (37–43)   arterial pH = 7.4
venous P_{O_2} = 40 mm Hg   venous P_{CO_2} = 46 mm Hg   venous pH = 7.38

*Although we are considering arterial P_{O_2} to be equal to alveolar P_{O_2}, in reality the P_{O_2} drops slightly after leaving the pulmonary capillaries. This is because a small amount of deoxygenated venous blood from the nonexchange portions of the respiratory tract and from the coronary circulation combines with oxygenated blood as it returns to the left side of the heart. The actual arterial P_{O_2} value is closer to 95 mm Hg.

TRY IT

Demonstrations for Chemical Control of Ventilation

Using what you have learned about the chemical control of ventilation, answer the following questions.

In which case can you hold your breath the longest?

a.   after normal breathing
b.   after hyperventilating
c.   after breathing into a paper bag

Explain your reasoning:

Will your breathing rate *increase* or *decrease* after doing the following?

a.   hyperventilating _____

b.   breathing into a paper bag _____

Explain your reasoning:

You can hyperventilate by increasing breathing rate or by taking deeper breaths at your usual breathing rate. After which type of hyperventilation do you think that you will be able to hold your breath longer?

a.   increasing breathing rate
b.   taking deeper breaths at your usual breathing rate

Explain your reasoning:

With a partner, do the following exercises. If you have time, repeat the sequence three times and take the average. Compare your results with your answers above.

1. Normal breathing (*eupnea*)

    Breathe normally. Have partner count and record the number of breaths per minute for three one-minute intervals.

2. No breathing (*apnea*)

    Breathe normally for several minutes. After a normal inspiration, time how long you can hold your breath.

3. Increased ventilation with no change in metabolic rate (*hyperventilation*)

    a. increased rate

        Breathe normal volumes rapidly for two minutes. Record the rate. Immediately after hyperventilation, hold your breath from the end of a normal inspiration for as long as possible. Time yourself.

    b. increased volume

        Breathe maximum volumes for two minutes, trying to keep the rate as close to your normal rate as possible. *If you get dizzy,* **stop!** *Note the time and extrapolate to two minutes.* Record the rate. Immediately after hyperventilation, hold your breath from the end of a normal inspiration for as long as possible. Time yourself.

4. Depressed blood carbon dioxide levels (P_{CO_2})

    Breathe deeply and rapidly for one minute. Stop sooner if you start to get dizzy. At the end of the one-minute test, breathe naturally (normal volumes) for three to four minutes. Record the rate in the first 90 seconds of normal breathing. Try not to regulate the rate in any way.

5. Elevated blood carbon dioxide levels

    Breathe normally into and out of a paper bag held over your nose and mouth. After one minute, record the breathing rate for the second minute. Then hold your breath for as long as possible. Time how long you are able to hold your breath.

	Test #1	Test #2	Test #3	Average rate	Time for breathholding
Normal breathing					
Increased rate, normal volume					
Increased volume, normal rate					
Increased rate and volume					
Breathe into paper bag					

19 The Kidneys

LEARNING OBJECTIVES

When you complete this chapter, you should be able to:

▨ List and describe the six functions of the kidneys.

▨ Trace the path of a drop of urine from the kidney to the external environment.

▨ Trace a drop of blood from a renal artery to a renal vein.

▨ Diagram the anatomical relationship between the vascular elements of the nephron and the tubular elements of the nephron. What are the juxtaglomerular apparatus, macula densa, and granular cells?

▨ List the three processes of the kidney and for each, describe: the direction of fluid/solute movement, the location of the fluid/solute (internal vs. external), and whether the process is selective or nonselective.

▨ Diagram fluid volume and osmolarity modifications by the nephron and describe the adaptive significance of the final volume and concentration of the urine.

▨ Diagram the filtration barriers a water molecule will pass as it travels from the blood into the lumen of the nephron. Show the anatomical structures and mechanisms by which filtration can be controlled.

▨ Describe the hydrostatic and osmotic pressures that contribute to glomerular filtration. Which direction of fluid movement does the net pressure favor?

▨ Define GFR and diagram its component factors. What is the average value for GFR?

▨ Diagram how GFR can be influenced by:

  ▪ variable resistance in afferent and efferent nephron arterioles

  ▪ myogenic and tubuloglomerular autoregulatory mechanisms

  ▪ hormonal control

  ▪ neural control.

▨ Diagram specific examples of: active transport, secondary active transport, passive reabsorption, and transcytosis mechanisms used by the kidney to accomplish reabsorption. Also, distinguish between epithelial transport and paracellular pathways.

▨ Create a generalized graph of the reabsorption of glucose in order to demonstrate how protein-mediated renal transport can reach saturation. Mark on the graph where transport maximum and renal threshold occur.

▨ Create a generalized graph of the secretion of penicillin, both alone and in the presence of probenecid, in order to demonstrate how protein-mediated renal transport can be affected by competition.

▨ Determine GFR when given plasma inulin concentration and rate of inulin excretion. Make sure you understand why this relationship is valid.

▨ Analyze renal handling of a substance and indicate if it is reabsorbed or secreted:

  ▪ when given GFR, plasma concentration, and excretion rate of the substance.

  ▪ by comparing clearance of the substance to clearance of inulin or creatinine.

- Understand and apply the equation E = F – R + S to analyze renal handling of a substance.

- Diagram the involuntary micturition reflex and incorporate the voluntary control influence exerted by higher brain centers.

SUMMARY

The urinary system consists of the kidneys, bladder, and accessory structures. The system produces urine and eliminates it to help the body maintain fluid and electrolyte balance. The kidneys have six functions: regulation of extracellular fluid volume, regulation of osmolarity, maintenance of ion balance, homeostatic regulation of pH, excretion of wastes and foreign substances, and production of hormones. There are four basic processes in the urinary system: filtration, reabsorption, secretion, and excretion.

The nephron is the functional unit of the kidney; each nephron is composed of vascular and tubular elements. Following the path of blood through the nephron, the anatomy is as follows: renal arteries, afferent arteriole, glomerulus, efferent arteriole, peritubular capillaries. Fluid filters out of the glomerulus into Bowman's capsule. It then moves into the proximal tubule, loop of Henle, distal tubule, collecting duct, and renal pelvis. After this point, the fluid will not be altered again, and it can be called urine. Urine drains into the urinary bladder via the ureters and is then eliminated in a process called urination (micturition).

One of the most important concepts to take from this chapter is that the amount of fluid excreted (E) is equal to the amount filtered (F), minus the amount reabsorbed (R), plus the amount secreted (S). Expressed mathematically, this becomes: E = F – R + S. Remembering this equation can often help you if you're in a bind trying to solve a problem.

Filtration is the movement of fluid from the blood of the glomerulus into the nephron lumen at Bowman's capsule. Filtered fluid composition is equal to that of plasma, minus blood cells and most proteins. Filtration occurs because the hydrostatic pressure exceeds the osmotic pressure, and the net driving force is 10 mm Hg in favor of filtration.

Specialized epithelial cells in the capsule, called podocytes, and mesangial cells of the glomerulus form slits that can be manipulated to change the glomerular filtration rate (GFR). GFR is the amount of fluid that filters into Bowman's capsule per unit time; the average GFR is 125 mL/min (180 L/day).

Control of GFR includes the myogenic response, tubuloglomerular feedback, and reflex control. In the myogenic response, smooth muscle of the arteriole stretches as blood pressure increases, and this ultimately causes vasoconstriction. Tubuloglomerular feedback regulates arteriolar diameter by means of chemical communication between the macula densa cells and JG cells (together called the juxtaglomerular apparatus). Reflex control involves sympathetic neurons and alpha receptors on afferent and efferent arterioles.

Reabsorption is the movement of filtered material from the nephron lumen back into the blood supply. Bulk reabsorption takes place in the proximal tubule, but regulated reabsorption takes place in later tubule segments. Most reabsorption involves transepithelial transport (movement across the apical and basolateral membranes). Reabsorption of water and solutes depends on both active and passive transporting mechanisms. The transport of Na^+ into the extracellular fluid creates concentration gradients that allow movement of other substances. Transport involves protein-substrate interaction, so saturation, competition, and specificity are also involved. The renal threshold for a substance is the plasma concentration of that substance at which saturation occurs. At concentrations above renal threshold, substances that are normally reabsorbed are excreted in the urine.

Secretion is the transfer of molecules from the extracellular fluid into the nephron lumen. Secretion depends mostly on active membrane transport. As with reabsorption, secretion shows saturation, competition, and specificity.

Excretion is the result of the other three processes. It depends on filtration rate and on whether reabsorption or secretion is involved. Clearance is an abstract concept describing how many milliliters of plasma passing through the kidneys have been totally cleared of a substance in a given period of time. Clearance is used clinically to assess renal handling of a substance, based only on the analysis of the blood and urine. Spend some valuable time working through the examples in the chapter to get a grasp of this concept mathematically. It can be a little tricky, so be sure you give yourself plenty of time to work through it.

Micturition is the elimination of urine from the bladder. Two sphincters close off the opening between the ureters and the bladder. The external sphincter is skeletal muscle and can be consciously controlled. Micturition is a simple spinal reflex initiated by stretch in the bladder wall.

TEACH YOURSELF THE BASICS
FUNCTIONS OF THE KIDNEYS

1. State the law of mass balance. (See Ch. 5.)

2. List the six functions of the kidneys:

   1. _____   4. _____

   2. _____   5. _____

   3. _____   6. _____

ANATOMY OF THE URINARY SYSTEM

See Anatomy Summary, Fig. 19-1.

IP *Urinary System: Anatomy Review*

The Urinary System Consists of Kidneys, Ureters, Bladder, and Urethra

3. Starting at the kidneys, follow a drop of urine to the external environment. (Fig. 19-1a)

4. Explain the term *retroperitoneal*.

5. Describe the vascular supply to the kidneys.

6. What is the function of the urinary bladder?

The Nephron Is the Functional Unit of the Kidney

7. The medulla is the (outer/inner?) layer and the cortex is the (outer/inner?) layer of the kidney. (Fig. 19-1c)

8. What is a nephron? Distinguish between cortical nephrons and juxtamedullary nephrons.

Vascular Elements of the Nephron

9. Trace a drop of blood through the nephron from a renal artery to a renal vein. (Fig. 19-1d–g)

10. Describe the portal system of the nephron. What is its function?

Tubular Elements of the Nephron

11. Trace a drop of fluid through the tubule of the nephron, ending in the renal pelvis. (Figs. 19-1c, j, 19-2)

12. What is the renal corpuscle? What happens here?

13. Diagram the relationship between the different segments of the nephron to the cortex and medulla of the kidney. In other words, which parts of the nephron are found in the renal cortex and which are found in the renal medulla?

14. Describe the juxtaglomerular apparatus. (Fig. 19-1g)

15. Fluid is considered urine when it enters the _____.

OVERVIEW OF KIDNEY FUNCTION

16. How much plasma on average enters the nephrons per day?

17. How much urine on average leaves the body per day as urine?

18. What happens to the fluid that doesn't leave in the urine?

Kidneys Filter, Reabsorb, and Secrete

19. List the three processes of the kidney and describe them. (Fig. 19-2)

20. In which of these processes is fluid entering the external environment? (Fig.1-2)

21. Which of these processes occurs by means of bulk flow?

22. Which of these processes uses transporting epithelia? (See Ch. 3.)

The Nephron Modifies Fluid Volume and Osmolarity

23. Fluid entering Bowman's capsule is nearly _____ osmotic with plasma. (See Ch. 5.)

24. Fill in the blanks: (Fig. 19-3)

Location in nephron	Volume of fluid	Osmolarity of fluid	What happens to the fluid in this nephric element?
Bowman's capsule	180 L/day	300 mOsM	
End of proximal tubule	54 L/day		
End of loop of Henle	18 L/day		
End of collecting duct (final urine)	1.5 L/day		

25. The final volume and concentration of the urine reflect what need(s) of the body?

26. What is excretion, and how is it different from secretion?

27. Describe the expression that relates excretion to the three processes of the nephron. (Fig. 19-3)

FILTRATION

28. Describe the composition of the filtrate that enters the lumen of the nephron. (Fig. 19-4)

The Renal Corpuscle Contains Filtration Barriers (Fig. 19-5)

IP *Urinary System: Glomerular Filtration*

29. List the three layers a water molecule will pass as it travels from the blood into the lumen of Bowman's capsule.

30. Describe the structure of glomerular capillaries and their pores. What are mesangial cells? (Fig. 19-5a–d)

31. Describe the structure and function of the basal lamina. (Fig. 19-5d)

32. Describe the filtration barrier created by the epithelium of Bowman's capsule. What are podocytes, fast processes, and filtration slits? (Fig. 19-5a–c)

Capillary Pressure Causes Filtration

33. Glomerular filtration occurs because (fill in the blanks in the chart): (Fig. 19-6)

Pressure type	Average pressure measurement	Favors movement of fluid from where to where
Glomerular capillary hydrostatic pressure (P_H)		
Colloid osmotic pressure (π)		
Bowman's capsule hydrostatic pressure (P_{fluid})		
Net pressure is:		

34. Compare the pressures involved in glomerular filtration to those involved in filtration and absorption in systemic capillaries. (See Ch. 15.)

35. Define GFR.

36. What two factors influence GFR the most?

37. Name the components of the filtration coefficient.

38. An average value for GFR is _____ L/day or _____ mL/min.

39. The total body plasma volume is _____ L, which means that the kidneys filter the plasma

_____ times each day.

GFR Is Relatively Constant
40. What is the relationship between blood pressure and GFR? (Fig. 19-7)

41. Diagram how GFR is controlled by the resistance at the renal arterioles. What happens to GFR when resistance increases or decreases at afferent arteriloes? Efferent arterioles? (Fig. 19-8)

GFR Is Subject to Autoregulation
42. What is meant by autoregulation of GFR? By what two mechanisms does your body accomplish autoregulation of GFR?

Myogenic Response

43. What is a myogenic response?

44. Describe the myogenic response to increased blood pressure in afferent arterioles. Include the mechanism.

45. Why is vasodilation not as effective as vasoconstriction in controlling GFR?

46. Why is a decrease in GFR when blood pressures fall below normal an adaptive response?

Tubuloglomerular Feedback

47. How does the distal tubule communicate with arterioles? (Fig. 19-9)

48. Describe tubuloglomerular feedback as a result of increased GFR. (Fig. 19-10)

49. What role does NaCl play in the tubuloglomerular feedback mechanism?

Hormones and Autonomic Neurons Also Influence GFR

50. In neural control of GFR, (sympathetic/parasympathetic?) neurons release (ACh/norepi?) onto (α, β_1, β_2?) receptors, causing (vasodilation/vasoconstriction?) of renal arterioles.

51. Vasoconstriction of the afferent arteriole will (increase/decrease?) its resistance, will (increase/decrease?) hydrostatic pressure in the glomerular capillaries, and will (increase/decrease?) GFR.

52. Vasoconstriction of the efferent arteriole will (increase/decrease) its resistance, will (increase/decrease?) hydrostatic pressure in the glomerular capillaries, and will (increase/decrease?) GFR.

53. Hormones that influence arteriolar resistance and GFR include _____,

a potent vasoconstrictor, and _____, which are vasodilators.

54. What are other regulatory actions of hormones in the nephron? Where in the kidney do they exert effect? What is the result on GFR?

REABSORPTION

IP *Urinary System: Early Filtrate Processing*

55. Bulk reabsorption in the nephron takes place in the _____.

56. Why does the kidney filter 180 L/day if 99% of what is filtered is reabsorbed?

Reabsorption May Be Active or Passive

57. To reabsorb molecules against their concentration gradient, the transporting epithelia of the nephron must use what process? (See Ch. 5.)

58. Which ion plays a key role in bulk reabsorption in the proximal tubule? (Fig. 19-11)

59. Distinguish between the roles of transepithelial transport and the paracellular pathway in reabsorption.

Active Transport of Sodium

60. Filtrate entering proximal tubule has [Na^+] similar to that of _____ and (higher/lower?) than the [Na^+] inside cells.

61. Diagram the transepithelial movement of Na^+ across proximal tubule cells. Include transporter proteins involved. (Fig. 19-12)

Secondary Active Transport: Symport With Sodium

62. List some molecules that are transported using Na^+-linked secondary active transport. (See Ch. 5.)

63. Diagram the transepithelial transport of glucose in the nephron. Include transporter proteins. (Figs. 19-13 and 5-15)

Passive Reabsorption: Urea Reabsorption

64. Urea in the proximal tubule can only move by diffusion. If the urea concentration of filtrate is equal to the urea concentration of plasma, what creates the urea concentration gradient needed for diffusion? (Fig. 19-11)

Transcytosis: Plasma Proteins

65. Some of the smaller protein hormones and enzymes can cross the filtration barrier at the glomerulus. By what mechanism are the filtered proteins reabsorbed?

66. For each of the following substances, tell how it crosses the apical and basolateral membranes of the proximal tubule cell and what form of transport it uses at each membrane (primary active, secondary active, facilitated diffusion, simple diffusion).

Molecule	Apical membrane	Type of transport?	Basolateral membrane	Type of transport?
Na^+				
glucose				
water				
urea				
proteins				

Renal Transport Can Reach Saturation

67. List the three characteristics of mediated transport seen in all protein-substrate interactions. (See Ch. 5.)

68. Briefly recap the molecular explanation of saturation.

69. Below saturation, the rate of transport is proportional to _____.

70. The rate of transport at saturation is also known as the _____. (Fig. 19-14)

71. For a particular substance, the plasma concentration at which that substance first appears in

    the urine is known as the _____. (Fig. 19-15)

72. Filtration (does/doesn't?) exhibit saturation. (Fig. 19-15a)

73. Normal plasma glucose concentrations are (less than, equal to, greater than?) the renal threshold for glucose. Therefore, normally all glucose filtered is (excreted/secreted/reabsorbed?).

74. What happens when the concentration of a substance in the plasma exceeds its renal threshold? (Fig. 19-15c, d)

75. What is the term for "glucose in the urine"?

76. If filtration of a substance is greater than reabsorption, then the excess substance is

    _____. (Fig. 19-15d)

   ▶ Remember! E = F − R + S

Peritubular Capillary Pressures Favor Reabsorption

77. How does fluid reabsorbed from the lumen of the nephron tubule system get reabsorbed into the peritubular capillaries?

SECRETION

78. In renal secretion, molecules move from the _____ to the _____.

79. By what means of transport is most secretion accomplished?

80. What is the point of secreting a substance in addition to filtering it?

Competition Decreases Penicillin Secretion

81. Describe competition in a mediated transport system, using the renal secretion of penicillin and probenecid as your example.

82. When probenecid is given at the same time as penicillin, what happens to the excretion rate of penicillin?

EXCRETION

83. Does looking at the composition of urine tell us if a substance has been filtered? _____

    reabsorbed? _____ secreted? _____ excreted? _____

Clearance Is a Noninvasive Way to Measure GFR

84. How can you determine GFR from measuring only plasma concentration and the excretion rate of inulin?

85. Define clearance and give its units. (Fig. 19-16)

86. Write the general equation for clearance.

87. Why is inulin used to estimate GFR? Why is it not used more often in clinical settings? What substance is used instead?

88. Take some time to review the derivation of the concept of clearance. How do you calculate the filtration rate (filtered load)? When can you use a clearance rate to estimate GFR?

89. If you haven't done so yet, work the example in Concept Check question number 12. Clearance can be tricky at first, so the more practice, the better!

Clearance Helps Us Determine Renal Handling

90. Once a GFR is known, what two factors must be measured to allow you to analyze renal handling of a substance? How would you use these measurements to determine renal handling?

91. How can you determine renal handling of a substance by comparing its clearance rate to the clearance rate of inulin or creatinine? (Table 19-2)

92. What simple equation can you remember to help you analyze renal handling?

93. What information about renal handling can clearance provide? What information about renal handling can clearance not provide? (Fig. 19-17)

94. Spend some time analyzing Fig. 19-17 to become comfortable with the concept of clearance.

MICTURITION

95. The urinary bladder can hold about _____ mL.

96. What prevents urine from leaving the bladder?

97. Describe the tonic control exerted by the CNS over the bladder. (Fig. 19-18a)

98. Fill in the steps of the involuntary micturition reflex. (Fig. 19-18b)

stimulus _____   receptor _____

afferent pathway _____   integrating center _____

efferent pathway(s) _____   effector(s) _____

tissue response(s) _____   systemic response _____

99.   What role do higher brain centers play in micturition? (Fig. 19-18b)

TALK THE TALK

afferent
afferent arteriole
angiotensin II (ANGII)
ascending limb
basal lamina
benzoate
blackwater fever
Bowman's capsule
catheter
clearance
collecting duct
colloid osmotic pressure
cortex
creatinine
creatinine clearance
descending limb
distal nephron
distal tubule
efferent
efferent arteriole
erythropoietin
excretion
external sphincter
fenestrated capillaries
filtrate
filtration
filtration barriers
filtration coefficient
filtration fraction
filtration slits
fluid pressure (P_{fluid})
foot processes

functional unit
glomerular filtration rate
   (GFR)
glomerulus
glucosuria (glycosuria)
granular cells
honey-urine disease
hydrostatic pressure (P_H)
inulin
inulin clearance
internal sphincter
juxtaglomerular apparatus
juxtaglomerular cells (JG cells)
juxtamedullary
kidneys
loop of Henle
macula densa
medulla
megalin
mesangial cells
micturition
myogenic response
nephrin
nephron
organic anion exchange (OAT)
paracellular pathway
penicillin
peritoneum
peritubular capillaries
podocin
podocytes
portal system

potassium benzoate
probenecid
proximal tubule
reabsorption
renal arteries
renal corpuscle
renal pelvis
renal physiology
renal threshold
renal veins
renin
retroperitoneal
saccharin
saturation
secretion
sodium-hydrogen antiporter
sphincters
transcellular transport
transepithelial transport
transport maximum (T_m)
tubuloglomerular feedback
urate transporter (UAT)
urate transporter 1 (URAT1)
urea
ureter
urethra
urinalysis
urinary bladder
urinary system
urine
urobilinogen
vasa recta

QUANTITATIVE THINKING

USEFUL FORMULAS FOR RENAL PHYSIOLOGY

excretion = filtration − reabsorption + secretion

filtered load of a substance (χ) = GFR × plasma concentration of χ

clearance of χ = excretion rate of χ/plasma concentration of χ

excretion rate of χ = urine flow rate × urine concentration of χ

1. Tamika goes in for a routine physical examination. Her urinalysis shows proteinuria but she has no other abnormalities upon physical exam. She weighs 60 kg and is 159 cm tall. Her lab data show:

   serum creatinine: 1.8 mg/dL   urine creatinine: 276 mg/dL
   urine volume: 1100 mL in 24 hours

   Calculate Tamika's creatinine clearance and GFR.

2. Plasma concentration of inulin: 1 mg/mL
   Plasma concentration of X: 1 mg/mL
   GFR 125 mL/min

   What is the filtration rate of inulin? of X?
   What is the excretion rate of inulin? of X?

3. An alien willingly allows you to test its renal function. Answer the following questions about the alien kidney function. (Remember: alien kidneys don't necessarily follow the same rules as human kidneys.)

   a. Tests show that the alien kidney freely filters glucose. Once in the lumen of the nephron, glucose is reabsorbed but not secreted. The renal threshold is determined to be 500 mg glucose/100 mL plasma, and the alien's glucose transport maximum is 90 mg/min. Can you calculate the alien's GFR from this information? If so, what is it? If not, what other information do you need?

b. Additional tests show that creatinine gives an accurate GFR in this alien species. The alien transport maximum for the reabsorption of phenol red is 40 mg/min. Look at the following data:

GFR: 25 mL/min
urine: creatinine = 5 mg/mL; phenol red = 5 mg/mL
plasma: phenol red = 2 mg/mL
urine flow: 2 mL/min

What is this alien's creatinine clearance? phenol red clearance?

4. Graphing question: You are given a chemical Z and told to determine how it is handled by the kidneys of a mouse. After a series of experiments, you determine that (a) Z is freely filtered; (b) Z is not reabsorbed; (c) Z is actively secreted; and (d) the renal threshold for Z secretion is a plasma concentration of 80 mg/mL plasma, and the transport maximum is 40 mg/min. The mouse GFR is 1 mL/min. On the graph below, show how filtration, secretion, and excretion are related. One axis will be plasma concentration of Z (mg/mL) with a range of 0–140, and the other axis will show rates of kidney processes (mg/min) with a range of 0–140.

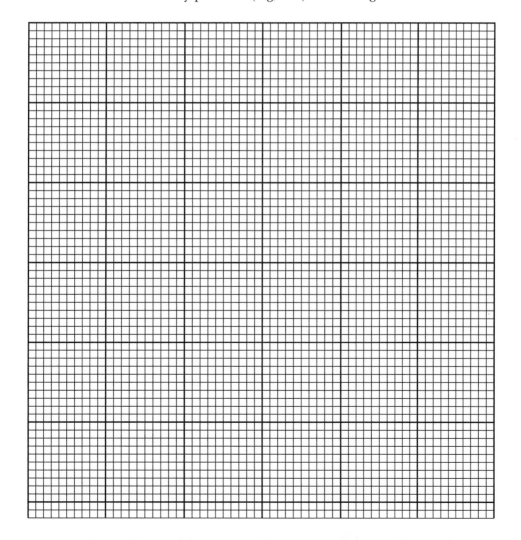

PRACTICE MAKES PERFECT

1. What has gone wrong with the nephron if a person has proteinuria (protein in the urine)?

2. Glucose is easily filtered by the kidneys. Why is glucose normally absent in the urine?

3. Diagram a nephron and label it. Next to each part name, write the processes or functions of that part.

4. Compare hydrostatic pressure, fluid pressure, osmotic pressure, and net direction of fluid flow in glomerular, peritubular, systemic, and pulmonary capillaries.

	Hydrostatic pressure	Fluid pressure	Osmotic pressure	Net direction of fluid flow
Glomerular capillaries				
Peritubular capillaries				
Systemic capillaries				
Pulmonary capillaries				

5. Define and give units for:

   a.   renal threshold

   b.   clearance

   c.   GFR

   spell out "GFR" _____

6. Compare and contrast the following concepts:

   a.   clearance and glomerular filtration rate

   b.   renal threshold and transport maximum

7. Using what you know about K^+ concentrations in the filtrate and proximal tubule cells, figure out how the cell could reabsorb K^+ from the tubule lumen. See Tables 5-2 and 5-3 for a list of common transporters.

	Apical membrane	Type of transport?	Basolateral membrane	Type of transport?
K^+				

8. Trace a water molecule from a capillary in the hand directly to the kidney and out through the urine. Name all anatomical structures through which the water molecule will pass. Do not write sentences. Use only names connected by arrows.

9. Inulin clearance when mean arterial pressure = 100 mm Hg is (greater than/less than/the same as?) inulin clearance when mean arterial pressure = 200 mm Hg.

10. Below is a graph of phenol red excretion in the bullfrog. You can assume that the structure of the frog kidney is like that of humans and that all four basic processes of human kidney function can occur. *Based on this graph*, answer the following questions:

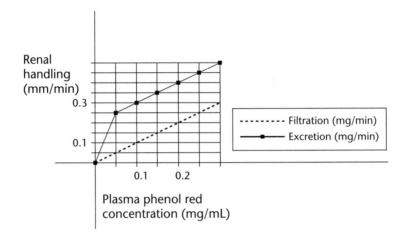

a. Circle the process or processes that allow excretion of phenol red by the bullfrog kidney:

    filtration   reabsorption   secretion

    On the graph above, draw in lines for any missing processes. (These lines should not be estimated but rather calculated based upon the given data.)

b. What causes the slope of the excretion line at a plasma phenol red concentration of 0.05 mg/mL to change?

c. If a phenol red transport *inhibitor* was administered concurrently with phenol red to this bullfrog, the clearance rate of phenol red compared to normal would:

    increase   decrease   stay the same

BEYOND THE PAGES

You can see a retroperitoneal kidney the next time you buy a whole chicken at the grocery. The abdominal organs are removed before the chicken is sold, but if you separate the chicken into breast and back halves by cutting through the ribs, you can see the paired kidneys lying alongside the backbone underneath the membranous lining of the abdomen.

ETHICS IN SCIENCE

Kidney transplants are now a very common procedure. How does organ availability affect our lives? Think about the urban myth that has spread through the Internet: a man visiting New Orleans was drugged following a wild party. The next morning he was found unconscious and bleeding, and doctors discovered that he was missing a kidney, removed for sale on the black market. Although this is just an urban legend, it brings up the question of whether people should be allowed to sell their organs. There have been reports of prison officials in China selling organs "to order" from prisoners who are to be executed. In the U.S., another ethical question is whether young women should be allowed to sell their ova (eggs) to infertile couples.

20 Integrative Physiology II: Fluid and Electrolyte Balance

LEARNING OBJECTIVES

When you complete this chapter, you should be able to:

- Map an overview of the body's integrated responses to blood volume and blood pressure changes.

- Diagram the tubular and vascular anatomy as well as the transport mechanisms (and solutes) that make the loop of Henle a countercurrent multiplier. Explain how this is the key to the regulation of urine concentration in response to the body's water needs.

- Diagram the mechanisms of vasopressin and aquaporins in water reabsorption.

- Map the stimuli and integrated responses that affect vasopressin release.

- Map the homeostatic responses to salt ingestion.

- Diagram aldosterone's cellular mechanisms of action at the P cells. What is the net result of aldosterone action? Where is aldosterone synthesized, how is it transported?

- Map how low blood pressure causes renin secretion.

- Map the renin-angiotensin-aldosterone (RAAS) pathway and show the multiple blood pressure-raising effects exerted by ANGII.

- Map the stimuli that result in formation of natriuretic peptides and show the homeostatic effects exerted by these NPs on target tissues. What is the net result of NP action?

- Explain why the regulation of body K^+ levels is essential to maintaining a state of well-being.

- Describe behavioral mechanisms involved in salt and water balance.

- Re-create Fig. 20-16 to show the different combinations of volume and osmolarity disturbances. For each combination of volume/osmolarity disturbance, diagram the appropriate homeostatic compensation.

- Map the homeostatic compensation mechanisms for severe dehydration.

- Compare and contrast the three mechanisms by which the body copes with minute-to-minute changes in pH. Which is the first line of defense? Second? Third?

- Using the law of mass action and the equation that describes how $CO_2 + H_2O$ create H_2CO_3 and $H^+ + HCO_3^-$, describe how buffers moderate pH fluctuations.

- Diagram the reflex pathways and cellular mechanisms of action that are involved in respiratory compensation of pH changes.

- Diagram the mechanisms by which the kidneys compensate (directly and indirectly) for pH changes. Identify the location (proximal or distal tubule) of the response mechanisms, identify specialized cell types and under which conditions they are active, identify all membrane transporters and whether they are found on the apical or basolateral membrane, and identify the solutes transported and the direction of transport. Finally, identify the net result of each mechanism and the overall net results of renal compensations in pH fluctuations.

- Map the causes and compensations involved in each of the four classes of acid-base disturbances (respiratory acidosis, metabolic acidosis, respiratory alkalosis, metabolic alkalosis). For each type of disturbance, show the changes in the equation for CO_2, H_2O, H_2CO_3, H^+, and HCO_3^- balance and note the factors (hallmarks) that are used to identify the nature of a particular disturbance.

SUMMARY

Fluid and electrolyte balance are under homeostatic control involving the renal, respiratory, and cardiovascular systems. Respiratory and cardiovascular mechanisms are primarily under neural control and are therefore more rapid than renal mechanisms, which are primarily under hormonal control.

Body osmolarity is homeostatically maintained at around 300 mOsM. Hypothalamic osmoreceptors sense changes in osmolarity and signal the kidneys to adjust urine concentration to correct for the changes. Kidneys must reabsorb water to make concentrated urine, and they must reabsorb Na^+ and not reabsorb water to make dilute urine. Water reabsorption takes place in the collecting duct of the neuron and is under the control of vasopressin released from the posterior pituitary. The presence of vasopressin increases the permeability of the collecting duct to water by inserting water pores into the apical membrane of the duct cell. Sodium reabsorption takes place in the distal tubule and collecting duct, and it is under the control of aldosterone from the adrenal cortex. Aldosterone initiates synthesis of Na^+-K^+-ATPase pumps on the basolateral membranes of P cells and thus allows Na^+ to be reabsorbed into the ECF.

Control of aldosterone release is complex. Secretion is controlled directly at the adrenal cortex by increased K^+ levels; secretion is inhibited by high osmolarity. Indirect control of aldosterone secretion involves renin produced in the juxtaglomerular cells of the kidney nephron. Renin secretion is directly or indirectly triggered by low blood pressure.

When secreted, renin converts angiotensinogen into angiotensin I (ANGI). ANGI is converted by angiotensin-converting enzyme (ACE) into angiotensin II (ANGII). ANGII causes release of aldosterone and widespread vasoconstriction, intended to contribute to an increase in blood pressure.

Changes in salt and water balance must be corrected, but it isn't always possible to correct the imbalances perfectly. In some instances, such as dehydration, integrating centers receive conflicting signals from different sensory pathways. When the body is faced with conflicting signals, the rule of thumb is to correct osmolarity imbalances first. Work through the "Integrated Control of Volume and Osmolarity" section in the chapter to get a full understanding of integrated salt and water balance. You should learn the integration of control well enough to be able to draw a map similar to the one in Fig. 20-17 without looking at any references.

Potassium balance is important to cell function, and is closely associated with Na^+ balance as well as with pH balance. Hyperkalemia and hypokalemia can cause problems with excitable tissues, especially the heart.

Acid-base balance is another homeostatic parameter that is crucial to body function. Remember that enzymes and other proteins are sensitive to pH changes. Acidosis is more common than alkalosis, but each requires compensation to restore homeostasis. The most important contributor to acidosis is CO_2 from respiration. Remember the equation: $CO_2 + H_2O \rightleftharpoons H_2CO_3 \rightleftharpoons H^+ + HCO_3^-$. There are three compensations that the body uses to combat acid-base imbalances: buffers, ventilation, and renal excretion of H^+ and HCO_3^-. Ventilation removes CO_2 and therefore decreases H^+ production. The kidneys remove the excess acid or base directly by excreting H^+ or HCO_3^- into the urine. Intercalated cells of the collecting duct are responsible for the excretion of ions: type A cells secrete H^+ and reabsorb HCO_3^- during acidosis; type B cells secrete HCO_3^- and reabsorb H^+ during alkalosis.

Acid-base disturbances are classified according to the pH change they induce and by how the imbalance originated. There are four possibilities: respiratory acidosis, respiratory alkalosis, metabolic acidosis, and metabolic alkalosis. When learning about acid-base disturbances, pay careful attention to how each of the four possibilities change the equation: $CO_2 + H_2O \rightleftharpoons H_2CO_3 \rightleftharpoons H^+ + HCO_3^-$. Also, learn the hallmarks associated with each condition as well as its compensations. If you can remember how the equation is manipulated, then you will have an easier time working problems concerned with acid-base balance. Make charts, diagrams, maps, or whatever helps you the most.

TEACH YOURSELF THE BASICS

FLUID AND ELECTROLYTE HOMEOSTASIS

1. What role does each of the following organs play in maintaining mass balance?

    kidneys

    lungs

2. How does behavior factor into the maintenance of fluid/electrolyte homeostasis?

3. For that matter, why are we concerned at all with fluid/electrolyte homeostasis? Why is it important? Include a discussion of some of the regulated substances and what happens if they get out of balance.

ECF Osmolarity Affects Cell Volume

 ▶ If you need a refresher on osmolarity, now's the time to go back to Ch. 5 and review.

4. Give a few examples of how osmolarity changes can affect cells and how specific cell types address osmolarity challenges.

Multiple Systems Integrate Fluid and Electrolyte Balance

5. Identify the differences between respiratory responses, cardiovascular responses, and renal responses. (Use a separate sheet of paper if necessary.) Compare response speed and reflex control. (See Ch. 6.)

6. Cite some instances of overlap in the regulation of fluid/electrolyte balance. Why do you think such an extensive network exists for maintaining these parameters? (Fig. 20-1)

WATER BALANCE

IP *Urinary System: Early Filtrate Processing*

7. In a 70-kg man, how much water is in his entire body? _____

    intracellular fluid (ICF)? _____ extracellular fluid (ECF)? _____

    plasma? _____ interstitial fluid? _____ (Fig. 5-25)

8. Now describe a woman's body water distribution.

Daily Water Intake and Excretion Are Balanced

9. List the normal routes of water input and water loss for the body. (Fig. 20-2)

10. Which are the most significant? Put a star next to them.

11. What is insensible water loss?

12. Give some examples of pathological water loss.

The Kidneys Conserve Water

13. TRUE/FALSE? Defend your answer.

    When body osmolarity goes up, the kidneys reabsorb water and bring osmolarity back to normal. (Fig. 20-3)

14. When water is lost from the body, how can it be restored?

The Renal Medulla Creates Concentrated Urine

15. By what process do the kidneys eliminate excess water?

16. By what process do the kidneys conserve water?

17. Diagram how the concentration gradient in the medullary interstitium is established. Include ions, osmolarities, and tubule anatomy. (Fig. 20-4)

Vasopressin Controls Water Reabsorption

18. How does the collecting duct alter its permeability to water and determine final urine concentration? (Fig. 20-5)

19. Is permeability absolute or graded? What does this accomplish?

Vasopressin and Aquaporins

20. What are aquaporins? Where are they stored when not influencing water movement?

21. Which is the aquaporin isoform regulated by vasopressin? Is this vasopressin-sensitive isoform inserted into the apical or basolateral membrane of a collecting duct cell? What is the significant of that?

22. Diagram vasopressin regulation of AQP2. What happens in low vasopressin concentrations? What is vasopressin's cellular mechanism of action? (Fig. 20-6)

23. What is the net result of vasopressin action?

Blood Volume and Osmolarity Activate Osmoreceptors

24. What stimuli control vasopressin secretion? (Fig. 20-7) Which of these is the most potent stimulus for vasopressin release?

25. Describe the current model of osmoreceptor mechanism of action.

26. Where are the primary osmoreceptors located? Diagram how they influence vasopressin release in response to a plasma osmolarity increase above 280 mOsM. (Fig. 20-8)

27. Diagram how a decrease in blood pressure or volume would influence vasopressin release.

28. Describe the circadian rhythm of vasopressin release in adults.

► Fig. 20-7 shows integrated responses to blood volume/pressure changes.

The Loop of Henle Is a Countercurrent Multiplier

29. What is the key to the kidney's ability to produce concentrated urine? (Fig. 20-4)

Countercurrent Exchange Systems

30. Describe a countercurrent exchange system. What are the requirements for such a system? What is the purpose of a countercurrent exchanger?

31. Diagram a countercurrent heat exchanger. (Fig. 20-9)

The Renal Countercurrent Multiplier

32. Diagram how the loop of Henle is a countercurrent exchange system and a countercurrent multiplier. What are some of the membrane transport proteins involved? (Fig. 20-10)

33. The descending limb of the loop is permeable to (water/solutes?) and impermeable to (water/solutes?). The ascending limb is permeable to (water/solutes?) and impermeable to (water/solutes?). As a result of the countercurrent system, fluid leaving the loop is (hypo-/iso-/hyper-?) osmotic to the blood.

The Vasa Recta Removes Water

34. Describe or diagram solute/water movement in the descending and ascending components of the vasa recta. What factors contribute to solute/water movement? (Fig. 20-10a)

35. How do the vasa recta participate in the countercurrent exchange/multiplier system?

Urea Increases Osmolarity of the Medullary Interstitium

36. What percentage of the solute in the medullary interstitial fluid is urea?

37. How does urea cross cell membranes in the collecting duct? Does it move via active or passive transport?

SODIUM BALANCE AND ECF VOLUME

IP *Urinary System: Late Filtrate Processing*

38. What is normal total body osmolarity? What is normal plasma Na^+ concentration measured from a venous blood sample?

39. The addition of NaCl to the body raises osmolarity. This stimulus triggers two responses: what are they?

40. Bringing increased osmolarity back into normal range often involves disrupting another set of homeostatic parameters. What are these parameters? How are they brought back into normal range? (Fig. 20-11)

41. Sodium excretion is a function of the _____. What are other ways in which sodium leaves the body?

Aldosterone Controls Sodium Balance

42. If aldosterone behaves like a typical steroid hormone, then describe how it is made; if aldosterone is stored, how is it transported to its target; and describe its cellular mechanism of action. (See Ch. 7 for more details.)

43. What are the target cells for aldosterone, and where are they located? Diagram these cells and their apical and basolateral membranes. (Fig. 20-12)

44. What effect does aldosterone have on its target cells? There's a faster early response and a slower response that comes later. Diagram both. What's the net result of aldosterone action? (Fig. 20-12)

45.   Compare water and sodium handling in the distal nephron versus the proximal tubule.

Low Blood Pressure Stimulates Aldosterone Secretion

46.   What are the direct and indirect stimuli for aldosterone secretion? (Table 20-1)

The Renin-Angiotensin-Aldosterone Pathway (RAAS)

47.   Arrange the following terms into a map of the RAAS pathway. (Fig. 20-13)

active	angiotensin I (ANGI)	liver
adrenal cortex	angiotensin II (ANGII)	plasma protein
afferent arteriole	angiotensinogen	renin
aldosterone	endothelium	
angiotensin converting	inactive	
enzyme (ACE)	JG cells	

48.   The stimuli that begin the RAAS pathway are all related directly or indirectly to what?

49.   What is renin, and where is it secreted? (Fig. 20-13)

50.   List three stimuli for renin secretion. (Fig. 20-14)

51.   TRUE/FALSE? Defend your answer.

   Na^+ retention immediately raises low blood pressure. (Fig. 20-11)

ANGII Has Many Effects

52.   List the effects of ANGII beyond stimulating aldosterone secretion. (Fig. 20-13)

53.   What are ACE inhibitors? What is their clinical significance? What are their side effects? What new therapies were developed as a result? How do they accomplish the same goals as ACE inhibitors?

ANP Promotes Na⁺ and Water Excretion

54. What is ANP, and where is it produced?

55. How is ANP made, stored, transported, and what is its cellular mechanism of action? (See Ch 7 for a review.)

56. What is the stimulus for NP secretion?

57. What is the net effect of natriuretic peptide action? What are the direct actions that accomplish this? (Fig. 20-15)

58. What are some of the indirect effects of NPs? Do these enhance or inhibit the natriuretic effects?

59. What is brain natriuretic peptide (BNP)? Describe its clinical utility.

POTASSIUM BALANCE

60. K⁺ balance is important despite its relatively (high/low?) ECF concentration. (See Ch. 8.)

61. Hyperkalemia (increases/decreases?) the K⁺ concentration gradient across cell membranes and (depolarizes/hyperpolarizes?) cells. This leads to (increased/decreased?) excitability in excitable tissues and can lead to cardiac arrhythmias.

62. Compare this to hypokalemia.

63. How does the body compensate for increased K⁺ levels?

64. Cite some possible sources for K⁺ disturbances.

BEHAVIORAL MECHANISMS IN SALT AND WATER BALANCE

65. Drinking water is normally the only way to _____, and eating salt is

   the only way to _____.

Drinking Replaces Fluid Loss

66. Which receptors trigger thirst, and what is their threshold osmolarity?

67. How does the act of drinking alleviate thirst? Which receptors are involved? What is the adaptive advantage of this?

Low Na⁺ Stimulates Salt Appetite

68. Define salt appetite.

69. Where are the human salt appetite centers?

Avoidance Behaviors Help Prevent Dehydration

70. List some other behaviors that help prevent dehydration. What is the adaptive significance of each?

INTEGRATED CONTROL OF VOLUME AND OSMOLARITY

IP *Fluids & Electrolytes: Water Homeostasis*

Osmolarity and Volume Can Change Independently

71. What are some examples of extreme fluid loss?

72. Re-create the chart in Fig. 20-16 that represents the possible disturbances of volume and osmolarity.

73. For each of the situations below, describe possible causes and in general terms explain the appropriate homeostatic response.

    increased volume and increased osmolarity

    increased volume, unchanged osmolarity

    increased volume and decreased osmolarity

    normal volume with increased osmolarity

    normal volume with decreased osmolarity

    decreased volume and increased osmolarity

    decreased volume, unchanged osmolarity

    decreased volume and decreased osmolarity

Dehydration Triggers Homeostatic Responses

▶ Table 20-1 summarizes pathways involved in volume/osmolarity homeostasis.

74. When integrating centers receive conflicting information about regulation of volume and

    osmolarity, correction of _____ has priority.

75. What are the compensatory mechanisms involved in severe dehydration? (Fig. 20-17)

76. What are the homeostatic challenges caused by dehydration?

77. Carotid, aortic _____ respond to BP changes. The cardiovascular control center (increases/decreases?) parasympathetic output and (increases/decreases?) sympathetic output in an effort to (increase/decrease?) blood pressure. Decreased volume is sensed by _____.

78. Autonomic influence is exerted on a variety of different effectors in response to severe dehydration. For each of these pathways, identify the autonomic branch that exerts influence and name the neurotransmitters involved, the receptors, and the specific target tissues. (You might need to review details in Chs. 11, 14, 15, 19, and 20.)

79. What effect does severe dehydration have on GFR?

80. List all the stimuli that will cause renin release in this pathway.

81. What behavioral pathway(s) will be initiated?

▶ Redundancy in control pathways ensures that four main compensatory mechanisms are activated.

82. Describe the four main compensatory mechanisms in the severe dehydration example presented in the chapter.

83. What is the net result of these compensatory measures?

84. If you haven't yet done it, make reflex pathways similar to Fig. 20-17 for the volume/osmolarity disturbances shown in Fig. 20-16. Use the pathways listed in Table 20-1.

ACID-BASE BALANCE

IP *Fluids & Electrolytes: Acid/Base Homeostasis*

85. Define pH. (See Ch. 2 and Appendix B.)

86. A change of 1 pH unit = a _____-fold change in H^+.

87. An alkaline solution has a (higher/lower?) H^+ concentration and a (higher/lower?) pH than an acidic solution.

pH Changes Can Denature Proteins

88. What are both the normal average pH and the normal range of pH in the human body?

89. Which body fluid is used clinically to indicate the body pH?

90. If the pH range compatible with life is 7.0–7.7, how can we survive stomach juices with a pH as low as 1 or urine with a pH ranging from 4.5–8.5?

91. If pH falls outside the normal range, what kinds of things go wrong?

92. Acidosis (increases/decreases?) neuron excitability; alkalosis (increases/decreases?) neuron excitability.

93. Disturbances in pH balance are associated with changes in K^+ balance. Name the membrane transporter that links movement of H^+ and K^+ in the kidney.

94. This transporter moves the ions in (the same/opposite?) directions, so that in acidosis, when H^+ is excreted, K^+ is _____. In alkalosis, when H^+ is (excreted/reabsorbed?), K^+ is (excreted/reabsorbed?).

Acids and Bases in the Body Come from Many Sources

95. In day-to-day functioning, is the body more challenged by intake and production acids or bases?

96. Maintaining mass balance requires what? (Fig. 20-19)

Acid Input

97. Cite examples of organic acids that contribute to H^+.

98. What is lactic acidosis? What is ketoacidosis? How can these states develop?

99. The biggest daily source of acid production is _____. How is this so?

100. Write the reaction for the production of H_2CO_3 from $CO_2 + H_2O$, and the subsequent production of H^+ and HCO_3^-.

101. What enzyme catalyzes the reaction in the previous question? (See Ch. 18.)

Base Input

102. There are few significant sources of bases in the diet and metabolism. List a couple.

▶ Because of this, the body spends more time correcting acid disturbances.

pH Homeostasis Depends on Buffers, the Lungs, and the Kidneys

103. Name the three mechanisms used by the body to cope with minute-to-minute changes in pH. List the order in which these mechanisms are employed—which is the first line of defense, second, third ... and explain why.

Buffer Systems Include Proteins, Phosphate Ions, and Bicarbonate Ions

104. Define the term buffer. (See Ch. 2.)

105. List some important buffers found in the human body.

106. Which is the most important extracellular buffer? Where does it come from, and how does it function? Include the equation. (See Ch. 18.)

107. According to the law of mass action, an increase in CO_2 will cause a/an (increase/decrease?) in H^+ and HCO_3^-.

108. What effect does an increase in H^+ from a metabolic source have on pH, HCO_3^-, and CO_2?

109. Sometimes changes in HCO_3^- concentrations aren't observed clinically. Explain why this is TRUE.

Ventilation Can Compensate for pH Disturbances

110. Because CO_2 and H^+ are linked in a dynamic equilibrium, P_{CO_2} changes will affect both H^+ and HCO_3^- levels.

    If ventilation increases, plasma P_{CO_2} (↑/↓?), H^+ (↑/↓?), and HCO_3^- (↑/↓?).

    If ventilation decreases, plasma P_{CO_2} (↑/↓?), H^+ (↑/↓?), and HCO_3^- (↑/↓?).

111. The body uses ventilation as a method for adjusting pH only if a stimulus associated with pH triggers the reflex response. Name the two stimuli that can do this: _____.
    (Fig. 20-20)

112. The carotid and aortic chemoreceptors respond to changes in _____ and initiate the following pathway (Fig. 20-19):

113. The central chemoreceptors respond to changes in _____ and initiate the following pathway: (Fig. 18-19)

Kidneys Use Ammonia and Phosphate Buffers

114. Describe the two ways that kidneys can alter pH.

115. How do the kidneys respond in times of acidosis? (Fig. 20-20)

116. What substances act as buffers in the kidneys?

117. Even with buffers, urine can become quite (acidic/basic?).

118. How do the kidneys respond in times of alkalosis?

119. Renal compensations are (faster/slower?) than respiratory compensations.

120. Diagram the membrane transporters involved in renal compensation of alkalosis. Show cell membranes crossed, ion movement, and concentration gradients.

The Proximal Tubule Secretes H⁺ and Reabsorbs HCO₃⁻

121. The proximal tubule reabsorbs most filtered HCO_3^- by indirect methods because there are no apical transporters to do what?

122. Looking at Fig. 20-21, you'll see two pathways where the transporters function together to achieve HCO_3^- and H^+ movement. Diagram the mechanisms by which bicarbonate is reabsorbed and H^+ secreted in the proximal tubule.

The Distal Nephron Controls Acid Excretion

123. The distal nephron plays a significant role in the _____ of acid-base balance.

124. How do I cells compare to P cells? What does "I" stand for here?

125. I cells are characterized by high concentrations of _____ in their cytoplasm.

126. Identify the differences between type A and type B cells: under which conditions is each type active, and what is the net result of ion movement in each?

127. Diagram how type A cells work. Include the transporters and ions involved. (Fig. 20-22a)

128. Describe how type B cells work. Include the transporters and ions involved. (Fig. 20-22b)

129. How are acid-base disturbances related to K^+ balance? Discuss transporters and acid-base states when answering this question.

Acid-Base Disturbances May Be Respiratory or Metabolic

130. Describe the classification of acid-base disturbances. (Table 20-2)

131. By the time an acid-base disturbance shows up as a change in plasma pH, what has happened?

132. If the problem is of respiratory origin, only which homeostatic compensation is available?

133. On the other hand, in metabolic acid-base disturbances, which homeostatic compensations are available?

Respiratory Acidosis

134. What alteration in ventilation will cause a respiratory acidosis?

135. Name some conditions/situations in which this might occur.

136. Fill in the equation with ↑ or ↓. _____ CO_2 + H_2O → _____ H^+ + _____ HCO_3^-

137. The hallmark of respiratory acidosis is:

138. What compensation mechanisms are available in this state?

139. As compensation occurs, the pH will (increase/decrease?) as H^+ is (excreted/reabsorbed?) and HCO_3^- is (excreted/reabsorbed?).

Metabolic Acidosis

140. Name some causes of metabolic acidosis.

141. Fill in the equation with ↑ or ↓. _____ CO_2 + H_2O ← _____ H^+ + _____ HCO_3^-

142. The hallmark of metabolic acidosis is:

143. What compensation mechanisms are available in this state?

144. Why don't you see an increase in P_{CO_2} associated with metabolic acidosis? How does ventilation change in compensation?

145. What do the kidneys do to compensate for this condition?

Respiratory Alkalosis

146. Respiratory alkalosis occurs as a result of:

147. Fill in the equation with ↑ or ↓. _____ CO_2 + H_2O ← _____ H^+ + _____ HCO_3^-

148. The hallmark that alkalosis is respiratory in origin is:

149. What compensation mechanisms are available in this state?

Metabolic Alkalosis

150. What could cause metabolic alkalosis?

151. Fill in the equation with ↑ or ↓. _____ CO_2 + H_2O → _____ H^+ + _____ HCO_3^-

152. The hallmark of metabolic alkalosis is:

153. What compensation mechanisms are available in this state?

154. The change in P_{CO_2} in this condition does what to ventilation?

155. The respiratory compensation helps correct the pH problem but does what to HCO_3^- levels?

156. What do the kidneys do to compensate for this condition?

157. To distinguish a respiratory acidosis from a respiratory alkalosis, you should look primarily at the relative concentrations of (H^+/CO_2/HCO_3^-?).

158. To distinguish a respiratory acidosis from a metabolic acidosis, you should look primarily at the relative concentrations of (H^+/CO_2/HCO_3^-?).

159. To distinguish a respiratory alkalosis from a metabolic alkalosis, you should look primarily at the relative concentrations of (H^+/CO_2/HCO_3^-?).

TALK THE TALK

ACE inhibitors
acidosis
aldosterone
alkalosis
angiotensin converting
    enzyme (ACE)
angiotensin I (ANGI)
angiotensin II (ANGII)
angiotensinogen
antidiuretic hormone (ADH)
apical sodium-hydrogen
    antiport
aquaporin-2
aquaporins
AT receptors
atrial natriuretic peptide
    (ANP)
basolateral $Na^+-HCO_3^-$ symport
bradykinin
brain natriuretic peptide
carbonic anhydrase
cardiovascular control center
    (CVCC)

chloride shift
chronic obstructive
    pulmonary disease (COPD)
countercurrent exchange
    system
countercurrent multiplier
diarrhea
dilute
diuresis
diuretics
granular cells
H^+-ATPase
H^+-K^+-ATPase
hypokalemia
hyponatremia
insensible water loss
intercalated cells (I cells)
interstitium
intravenous (IV) injection
ketoacidosis
ketoacids
lactic acidosis
mass balance

medullary interstitial osmolarity
membrane recycling
Na^+-H^+ exchanger (NHE)
natriuresis
NKCC symporter
osmoreceptors
paracrine feedback
physiological psychology
principal cells (P cells)
renin-angiotensin-aldosterone
    system (RAAS)
respiratory compensation
salt appetite
sartans
sympathetic neurons
tetanus
thirst
vasa recta
vasopressin
vasopressin receptor
    antagonists
water pores

QUANTITATIVE THINKING

OSMOTIC DIURESIS

When a diabetic's plasma glucose concentration exceeds the renal threshold for glucose reabsorption, the unreabsorbed glucose is excreted in the urine. The presence of this excess solute in the tubule lumen will cause additional water loss and create an *osmotic diuresis*. In the clinics, it is sometimes said that "the glucose holds the water in the urine." Physiologically, osmotic diuresis is based on simple amount/concentration relationships. The nephron can concentrate urine only to a maximum of 1200 mOsM/L. If there is more solute in the lumen (i.e., greater amount of solute) but the osmolarity cannot increase above 1200 mOsM, there will be a greater urine volume. To understand why this is so, work through the following problem.

The following shows a diagram of a nephron. The numbers inside the nephron at points A, B, and C represent the concentration of the filtrate at those points. Assume for this example that vasopressin is present in amounts that allow maximal concentration of the urine and that there is no solute reabsorption in the distal parts of the nephron.

Assume that in a normal individual, 150 mOsM of NaCl per day pass from the loop of Henle into the urine. Given the concentrations shown at points A–C, what volume of fluid passes those points in one day?

Point A _____ Point B _____ Point C _____

Now suppose the person suddenly becomes diabetic and is unable to reabsorb all filtered glucose. The 150 mosmoles of NaCl in the distal nephron are joined by 150 mosmoles of unreabsorbed glucose. Now the *amount* of solute leaving the loop of Henle has increased from 150 to 300 mosmoles. Do you think the volume of fluid leaving the loop of Henle will be the same, greater, or smaller?

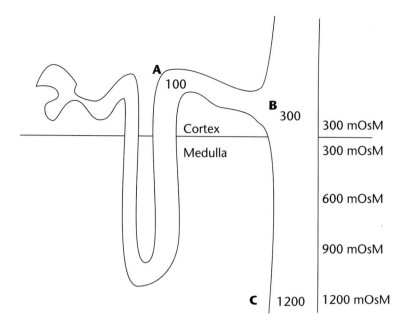

Now do the calculations:

	Osmolarity	Volume
Point A		
Point B		
Point C		

What happened to the volume of urine when the *amount* of solute entering the distal nephron doubled? This shows you the process behind fluid loss in osmotic diuresis.

PRACTICE MAKES PERFECT

1. Compare the following pairs of items. Put the symbols below in the space.

    greater than >   less than <   same as or equal =

    a ____ b   a.   urine osmolarity in a normal person with maximal vasopressin
                 b.   urine osmolarity in a diabetic who is excreting glucose with maximal vasopressin

    a ____ b   a.   aldosterone secretion when osmolarity is high and blood pressure low
                 b.   aldosterone secretion when osmolarity and blood pressure are both low

    a ____ b   a.   plasma P_{CO_2} in respiratory acidosis
                 b.   plasma P_{CO_2} in respiratory alkalosis

    a ____ b   a.   renal reabsorption of HCO_3^- in acidosis
                 b.   renal reabsorption of HCO_3^- in alkalosis

    a ____ b   a.   ventilation in metabolic alkalosis
                 b.   ventilation in metabolic acidosis

    a ____ b   a.   renin secretion when blood pressure is high
                 b.   renin secretion when blood pressure is low

    a ____ b   a.   renal filtration of HCO_3^- in acidosis
                 b.   renal filtration of HCO_3^- in alkalosis

2. What effect does a decrease in mean arterial blood pressure have on each item below?

    a. Na^+ reabsorption in the proximal tubule

    b.   blood pressure in the afferent arteriole

    c.   atrial stretch

    d.   aldosterone secretion

    e.   angiotensinogen levels in the plasma

    f.   blood volume

3. The labels were left off the axes on the following graph. One axis is vasopressin concentration and one axis is plasma osmolarity. Which is which? Defend your answer.

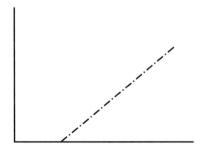

4. It's the night before your physiology test, and you awake from a crazy dream to find yourself still at your desk in a pile of books and papers. You remember that in your dream, you were walking through a peach orchard. As you were walking down the path, you noticed that all the trees on your left were turning into CO_2 and H_2O. Looking to your right, you noticed that all the trees on that side were turning into H^+ and HCO_3^-. Having thoroughly studied acid-base balance, you recognized that you were standing in the middle of the important acid-base equation. As you realize this, a funny new tree pops up. On the right side of the tree there are green fruit, and on the left side there are red fruit. You are hungry from all your studying, so your subconscious picks a green fruit from the tree.

    As you eat the green fruit, something strange happens (as if this isn't strange enough already!). The left side of the path (the $CO_2 + H_2O$ side) grows and grows until it looks as if it will topple over you. You suddenly realize that you have a touch pad in your hand that will enable you to set the proper compensatory mechanisms in action. What do you do to prevent your impending doom?

    Write the equation of interest, show the imbalance that occurred, and describe what physiological mechanisms you employed to compensate for the imbalance. Why did you choose the way you did?

5. In a patient with an acid-base disturbance characterized by a plasma P_{CO_2} of 80 mm Hg and a plasma bicarbonate concentration of 33 mEq/L (normal = 24), you would expect to find (circle all that apply):

   a. virtually complete renal bicarbonate reabsorption despite the elevated plasma bicarbonate
   b. urine pH less than 6.0
   c. stimulation of distal tubule H^+ secretion due to elevated plasma P_{CO_2}
   d. hypokalemia

6. Mr. Osgoode arrives at the hospital hyperventilating and disoriented. You order tests and receive these results: plasma pH 7.31; arterial P_{CO_2} = 30 mm Hg; plasma HCO_3^- = 20 mEq/L (normal = 24). What are the normal values for pH and P_{CO_2}? What is Mr. Osgoode's acid-base state, and how did you come to that conclusion?

7. Ari wants to play a trick on his friends, so he gets a long tube and hides in the bottom of a pond, breathing through the tube. After a few minutes, he feels that he is having trouble getting air and considers surfacing. If you tested his plasma pH, plasma HCO_3^-, and arterial P_{CO_2} at this time, would you expect each of them to be normal, above normal, or below normal? Explain your answers.

MAPS

1. Create a summary map showing the compensations for acid-base disturbances, using the following terms and any others you wish to add.

acidosis	filtration	metabolic alkalosis
alkalosis	H⁺-ATPase	Na⁺-HCO₃⁻ symport
ammonia	H⁺-K⁺-ATPase	Na⁺-H⁺ antiport
bicarbonate	HCO₃⁻ –Cl⁻ antiport	Na⁺-K⁺-ATPase
buffer	hyperkalemia	pH
carbonic anhydrase	hyperventilation	phosphate ion
carotid and aortic chemore-ceptor	hypokalemia	reabsorption
	hypoventilation	respiratory acidosis
central chemoreceptor	intercalated cell (I cell)	respiratory alkalosis
chloride shift	law of mass action	respiratory compensation
CO₂	medullary chemoreceptor	secretion
excretion	metabolic acidosis	

2. Use the map on the renin-angiotensin pathway (Fig. 20-13) as the basis for a large map. Fill in mechanisms and receptors and neurotransmitters—and anything else you can think of. Here's your chance to demonstrate mastery of physiology!

3. Take each of the following situations and

    a. figure out what changes will occur in the volume, osmolarity, and sodium concentration of the body.
    b. figure out what homeostatic responses these changes will trigger and how.
    c. describe *in as much detail as possible* the homeostatic response.
    d. explain how the homeostatic response will bring the altered condition(s) back to normal.

    On large pieces of paper, map the detailed homeostatic pathways for each situation below.

    Example 1: Person ingests a large amount of salt and drinks enough water so that the result is as if he/she drank a large volume of isotonic saline.

    Example 2: Person drinks a large volume of pure water.

    Example 3: Person loses water and salt, but more water than salt through sweating (loses hyposmotic fluid).

    Example 4: Person loses a large volume of blood (hemorrhage).

HINTS FOR MAPPING VOLUME AND OSMOLARITY PROBLEMS

1. What are the possible stimuli? The primary stimuli are blood pressure and osmolarity (increased, decreased, or no change ... see the matrix). You should also consider Na^+, K^+, and blood volume.

2. Make a list of all receptors/tissues that respond to the stimuli in #1. Don't forget tissues that respond directly to the stimuli, such as the atrial cells that secrete ANP and the aldosterone-secreting cells of the adrenal cortex. Don't forget GFR and blood pressure homeostasis via the cardiovascular control center.

3. Once you have established which pathways will be stimulated/inhibited, draw out all reflex pathways through to the desired response, which should be the opposite of the stimulus. If you get to the end and have a response that is the same as the stimulus (i.e., stimulus was increased BP but the response increases BP), go back ... you've done something wrong or forgotten a link. Be sure to look for places where pathways intersect.

4. Sometimes two pathways will have opposite effects on a single integrating center (i.e., one stimulates aldosterone, the other inhibits). Assume that the response that opposes the stimulus is dominant. Alternatively, remember that the body defends osmolarity before it defends volume.

BEYOND THE PAGES

TRY IT

Kitchen Buffers

Your kitchen contains the necessary ingredients for a simple demonstration of the bicarbonate buffer system. Pour a couple of tablespoons of white or cider vinegar into a small bowl. Taste the vinegar and notice the sour or tart taste. The sensation of sour taste is directly proportional to the acidity of a solution. Now put a tablespoon of baking soda into another small bowl. Baking soda is sodium bicarbonate, the sodium salt of the bicarbonate buffer.

Remember the equation in which bicarbonate buffers acid:

$$CO_2 + H_2O \rightleftharpoons H_2CO_3 \rightleftharpoons H^+ + HCO_3^-$$

Based on this equation, what do you predict will happen when you pour the vinegar into the baking soda? If you taste the resulting solution, do you think the taste will be the same as it was before? Explain.

Now conduct your experiment. What happens when the vinegar and baking soda mix? This reaction has been used in baking for centuries. Allow the excess baking soda to settle to the bottom of the bowl and taste the remaining vinegar solution. Is it as sour as before?

21 The Digestive System

LEARNING OBJECTIVES

When you complete this chapter, you should be able to:

- Trace the path a piece of undigested food would travel from the mouth to the anus and name all the anatomical structures it might encounter.

- Diagram the layered structure of the GI tract wall and explain structures within and functions of each layer.

- Diagram how slow wave potentials are created in the intestines and how this can lead to smooth muscle contraction.

- Distinguish and describe the different patterns of smooth muscle contraction observed in the GI tract.

- Diagram the location and the cellular processes by which the GI tract secretes acid, bicarbonate, and NaCl.

- Explain the adaptive significance of why the body secretes many digestive enzymes as zymogens.

- Contrast long reflexes, short reflexes, and reflexes involving GI peptides.

- Describe the anatomy and physiology of the ENS and compare/contrast it with the CNS.

- Identify the three families of GI hormones and give examples of each.

- Explain how pH can be used to predict the location where a particular digestive enzyme might be most active.

- Diagram the absorption mechanisms for carbohydrates, proteins, fats, nucleic acids, ions, vitamins, minerals, and water.

- Map the key processes and control pathways associated with the cephalic phase, the gastric phase, and the intestinal phase of digestion. Where possible, include diagrams of cellular mechanisms and all enzymes, hormones, paracrines, ions, or other substances that might be involved.

- Compare and contrast small and large intestines in terms of anatomy, digestion, absorption, secretion, and motility.

- Describe the GALT and explain how M cells provide information about the contents of the lumen.

SUMMARY

Digestion concepts and details would work well as a series of charts connected by pathway maps. In other words, start a large concept map and incorporate charts as well as the more standard concept map elements.

If you trace food through the digestive system, it passes through the following structures: mouth, pharynx, esophagus, stomach (fundus, body, antrum), small intestine (duodenum, jejunum, ileum), large intestine, rectum, and anus. Exocrine secretions are added by the salivary glands, pancreas, and liver. The GI tract contains the largest collection of lymphoid tissue, called the gut-associated lymphoid tissue (GALT). The enteric nervous system (ENS) integrates and initiates GI reflexes. The ENS can act in coordination with or completely separate from the cephalic brain.

There are three phases of food processing: cephalic, gastric, and intestinal. There are four digestive processes that can take place during food processing: digestion, absorption, motility, and secretion.

Digestion of biomolecules takes place at various locations throughout the GI tract. Carbohydrates are digested in the mouth and small intestine by amylases and disaccharidases. Proteins are digested in the stomach and small intestine by endopeptidases and exopeptidases. Fat digestion begins in the mouth and small intestine with the help of lipases, but most fat digestion takes place in the small intestine via pancreatic lipase and colipase. Bile from the liver emulsifies fats, increasing the surface area exposed to enzyme action.

Absorption takes place mostly in the small intestine. Because the intestinal wall is composed of transporting epithelium, transepithelial transport mechanisms similar to those in the kidneys are observed. Glucose is transported across the apical membrane by the Na^+-glucose SGLT symporter and across the basolateral membrane by the GLUT2 transporter. Amino acids are absorbed by a Na^+-dependent cotransporter, while small peptides are absorbed by the H^+-dependent oligopeptide transporter (PepT1) or by transcytosis. Fat absorption is primarily by simple diffusion. In epithelial cell cytoplasm, monoglycerides and fatty acids are moved to the smooth ER where they combine with cholesterol and proteins to form chylomicrons. Chylomicrons are then transported out of the cells by exocytosis and moved into the lymphatic system—bypassing the liver and entering the venous blood just before it reaches the heart. Most nutrients are absorbed into the hepatic portal system by which they are delivered into the liver.

Motility refers to the movement of material through the digestive system. The muscles of the GI tract are single-unit smooth muscle whose cells are connected by gap junctions. Interstitial cells of Cajal generate spontaneous slow-wave potentials that then spread to nearby cells via gap junctions. The wall of the intestinal tract is composed of an outer layer of longitudinal muscle and an inner layer of circular muscle. Peristaltic contractions are progressive waves of contraction that propel material from the esophagus to the rectum. Segmental contractions are mixing contractions that churn the material without propelling it forward. Material in the large intestine is moved forward by mass movement and is removed from the body by means of the defecation reflex. Each segment of the GI tract is separated by muscular sphincters that are tonically contracted except when food must move into the next segment or out of the body. Motility in each section of the GI tract is under multiple controls—both neural and hormonal.

Various components of the digestive system contribute mucus, enzymes, hormones, and paracrines that make up the 7 L secreted by the body into the GI lumen. Digestive enzymes are secreted by either exocrine glands or epithelial cells in the mucosa of the stomach or small intestine. Because enzymes are proteins, they are synthesized ahead of time, stored in secretory vesicles, and released on demand. To prevent autodigestion and provide an additional level of control, some enzymes are secreted as inactive zymogens that must be activated in the GI lumen. Mucus forms a protective coating over the GI mucosa and also provides lubrication for food movement. Large amounts of water and ions (Na^+, K^+, Cl^-, H^+, and HCO_3^-) are also secreted. For example, the stomach secretes H^+-rich fluid while the pancreas secretes HCO_3^--rich fluid.

GI tract processes are governed by two types of neural reflexes: short reflexes, which take place entirely in the enteric nervous system; and long reflexes, which are integrated in the CNS. Stimuli for long reflexes may originate either inside or outside the GI tract. Peptides controlling GI peptides can be grouped into three families: gastrin family, secretin family, and those not belonging to either. GI peptides can act locally or affect brain regions to influence behavior.

TEACH YOURSELF THE BASICS

FUNCTION AND PROCESSES OF THE DIGESTIVE SYSTEM

1. What is the function of the GI tract?

2. List challenges the GI tract must overcome to carry out this function. (Fig. 21-1)

3. List and define the four basic processes of the digestive system. (Fig. 21-2)

4. What happens to nutrients brought into the body through the digestive system?

5. Why does the GI tract have the largest collection of lymphoid tissue of any organ?

6. What is the name given to the GI system's immune tissue?

ANATOMY OF THE DIGESTIVE SYSTEM

IP *GI—Anatomy Review*
7. What are the first steps in digestion? Where do these steps take place? What glands and secretions are involved?

8. The GI tract is a long tube separated into segments by _____. Along the way, food becomes

   chyme through the addition of secretions from _____. (Fig. 21-3)

9. Digestion takes place mainly in the _____ of the GI tract, and digestive products are

   _____ across the epithelium and pass into the _____ compartment.

The Digestive System Consists of Oral Cavity, GI Tract, and Accessory Glandular Organs
10. Trace a piece of food that enters the digestive system through the mouth, and follow its undigested portion until it is excreted. Include the sphincters the food passes. (Fig. 21-3)

11. List the three sections of the stomach.

12. List the three sections of the small intestine.

13. The digestive waste that leaves the body is called _____.

The GI Tract Wall Has Four Layers

14. List and briefly describe the four layers of the GI wall in the stomach and intestines, from inner to outer layers. (Fig. 21-3c–e)

The Mucosa

15. List and describe the three layers of the mucosa.

16. What anatomical modifications increase surface area facing the lumen of the stomach and intestine? (Fig. 21-3d)

17. The tubular invaginations of the lumen that extend into supporting connective tissue are called

    _____ in the stomach and _____ in

    the small intestine.

18. List the types of epithelial cells found in the lumen of the GI tract. What is the function of each?

19. Compare cell-to-cell junctions of the stomach and intestines.

20. Why are GI stem cells constantly replacing epithelial cells in the crypts and gastric glands?

21. What is found in the layer known as the lamina propria? (Fig. 21-3c)

22. What are Peyer's patches? (Fig. 21-3e)

23. Describe the muscularis mucosae.

The Submucosa

24. What structures are found in the submucosa? (Fig. 21-3c, e)

The Muscularis Externa and Serosa

25. The outer intestinal wall consists of two smooth muscle layers. (Fig. 21-3d, e) Contraction of

the _____ layer decreases lumen diameter, and contraction of the _____

layer shortens the length of the tube. The stomach has an incomplete third muscle layer,

arranged _____.

26. What is the myenteric plexus and where is it found?

27. What is the serosa and where is it found?

28. What are the peritoneum (peritoneal membrane) and mesentery?

MOTILITY

IP GI—Motility

29. What are the two purposes served by GI motility?

30. Gastrointestinal motility is determined by:

GI Smooth Muscle Contracts Spontaneously

31. Most of the intestinal tract is composed of _____-unit smooth muscle whose

cells are electrically connected by _____ junctions. (See Ch. 12.)

32. Distinguish between tonic and phasic contractions and tell where in the GI tract each contraction type can be found.

33. Describe how a slow-wave potential generates smooth muscle contraction in the GI tract. What roles do interstitial cells of Cajal and gap junctions play in this process? (Fig. 21-4)

GI Smooth Muscle Exhibits Different Patterns of Contraction

34.   Describe the three general patterns of muscle contraction in the gut. What is the function of each? (Fig. 21-5)

35.   What is a bolus?

36.   What are some common motility disorders?

SECRETION

IP GI—Secretion

37.   List the sources and volumes of fluid input into the GI tract. (Fig. 21-1)

38.   What is the significance of this secreted volume?

The Digestive System Secretes Ions and Water

39.   List the five major ions found in digestive secretions.

40.   Movement of water and ions in the GI tract is very similar to water and ion movement in what other anatomical location?

41.   List some of the transporters found on apical and basolateral membrane surfaces in GI epithelial cells.

Acid Secretion

42. Diagram how the parietal cells secrete hydrochloric acid into the stomach lumen. Include all ions involved. (Fig. 21-6)

Bicarbonate Secretion

43. What are the sources of duodenal bicarbonate?

44. Diagram the production of pancreatic bicarbonate. Include all ions involved. (Figs. 21-7, 21-8)

45. How does the process of pancreatic bicarbonate production result in secretion of a watery sodium bicarbonate solution?

46. What is cystic fibrosis? What are the pathological elements of the condition?

NaCl Secretion

47. Diagram the production of the isotonic NaCl solution secreted by crypt cells in the small intestine and the colon. (Fig. 21-9)

Digestive Enzymes Are Secreted into the Lumen

48. List the structures that secrete digestive enzymes.

49. Describe the synthesis, storage, and release of digestive enzymes. (See Chs. 3 and 7.)

50. Enzymes secreted in inactive form are known collectively as _____.

51. What types of pathways control enzyme release?

Specialized Cells Secrete Mucus

52. Mucus is composed of glycoproteins called _____.

53. Name two functions of mucus.

54. Mucus is made by specialized cells: _____ cells in the stomach, _____ cells in

    the intestine, and _____ in the mouth. (Fig. 3-27)

55. List signals for mucus release.

Saliva Is an Exocrine Secretion

56. Describe saliva.

57. Diagram how saliva is produced.

58. Describe the regulatory control of salivation.

The Liver Secretes Bile

59. What is bile? Which cells secrete it? (Fig. 21-10)

60. What are the key components of bile? What is the relevance of each component?

61. What else is found in bile?

62. How are bile salts made?

63. What is the role of the gallbladder?

REGULATION OF GI FUNCTION

IP *GI—Control of the Digestive System*

Overview of GI control mechanisms:

64. Long reflexes integrated in the CNS: (Fig. 21-11)

    Distinguish between long reflexes and cephalic reflexes. Give examples of each.

    Describe emotional influence on the long reflexes.

    Describe the autonomic influence in long reflexes.

65. Short reflexes integrated in the enteric nervous system (ENS): (Fig. 21-11)

    Contrast short reflexes with long reflexes.

    The primary responses controlled by the ENS are related to:

Describe the functions of:

submucosal plexus

myenteric neurons

66. Reflexes involving GI peptides:

Describe the location and nature of action for GI peptide hormones and paracrines.

Which GI functions are under GI peptide control? Give examples. (Fig. 21-12)

Describe ways in which GI peptides act on the brain.

The Enteric Nervous System Can Act Independently

67. What is one evolutionary link between the ENS and nerve networks in other species?

68. Describe the similarities between the ENS and the CNS.

GI Peptides Include Hormones, Neuropeptides, and Cytokines

69. Name the three families of GI peptides and give examples of each.

70. What do the following abbreviations stand for? (Table 21-1)

CCK _____   GIP _____

GLP-1 _____   VIP _____

▶ Table 21-1 summarizes the sources, targets, and effects of some GI hormones.

DIGESTION AND ABSORPTION

IP *GI—Digestion and Absorption*

71. How is digestion accomplished?

72. What is the role of bile?

73. The pH at which different enzymes best function reflects what?

74. Where does most nutrient absorption take place?

75. What is the brush border of epithelial cells? What is its purpose? (Fig. 21-13)

Carbohydrates Are Absorbed as Monosaccharides

76. Name the enzyme that is responsible for the following reactions. (Figs. 21-14 and 2-7)

maltose → monosaccharides _____

starch → maltose _____

sucrose → monosaccharides _____

lactose → monosaccharides _____

77. Into what monosaccharides are the following disaccharides digested? (Fig. 21-14)

maltose → _____

sucrose → _____

lactose → _____

78. Diagram the processes for intestinal absorption of glucose, galactose, and fructose. (Fig. 21-15)

79. Why don't enterocytes consume the glucose they absorb?

Proteins Are Digested into Small Peptides and Amino Acids

80. Name the type of enzyme that is responsible for each of the following reactions, and give some examples of each. (Figs. 21-16 and 2-9)

_____ breaks interior peptide bonds to make smaller peptides.

Examples:

_____ breaks exterior peptide bonds to make single amino acids.

Examples:

81. What are the primary products of protein digestion? How is each product absorbed? (Fig. 21-17)

82. Between 30%–60% of the protein found in the intestinal lumen comes from what source?

Some Larger Peptides Can Be Absorbed Intact

83. For those larger peptides for which it is possible, how are they absorbed? (See Ch. 5.)

84. What are the implications and medical applications of larger peptide absorption? (Fig. 21-13)

Bile Salts Facilitate Fat Digestion

85. Name five common forms of fat or fat-related molecules in the Western diet. (Fig. 2-8)

86. List and describe some of the enzymes involved in fat digestion? (Fig. 21-18)

87. Why is it necessary to emulsify fats? How is this emulsification carried out and what substances are involved? (Figs. 21-19, 21-20; also see Ch. 3)

88. Why is colipase necessary?

89. Why are fats absorbed primarily by simple diffusion?

90. How is cholesterol absorbed?

91. Diagram the process of fat digestion and absorption, beginning at the stomach and ending at the heart. (Fig. 21-20)

Nucleic Acids Are Digested into Nitrogenous Bases and Monosaccharides

92. How are nucleic acids digested and absorbed? (Fig. 2-11)

The Intestine Absorbs Vitamins and Minerals

93. How are fat-soluble vitamins absorbed?

94. How are water-soluble vitamins absorbed? What's the major exception?

95. How are minerals absorbed?

96. What is significant about iron and calcium absorption?

97. Describe iron absorption. How is it regulated?

98. Describe calcium absorption.

The Intestines Absorb Ions and Water

99. Where does most water absorption take place? Where else is water absorbed?

100. Now describe how water is absorbed and how it is related to specific Na^+ transport processes. (Fig. 21-21)

▶ Fig 21-22 summarizes the main events that occur in each section of the GI tract.

101. Food processing is traditionally divided into three phases. Name those three phases.

THE CEPHALIC PHASE

102. What are some stimuli that cause anticipatory (feedforward) long reflexes during the cephalic phase? (Fig. 21-23)

103. What are the body's responses to these anticipatory stimuli?

Chemical and Mechanical Digestion Begins in the Mouth

104. Which division of the nervous system controls saliva secretion? What stimuli can trigger saliva secretion?

105. List the functions of saliva.

106. Saliva begins chemical digestion with the secretion of which two enzymes? In which two digestive functions do these enzymes participate?

107. What is the result of food processing (chemical and mechanical) in the oral cavity?

Swallowing Moves Food from Mouth to Stomach

108. What is the stimulus for the swallowing (deglutition) reflex? (Fig. 21-24)

109. Diagram the deglutition reflex. (Fig. 21-24)

110. Describe how food enters the stomach.

111. Describe the structure and function of the lower esophageal sphincter.

THE GASTRIC PHASE

112. How many liters of food, drink, and saliva enter the fundus of the stomach each day?

113. Name and describe the three general functions of the stomach.

114. When does the gastric phase of digestion begin? Describe the reflexes of the gastric phase.

115. What happens before food arrives in the stomach? To which phase does this reflex belong?

The Stomach Stores Food

116. Compare upper stomach action to lower stomach action upon the arrival of food.

117. What is the importance of food storage in the stomach?

118. Enhanced gastric motility during a meal is primarily under _____ control and is stimulated

by _____ of the stomach.

The Stomach Secretes Acid and Enzymes

119. Outline the processes of chemical digestion in the stomach. (Fig. 21-25)

120. Once food enters the stomach, which gastric-phase signal molecules promote gastric secretion and increased motility? What is the primary negative feedback signal for gastric phase secretion?

121. Diagram the coordinated function of gastric secretion. (Fig. 21-26)

The Stomach Balances Digestion and Protection

122. What normally protects the gastric mucosa from enzymes and acid? (Fig. 21-27) How is this protection compromised in Zollinger-Ellison syndrome?

123. What are other causes of excess acid secretion?

124. What are some therapies for dypepsia?

THE INTESTINAL PHASE

125. When does the intestinal phase begin? What was the net result of the gastric phase?

126. Reflexes during the intestinal phase serve to regulate what? Are they hormonal, neural, or otherwise in their action? (Fig. 21-28)

127. Take the following stimuli and describe the reflex initiated:

presence of acidic chyme in the duodenum

meal containing fats

meal containing carbohydrates

hyperosmotic solution in the intestine

128. How is food propelled through the small intestine? How is this process regulated? What are some of the factors the body must take into consideration while moving food through the small intestine?

Bicarbonate Neutralizes Gastric Acid

129. Name the secretions added to chyme as it passes through the small intestine. What cells/tissues/organs are secreting these substances, and what does each accomplish? (Fig. 21-1)

130. What are the signals for pancreatic enzyme release?

131. How are most pancreatic enzymes secreted? How are these enzymes activated? (Fig. 21-29)

Most Fluid Is Absorbed in the Small Intestine

132. What is the significance of the hepatic portal system in intestinal absorption? Describe its anatomy and physiology. (Fig. 21-30)

Most Digestion Occurs in the Small Intestine

133. Describe protein digestion in the small intestine. What digestion has already happened before arrival in the small intestine? What digestion takes place in the small intestine?

134. Describe carbohydrate digestion in the small intestine. What digestion has already happened before arrival in the small intestine? What digestion takes place in the small intestine?

135. Describe fat digestion in the small intestine. What digestion has already happened before arrival in the small intestine? What digestion takes place in the small intestine?

The Large Intestine Concentrates Waste

136. List the seven regions of the large intestine as they would be encountered by chyme entering from the small intestine. Include any valves that might be encountered. (Fig. 21-31)

137. The anus is closed by two sphincters. Describe those sphincters.

138. How does the wall of the large intestine differ from the wall of the small intestine? Describe the muscularis, the mucosa, and the luminal surface.

Motility in the Large Intestine

139. Chyme that enters the colon continues to be mixed by _____.

140. Describe mass movement, the unique colonic contraction. This type of contraction is associated with what reflex?

141. Diagram the defecation reflex, beginning with the stimulus for defecation.

142. Describe some of the results that emotional influences can have on defecation.

Digestion and Absorption in the Large Intestine

143. How do bacteria in the large intestine participate in digestion?

144. What is flatus?

Diarrhea Can Cause Dehydration

145. What is diarrhea? What are some causes of the condition?

146. What is osmotic diarrhea? How does it occur?

147. What is secretory diarrhea? How does it occur?

148. How can diarrhea cause dehydration? What are the appropriate therapies?

IMMUNE FUNCTIONS OF THE GI TRACT

149. What are the GI tract's first lines of defense in fighting off pathogens?

150. Pathogens or toxic materials in the small intestine trigger sensory receptors and the immune cells of the _____. What are two common responses?

M Cells Sample Gut Contents

151. Describe the components of the immune system of the intestinal mucosa.

152. How are antigens brought into the M cells? (See Ch. 5.) What do M cells do with the antigens they've ingested? (See Ch. 16.)

153. If the antigen is threatening, what responses does the immune system initiate?

154. What are inflammatory bowel diseases?

155. How do certain pathogenic bacteria cross the barrier created by the intestinal epithelium?

Vomiting Is a Protective Reflex
156. What is vomiting? What does it accomplish?

157. Excessive vomiting can cause what condition? (See Ch. 20.)

158. Diagram the vomiting reflex.

TALK THE TALK

absorption
accessory glandular organs
acid reflux
acini
alkaline tide
aminopeptideases
amphipathic
amylase
antacids
antigens
antrum
anus
appendix
ascending colon
aspiration pneumonia

bicarbonate secretion
bile
bile acids
bile salts
body
bolus
brush border
carbonic anhydrase
carboxypeptideases
cecum
cellulose
cephalic phase
cephalic reflexes
chief cells
cholecytokinin (CCK)

chylomicrons
chyme
chymotrypsin
colipase
colon
colonocytes
common bile duct
constipation
crypts
cystic fibrosis
cystic fibrosis transmembrane
    conductance regulator
    (CFTR chloride channel)
cytochrome
D cells

defecation reflex
deglutition
denatures
descending colon
digestion
digestive enzymes
disaccharidases
divalent metal transporter (DMT1)
duodenum
dyspepsia
emesis
endopeptidases
enteric nervous system (ENS)
enterochromaffin-like (ECL) cells
enterocytes
enteropeptidase
epiglottis
Escherichia coli (*E. coli*)
esophagus
exopeptidases
external anal sphincter
extrinsic neurons
ezetimbe
feces
feedforward reflexes
ferroportin
fiber (roughage)
fistula
flatus
fructose
fundus
G cells
gallbladder
gastric acid
gastric glands
gastric inhibitory peptide (GIP)
gastric lipase
gastric phase
gastrin
gastrin-releasing peptides
gastrocolic reflex
gastroileal reflex
gastrointestinal system
gastrointestinal tract (GI tract)
ghrelin
GI stem cells
glucagon-like peptide 1 (GLP-1)
glucose
glucose-dependent insulinotropic peptide (GIP)
glycogen
goblet cells
gut

gut-associated lymphoid tissue (GALT)
haustra
Helicobacter pylori
hepatic portal system
hepatocytes
hepcidin
hormone
ileocecal valve
ileum
immunoglobulins
incretin hormones
inflammatory bowel disease
internal anal sphincter
interstitial cells of Cajal
intestinal phase
intrinsic factor
intrinsic neurons
irritable bowel syndrome
jejunum
lacteals
lactose
lamina propia
large intestine
lipases
long reflexes
lysozyme
M cells
maltose
mass movement
mastication
mesentery
micelles
migrating motor complex
morbidity
mortality
motilin
motility
mucins
mucosa
mucous cells
muscularis externa
muscularis mucosae
myenteric plexus
NPC1L1 cholesterol transporter
nodules
oligopeptide
oligopeptide transporter (PepT1)
osmotic diarrhea
parietal cells
parotid glands
pepsin
pepsinogen
peptic ulcers
peristalsis

peritoneal membrane
pernicious anemia
Peyer's patches
phasic contractions
phospholipase
plicae
proenzymes
proton pump inhibitors (PPIs)
pyloric valve (pylorus)
receiving segment
receptive relaxation
receptors
rectum
rugae
saliva
salivary amylase
salivary gland
Salmonella
satiety
secretin
secretion
secretory diarrheas
segmental contractions
serosa
serosal
Shigella
short reflexes
sigmoid colon
slow wave potentials
small intestine
starch
stomach
sublingual glands
submandibular glands
submucosa
submucosal glands
submucosal plexus
sucralose
sucrose
tenia coli
tonic contractions
transferrin
transverse colon
trypsin
vagal reflex
valsalva maneuver
vasoactive intestinal peptide (VIP)
Vibrio cholerae
villi
vitamin B_{12} (cobalamin)
vomitus
Zollinger-Ellison syndrome
zymogens

PRACTICE MAKES PERFECT

1. Number the following structures of the gastrointestinal tract in the order in which food passes:

   _____ stomach   _____ ileum   _____ esophagus

   _____ ascending colon   _____ pyloric sphincter   _____ duodenum

2. Why doesn't salivary amylase work in the stomach?

3. Pepsin is produced by the _____, and trypsin is produced by the

   _____. Pepsin is active only at _____ H⁺ concentrations, while trypsin is active

   at _____ H⁺ concentrations.

4. How does the digestive system prevent autodigestion?

5. TRUE/FALSE? Defend your answers.

   a. The pathway from the intestinal lumen to the circulating blood for a short-chain fatty acid (<10 carbon atoms) is intestinal mucosa cell to chylomicrons to capillary to systemic venous blood in the hepatic portal vein.

   b. Most of the bile salts secreted by the liver into the lumen of the intestine are excreted in the feces.

   c. The amount of water that is reabsorbed from the intestinal tract in a day is equal to the amount of water that is ingested in a day.

d.   If the blood supply to the small intestine decreases dramatically so that the cells become hypoxic, the absorption of glucose will decrease.

6.   The parietal cells of the stomach secrete hydrochloric acid.

a.   Based on your knowledge of acid secretion in the kidney, draw the mechanism of HCl secretion in the parietal cell.

b.   Based on the model you just drew, what kind of acid-base disturbance would excessive vomiting of stomach contents cause? In one sentence, defend your answer.

7. Why is the enteric nervous system considered by some to be a third division of the nervous
    system that is equivalent to the CNS and peripheral nervous system?

8. Why do physicians recommend that parents delay feeding infants cereals containing glutens?

9. Fill in the reflex pathway below. The stimulus and response are given.

                    stimulus <u>test anxiety</u>

    receptors _____   afferent path _____

    integrating center _____   efferent path _____

    effector_____   cellular response _____

    tissue response _____

                    systemic response <u>acid stomach syndrome</u>

10. In no more than two phrases per pair, compare/contrast the four processes of the urinary
    system to the four processes of the digestive system.

11. You have been doing some experiments in lingual and gastric lipase, but someone marked the test tubes with a marker that rubbed off. You know that one tube has the gastric enzyme and one tube has the lingual enzyme. What test could you do with the enzymes to tell which is which?

MAPS

1. Start with some sucrose and starch molecules in a candy bar. Follow these molecules through ingestion, digestion, and absorption as they move through the digestive tract.

2. Map digestive smooth muscle contraction, using the following terms and any others you wish to add.

circular layer	paracrines in the GI tract	segmental contraction
gap junction	peristaltic contraction	short reflex
longitudinal layer	peristaltic reflex	single-unit smooth muscle
motility	pharmacomechanical	slow wave potential
muscularis mucosa	coupling	tonic contraction
myenteric plexus	receiving segment	

3. Map the long and short reflexes of the gastrointestinal system.

4. Create a map of stomach function.

5. Compile information from the entire chapter to create a map of small intestine function.

BEYOND THE PAGES

TRY IT

Digestion of Starch by Salivary Amylase

You can sense the changes that take place during starch digestion by chewing on a soda cracker and holding it in your mouth rather than swallowing it. For best results use a starchy, unflavored soda cracker like a Saltine®. Put it in your mouth and start chewing. What happens to the taste as you chew? Rather than swallowing the chewed cracker right away, keep it in your mouth for a minute or two and see if the taste continues to change. Your salivary amylase acts on the starch in the cracker, breaking it up into smaller glucose polymers and the disaccharide maltose. Maltose has about 40% of the sweetness of sugar, so you should notice the taste of the cracker changing as you chew.

Terminology

GI physiology is full of technical words for functions that go by much more common names. Impress your friends with these:

("Excuse me. I need to eructate.")
mastication = chewing
emesis = vomiting
flatulence = having intestinal gas (flatus)
borborygmi = rumbling noises in the GI tract from intestinal gas
deglutition = swallowing
eructation = burping

Organizing Study

This is another chapter that is top heavy with terminology and information that needs to be memorized. There are several ways that you can organize the information in the chapter for study. One is to go anatomically, section by section, and outline everything that happens in each section. Within a section, you can subdivide into the four processes of motility, secretion, digestion, and absorption. For the latter two sections, you can further subdivide according to nutrients. Another way to organize is to take the three major classes of biomolecules and follow them one at a time from ingestion through absorption. Whichever way you organize it, maps are a great method for putting lots of integrated information into a relatively compact space. You may want to buy some poster board for the maps in this chapter, though!

Here is an example of a chart to make:

	Carbohydrate	Fat	Protein
List *all* places in the digestive tract where significant digestion of this takes place. What enzymes?			
In what form is this group of foods absorbed? (i.e., What is the product of digestion?)			
Describe the process of absorption.			

22 Metabolism and Energy Balance

LEARNING OBJECTIVES

When you complete this chapter, you should be able to:

- Contrast the glucostatic and lipostatic theories of food intake regulation, and then describe the current model of how food intake appears to be regulated.

- Explain how we measure energy use and metabolic rate. What factors can affect metabolic rate?

- Determine the total caloric intake of a particular food based on the metabolic energy content of its components.

- Distinguish between anabolic and catabolic pathways.

- Distinguish between the fed state and the fasted state.

- List the three possible fates an ingested nutrient might meet, depending on the body's needs.

- Create a map that summarizes the balance of nutrient pools, nutrient storage, and intake/output for carbohydrates, proteins, and lipids.

- Explain the regulatory significance of push-pull control.

- Create a summary diagram of anabolic carbohydrate, protein, and lipid metabolism in the fed state.

- Create a summary diagram of catabolic carbohydrate, protein, and lipid metabolism in the fasted state.

- Explain how the ratio of insulin to glucagon determines the dominant metabolic state.

- Create a map for insulin. Show the stimuli that increase/enhance its secretion, the stimuli that decrease/inhibit its secretion, its systemic effects, and its effects at specific target tissues (include mechanisms of action where possible).

- Create a similar map for glucagon. Show the stimuli that increase/enhance its secretion, the stimuli that decrease/inhibit its secretion, its systemic effects, and its effects at specific target tissues (include mechanisms of action where possible).

- Contrast type I diabetes mellitus with type II diabetes mellitus.

- Create a map for type I diabetes to show the body's responses to elevated plasma glucose in absence of insulin. Make sure to cover protein metabolism, fat metabolism, glucose metabolism, brain metabolism, osmotic diuresis and polyuria, dehydration, and metabolic acidosis.

- Map the homeostatic control of body temperature. Demonstrate the normal physiological responses as the body is subjected to temperatures above and below the thermoneutral zone. Identify pathological conditions that can raise or lower body temperature.

SUMMARY

Metabolism is the sum of the chemical reactions in your body that extract and use energy or store it for later use. There are two general types of metabolic reactions: anabolic (synthesis) and catabolic (breakdown). Normally, these reactions are balanced so that energy intake equals energy output. Energy is used to fuel biological work such as transport work, mechanical work, and chemical work. Body temperature regulation is also closely linked with metabolism.

Metabolism is divided into two states: fed (absorptive) and fasted (postabsorptive). The fed state is predominantly anabolic and the fasting state is predominantly catabolic. During the fed state, energy storage compounds are made. These include glycogen (liver, skeletal muscles) and fat (adipose). When the body enters the fasting state, these compounds are broken down to provide energy. Reactions involved in glycogen processing are glycogenesis (glucose → glycogen) and glycogenolysis (glycogen → glucose). Glucose can also be synthesized from noncarbohydrate precursors. Fats and proteins are converted into glucose through a process called gluconeogenesis. Different enzymes control the forward and reverse reactions in these processes, adding an additional level of control to metabolism called push-pull control. The goal of all this is to maintain plasma glucose concentrations so that the brain receives enough glucose.

The hour-to-hour control of glucose concentration depends on the ratio of insulin to glucagon. These are hormones secreted by islets of Langerhans cells in the endocrine pancreas: beta cells produce insulin and alpha cells produce glucagon. Insulin and glucagon work antagonistically to maintain plasma glucose. Insulin dominates during the fed state, and glucagon dominates during the fasted state. Insulin lowers glucose levels by facilitating glucose uptake and utilization primarily in liver, adipose, and skeletal muscle cells. Glucagon has the opposite effect from that of insulin. It raises plasma glucose by initiating glycogenolysis and gluconeogenesis and increasing hepatic glucose output.

Diabetes mellitus is a collection of diseases marked by abnormal secretion or activity of insulin. *Beware:* This one topic integrates almost every control pathway you have learned in this course! Be sure to give yourself plenty of time to learn and understand every pathway involved. Once you have mastered the integration of diabetes, you can rest assured that you have reached a new level in understanding human physiology. Insulin-dependent diabetes (type 1) is the more severe of the diabetes mellitus conditions. With type 1 diabetes, despite high plasma glucose concentrations, lack of insulin prevents most cells from taking up glucose. Therefore, the body is tricked into thinking it is in a constant state of fasting. The cells are breaking down proteins and fats, leading to muscle wasting and ketoacidosis. Suprathreshold glucose concentrations lead to glucosuria, which in turn causes osmotic diuresis and polyuria. Fluid loss leads to dehydration, which triggers homeostatic renal and cardiovascular reflexes. If all these problems aren't resolved, coma and death are inevitable. However, insulin injections and electrolyte therapy can correct the imbalances and prevent death. Study Fig. 22-16 to see the integration of diabetes control. See if you can supply additional detail to this map (like receptor types and neurotransmitters).

The final topic in this chapter is temperature regulation. Humans are homeothermic, so our bodies maintain tight control over body temperature to ensure we stay in a thermoneutral zone that allows optimal functioning of enzymes, receptors, etc. The hypothalamus contains the thermoregulatory center. It's here that sensory information from peripheral and core thermoreceptors is integrated and physiological responses to temperature changes are initiated. Heat loss from the body is promoted by dilation of blood vessels in the skin and sweating. Heat gain is generated by shivering and possibly by nonshivering thermogenesis.

TEACH YOURSELF THE BASICS

The Brain Controls Food Intake

1. Our current model of the control of food intake is based on:

2. Distinguish between the feeding center and the satiety center:

3. Contrast the glucostatic theory and the lipostatic theory.

4. Briefly describe some of the chemical, neural, psychological, and external factors that are involved in regulation of food intake. (Fig. 22-1; Table 22-1)

ENERGY BALANCE

Energy Input Equals Energy Output

5. The first law of thermodynamics states that: (See Ch. 4.)

6. Energy intake consists of:

7. Energy output by the body takes one of two forms:

8. List the three kinds of biological work and give an example of each: (See Ch. 4.)

Energy Use Is Reflected by Oxygen Consumption

▶ We must be able to estimate energy intake (food) and energy output (heat loss, work).

9. What is direct calorimetry? How is the energy measured experimentally?

10. TRUE/FALSE? Defend your answer.

   The metabolic energy content of food is equal to the content measured by direct calorimetry.

11. List the metabolic energy content of 1g of carbohydrate, 1g of protein, and 1g of fat. How would you determine the total number of kilocalories for a particular food that contains a variety of biomolecules?

Oxygen Consumption Reflects Metabolic Rate

12. What are some ways to measure an individual's metabolic rate?

Many Factors Influence Metabolic Rate

13. What is the basal metabolic rate (BMR)?

14. What is the most accurate way to measure a person's BMR? What is the most practical way to measure a person's BMR? What is this alternative rate called?

15. What are some factors that influence metabolic rate?

16.   What is meant by diet-induced thermogenesis? What factors are involved?

17.   What behavioral changes can we make to influence our metabolic rate?

Energy Is Stored in Fat and Glycogen

18.   How is our daily energy requirement expressed?

19.   Compare glycogen versus fats for energy storage. Which holds more energy? Which is easier to access?

20.   Where is glycogen stored in the body? About how much is in reserve in these locations?

METABOLISM

21.   Define metabolism.

22.   Distinguish between anabolic and catabolic pathways.

23.   What are the two states of metabolism? Which is catabolic and which is anabolic?

Ingested Energy May Be Used or Stored

24.   What three possible fates do biomolecules meet in the body?

25.   What are nutrient pools? (Fig. 22-2)

26.   Give a brief overview of the body's needs and regulation of:

amino acid pool

free fatty acid pool

glucose pool

27. The plasma concentration of which biomolecule is most closely regulated?

28. The amino acid pool of the body is used primarily for:

29. Define the following terms:

    glycogenesis

    glycogenolysis

    gluconeogenesis

    lipogenesis

    lipolysis

Hormones Alter Enzyme Activity to Control Metabolism

▶ Review Fig. 22-3 to examine the biological pathways important for energy production and how they interact. (See Ch. 4 for detailed energy pathways.)

30. Describe push-pull control of metabolic reactions and give an example. Why is push-pull control important? (Fig. 22-4)

Anabolic Metabolism Dominates in the Fed State

▶ Review the fates of nutrients during the fed state as outlined in Table 22-2.

Carbohydrates Provide Energy

31. From a macro perspective, what is the anatomical path of glucose absorption?

32. On the cellular level, how/where is glucose absorbed? (See Ch. 5.)

33. What are the potential fates for absorbed glucose?

34. How is unused glucose stored?

Amino Acids Make Proteins

35. From a macro perspective, what is the anatomical path of amino acid absorption?

36. What are the potential fates of absorbed amino acids?

Fats Store Energy

37. How are fats absorbed? How do they enter into the bloodstream? (Fig. 22-5; also see Ch. 21)

38. Once in the bloodstream, how are triglycerides converted into free fatty acids and glycerol? What then happens to these molecules?

39. What happens to the chylomicron remnants that remain in circulation? (Fig. 22-5)

40. After processing in the liver, lipoprotein complexes re-enter the bloodstream containing varying amounts of lipids and apoproteins. How are these lipoprotein complexes targeted to different tissues in the body, and how are they brought into their target cells? (See Ch. 5.)

41. What's the difference between "good" cholesterol and "bad" cholesterol? What are the roles of apoA and apoB in the different lipoprotein-cholesterol?

Plasma Cholesterol Predicts Coronary Heart Disease

42. Which pathological states are predicted by plasma lipid levels? (Fig. 22-6)

43. How are plasma lipid levels measured?

44. What are the current desirable lipid values for cholesterol (total, LDL-C, and HDL-C)?

45. What are some lifestyle modifications that can improve lipid profiles?

46. What aspects of cholesterol absorption and metabolism are targeted by the pharmacological therapies for treating elevated cholesterol?

47. What are some of the pharmacological agents currently being used to treat elevated lipid levels?

Catabolic Metabolism Dominates in the Fasted State

48. What is the signal for a shift in metabolic state?

49. What is the goal of the fasted state?

50. Which organ is the primary source of glucose production during the fasted state? (Fig. 22-7) How does it contribute to available glucose levels?

51. What are other sources of energy during the fasted state? (Fig. 22-3)

52. What is β-oxidation? What are the potential dangers of β-oxidation?

HOMEOSTATIC CONTROL OF METABOLISM

53. Hour-to-hour regulation of metabolism depends on what ratio?

The Pancreas Secretes Insulin and Glucagon

54. Identify the peptide(s) secreted by each of these cell types within the islets of Langerhans: (Fig. 22-8)

     α cells

     β cells

     D cells

     PP cells or F cells

55. Like all endocrine glands, the islets are closely associated with _____.

56. Neural control of the endocrine pancreas involves which neurons? Would you expect antagonistic or tonic nervous control?

The Insulin-to-Glucose Ratio Regulates Metabolism

57. Insulin and glucagon are (synergistic/antagonistic/permissive?) hormones. (See Ch. 7.)

58. What factor determines whether insulin or glucagon dominates?

59. Which hormone is dominant during the fed state (Fig. 22-9a)? What is the net metabolic effect?

60. Which hormone is dominant during the fasted state (Fig. 22-9b)? What is the net metabolic effect?

61. Describe the fluctuations of glucose, glucagon, and insulin concentrations during the course of a day. (Fig. 22-10)

Insulin Is the Dominant Hormone of the Fed State

62. How does insulin behave like a typical peptide hormone? (Table 22-3; also see Fig. 7-4c)

63. List and briefly describe the factors that influence insulin secretion:

64. What is the mechanism by which increased glucose concentration causes insulin secretion? (Fig. 5-38)

Insulin Promotes Anabolism

65. Which receptor type is involved in the signal transduction pathway for insulin? (Fig. 22-11; also see Ch. 6)

66. What are the primary target tissues for insulin? What are the effects of insulin on its target cells?

67. Which tissues do not require insulin for glucose uptake?

68. What is the cellular mechanism for insulin activity:

    a.   in adipose and resting skeletal muscle tissues? (Fig. 22-12)

    b.   in exercising skeletal muscle?

    c.   in hepatocytes? What happens to glucose in hepatocytes in absence of insulin? (Fig. 22-13)

69. Insulin is considered to be (catabolic/anabolic?). Which metabolic pathways are activated and which are inhibited? (Fig. 22-14)

70. Summarize the four ways in which insulin lowers plasma glucose.

Glucagon Is Dominant in the Fasted State

71. How is glucagon antagonistic to insulin? (Table 22-5)

72. Is it always present, or is it secreted only upon demand? (Fig. 22-10)

73. What is the primary stimulus for glucagon release? What else can stimulate glucagon release?

74. What is glucagon's primary target tissue? (Fig. 22-15; Table 22-5)

75. What metabolic pathways does glucagon trigger?

Diabetes Mellitus Is a Family of Diseases

76. How is diabetes characterized? What are some of the causes of diabetes?

77. Distinguish between type 1 and type 2 diabetes mellitus.

Type 1 Diabetics Are Prone to Ketoacidosis

78. Explain the physiology behind each of the symptoms of untreated type 1 diabetes mellitus (Fig. 22-16):

glucosuria (see Ch. 19)

ketone production (see Ch. 20)

muscle wasting

excessive urination

metabolic acidosis (see Ch. 20)

polydipsia (see Fig. 20-17)

polyphagia

osmotic diuresis

hyperglycemia

increased ventilation (see Ch. 18)

hyperkalemia (see Ch. 8)

79. What are the treatment options for type 1 diabetes mellitus?

▶ Type 1 diabetes integrates just about everything you've learned so far in physiology. Spend some more time learning the pathways affected by this disease before you move on. See if you can add to the map in Fig. 22-16.

Type 2 Diabetics Often Have Elevated Insulin Levels

80. What are the potential reasons type 2 diabetics have elevated insulin levels?

81. Explain the glucose tolerance test and its role in diagnosing diabetes. (Fig. 22-17)

82. Why do patients with untreated type 2 diabetes usually not develop ketosis?

83. What are the therapy options for type 2 diabetes? (Table 22-6)

84. Diagram how, in normal physiology, the combined actions of amylin, GIP, and GLP-1 create a self-regulating cycle for glucose absorption and fed-state glucose metabolism.

Metabolic Syndrome Links Diabetes and Cardiovascular Disease

85. What is metabolic syndrome?

86. What are the criteria for classifying a person as having metabolic diesease?

87. What are PPARs? What roles do they play in metabolic syndrome?

REGULATION OF BODY TEMPERATURE

Body Temperature Balances Heat Production, Gain, and Loss (Fig. 22-18)

88. How is metabolic efficiency related to obesity?

89. What is meant by homeothermic?

90. What is the normal body temperature range of humans (in °C)?

91. List factors that can affect body temperature.

92. When during the day is body temperature generally highest?

Heat Gain and Loss Are Balanced

93. List two sources of internal heat production. (Fig. 22-19)

94. List the sources of external heat input.

95. Describe the four different kinds of heat loss.

Body Temperature Is Homeostatically Regulated

96. What is the thermoneutral zone?

97. What is the temperature range for the human thermoneutral zone? What happens to the body's net heat production in temperatures above or below the thermoneutral zone?

98. The greatest physiological challenge to homeostasis comes from (hot/cold?) temperatures.

99. The integrating center for control of body temperature is located in the _____. (Fig. 22-20)

100. Where in the body are temperatures monitored by thermoreceptors?

101. Heat loss is achieved by what two mechanisms?

102. Heat generation is achieved by what two mechanisms?

Alterations in Cutaneous Blood Flow Conserve or Release Heat

103. Does blood flow through cutaneous blood vessels increase or decrease to achieve heat loss? To achieve heat conservation?

104. Diagram the mechanisms for regulation of cutaneous responses to body temperature fluctuations (to both rising and falling temperatures).

Sweat Contributes to Heat Loss

105. Describe the anatomy of a sweat gland and the composition of sweat. What is a typical value for sweat production?

106. Which neurons regulate sweat production?

107. How does sweat contribute to surface heat loss?

Movement and Metabolism Produce Heat

108. What are the two broad categories of heat production in the body?

109. Describe shivering thermogenesis (include the mechanism).

110. Describe nonshivering thermogenesis (include the mechanism).

▶ Responses to high/low temperatures are shown in Fig. 22-21.

The Body's Thermostat Can Reset

111. List some physiological causes for temperature variation.

112. List some pathological causes for temperature variation.

113. What is fever? What are some of the ways fever is generated?

114. What is the mechanism behind malignant hyperthermia?

TALK THE TALK

absorptive state
adipocytokines
alpha cells
amylin
anabolic pathways
apoprotein
apoprotein A (apoA)
apoprotein B (apoB)
basal metabolic rate (BMR)
beta cells
bile acid sequestrants
bomb calorimeter
brain metabolism
catabolic pathways
chemical work
conductive heat gain
conductive heat loss

convective heat loss
cutaneous
D cells
dehydration
diabetes mellitus
diabetic ketoacidosis
dipeptidyl peptidase-4
direct calorimetry
DPP-4 inhibitors
energy intake
evaporative heat loss
ezetimibe
fasted state
fat metabolism
fed state
feedforward effects of GI
    hormones

feeding center
first law of thermodynamics
ghrelin
glucagon
gluconeogenesis
glucose metabolism
glucostatic theory
glucosuria
glycogenesis
glycogenolysis
heat exhaustion
heat index
heat stroke
high-density lipoprotein (HDL)
homeothermic
humidex
hyperthermia

hypothermia
increased amino
    concentrations
increased glucose
    concentrations
indirect calorimetry
insulin-receptor substrates
juvenile-onset diabetes
ketone bodies
leptin
lipogenesis
lipoprotein lipase
lipostatic theory
long-term energy storage
low-density lypoprotein (LDL)
malignant hyperthermia
mechanical work
metabolic acidosis
metabolic rate

metabolic syndrome
metabolism
mitochondrial uncoupling
neuropeptide
nonshivering thermogenesis
nutrient pools
obestatin
osmotic diuresis
osmotic diuresis and polyuria
oxygen consumption
parasympathetic activity
peroxisome proliferator-
    activated receptors (PPARs)
postabsorptive
protein metabolism
push-pull control
pyrogens
radiant heat gain
radiant heat loss

renal threshold
respiratory exchange ratio
    (RER)
respiratory quotient (RQ)
resting metabolic rate (RMR)
satiety center
shivering thermogenesis
short-term energy storage
sitagliptin
sympathetic activity
thermography
thermoneutral zone
thermoreceptors
transport work
type 1 diabetes mellitus
type 2 diabetes
very-low-density lipoprotein
    (VLDL)

QUANTITATIVE THINKING

Body Mass Index: The **body mass index** (BMI) has been shown to correlate well with how much body fat a person has, and it can be calculated without special equipment or testing. The 1995 NIH guidelines define a BMI below 25 as healthy.

To calculate: $BMI = w/h^2$

w = weight in kilograms or weight in pounds divided by 2.2
h = height in meters or height in inches divided by 39.4

Calculate your BMI: weight = _____ lbs. ÷ 2.2 = _____ kg

height = _____ in. ÷ 39.4 = _____ m

$BMI = weight (kg)/height (m)^2$

PRACTICE MAKES PERFECT

1. Circle the letter of each pair representing an *incorrect* cause/effect relationship:

    a. epinephrine/increased glycogenolysis in the liver
    b. insulin/increased protein synthesis
    c. glucagon/decreased gluconeogenesis

2. TRUE/FALSE? Defend your answer.

    The heat produced by an organism is one way of defining (or measuring) metabolism.

3. Classify each of the following hormones as anabolic (A) or catabolic (C).

    glucagon _____   insulin _____

4. If you go on a no-carbohydrate diet, why doesn't the brain starve to death for lack of glucose?

5. Compare insulin secretion when glucose is given orally to insulin secretion after the same amount of glucose is given intravenously.

6. Generally insulin and glucagon are released by opposing stimuli and have opposing effects on metabolism. However, *both* hormones are released by the stimulus of an increase in blood amino acids. Circle all the answers below that explain correctly why this occurs.

   a. Glucagon will prevent hypoglycemia following ingestion of a pure protein meal.
   b. Both insulin and glucagon promote amino acid absorption at the small intestine.
   c. Amino acids are present in the blood during both anabolism and catabolism.
   d. Glucagon release is part of a positive feedback loop.
   e. Amino acids stimulate the release of insulin that stimulates the release of glucagon.
   f. Glucagon stimulates transcription and translation of amino acid transporters at the cell membrane.

7. Analyze the food label below for fat content (% of total calories) and critique the labeling. Is this a "good" food according to current guidelines on recommended total fat intake (25–30% calories from fat)?

   Calories per serving   190
   Total fat   7 g
   Total carbohydrate   25 g
   Total protein   9 g

8. Mice with leptin deficiency become obese, but obese humans have elevated levels of leptin in their blood. Can you think of an alternate explanation for leptin-related obesity in humans that fits these findings?

9. TRUE/FALSE? Defend your answers.

   a. Glucagon promotes glycogenolysis, gluconeogenesis, and ketogenesis.

   b. Insulin promotes transport of glucose into liver.

   c. Glucose transport in all cells of the body is via a mediated transport system that exhibits saturation.

10. Why are body-builders wasting their money if they take amino acid supplements in addition to a balanced diet?

11. A patient comes in with a diagnosis of alpha-cell hyperplasia due to a pancreatic tumor. You draw a blood sample and send it out for analysis. Before it comes back, predict what changes you expect to see in the following parameters (up, down, no change) and explain your rationale.

Plasma concentration	Change: ↑, ↓, or N/C	Rationale for your answer
glucose		
insulin		
amino acids		
ketones		
K^+		

MAPS

Map in detail the physiological complications arising from insulin-dependent diabetes mellitus. Include all homeostatic controls, all hormones, neurotransmitters, receptors, integrating centers, effector tissues, etc. Be very *specific*. This map will require a large sheet of paper or a poster board.

23 Endocrine Control of Growth and Metabolism

LEARNING OBJECTIVES

When you complete this chapter, you should be able to:

- Describe the anatomy of the adrenal glands. List the steroid hormones produced by the adrenal cortex and describe where and how they are synthesized.

- Diagram the HPA pathway and include all feedback signals.

- Describe cortisol secretion patterns, how it is transported, its target cells, where the receptors are, how it creates a cellular response, and its metabolic effects.

- Describe the therapeutic uses of cortisol.

- Identify the hallmarks of hypercortisolism and hypocortisolism. Distinguish between potential causes (iatrogenic, primary, and secondary causes).

- Identify the additional physiology functions of CRH and ACTH.

- Describe the anatomy of the thyroid gland and identify the hormones secreted by the distinct cell types.

- Diagram the synthesis of the thyroid hormones (T_3 and T_4).

- Describe secretion of thyroid hormones, how they are transported, where the receptors are, how a cellular response is created, and their metabolic effects. Which is the more biologically active form of thyroid hormone?

- Diagram the thyroid hormone control pathway.

- Identify the hallmarks of hyperthyroidism and hypothyroidism. Distinguish between primary and secondary thyroid pathologies.

- List the factors that affect normal growth.

- Diagram the control pathway for growth hormone release.

- Describe growth hormone secretion patterns, how growth hormone is transported, where its receptors are, how it creates a cellular response, and its metabolic effects.

- Identify the hallmarks of hypersecretion and hyposecretion/no secretion of growth hormone.

- Describe tissue growth and the hormones involved.

- Describe the structure of bone and how bone is a dynamic tissue.

- Diagram the mechanisms by which bone adds diameter and length. Which cells and which hormones are involved?

- List and summarize the physiological functions of calcium.

- Diagram the general mechanisms that regulate plasma calcium concentration.

- Diagram the specific actions of parathyroid hormone, calcitriol, and calcitonin. What is the nature of each hormone, where is each hormone produced, what are the stimuli for hormone release, what are the cellular mechanisms of action, and how does each hormone affect calcium balance in the body?

SUMMARY

Recall from Chapter 22 that metabolism is the sum of the chemical reactions in your body that extract and use energy or store it for later use. There are two general types of metabolic reactions: anabolic (synthesis) and catabolic (breakdown). Insulin and glucagon are the primary hormones of minute-to-minute metabolic control.

Other hormones play roles in the long-term regulation of metabolism: cortisol (from the adrenal cortex), thyroid hormones (thyroid gland), and growth hormone (anterior pituitary).

Cortisol is essential for life; it's overall catabolic effects are aimed at preventing hypoglycemia. Cortisol and the other adrenal glucocorticoids are known as stress hormones because of their roles in mediating long-term stress. Cortisol also has immunosuppressant effects, making it a useful therapeutic drug.

Thyroid hormones affect the quality of life. Their overall effect in adults is to provide substrates for oxidative metabolism. In children, thyroid hormones are necessary for full-expression of growth hormone.

Growth hormone is secreted throughout our lifetime, but peak GH secretion is around the time of puberty. Normal growth requires adequate amounts of growth hormone, thyroid hormones, insulin, and sex hormones. Growth hormone controls the release of insulin-like growth factors (IGFs) from the liver, which in turn stimulate bone and soft tissue growth.

The extracellular matrix of bone contains large amounts of calcium phosphate. Calcium is extremely important to our physiology, so the calcium stored in bones serves as a reservoir that can be tapped when free plasma calcium levels drop. Bone is constantly being formed and resorbed depending on calcium needs. Hormone control of calcium levels involves parathyroid hormone, calcitriol, and calcitonin.

TEACH YOURSELF THE BASICS

REVIEW OF ENDOCRINE PRINCIPLES

1. Take some time to review the endocrine principles first presented in Ch. 7.

ADRENAL GLUCOCORTICOIDS

2. Distinguish between the adrenal cortex and the adrenal medulla. (Fig. 23-1)

The Adrenal Cortex Secretes Steroid Hormones

3. Which hormones does the adrenal cortex secrete? For each, indicate the generalized effect and the cortical zone (layer) where it is produced. (Fig. 23-1)

▶ All steroid hormones are synthesized from cholesterol. (Fig. 23-2)

4. What are crossover effects? Why are they observed with steroid hormones? Give one example.

Cortisol Secretion Is Controlled by ACTH

5. Diagram the cortisol control pathway (HPA pathway). (Fig. 23-3)

6. Describe the diurnal secretion pattern of cortisol. (Fig. 23-4)

7. If cortisol is a typical steroid hormone, answer the following: (Fig. 7-7)

    Where (which organelle) in the cell would you expect it to be produced?

    How is it stored? (explain)

    When/how is it secreted and transported?

    Where are the receptors at its target cell?

    How does it elicit its cellular action?

Cortisol Is Essential for Life

8. List the metabolic effects of cortisol action. (Fig. 23-3; Table 23-1)

9. What is cortisol's most important metabolic effect? How does it interact with glucagon? (See Ch. 7.)

Cortisol Is a Useful Therapeutic Drug

10. What are cortisol's immunosuppressant effects?

11. What group of drugs has diminished the use of glucocorticoids for treating minor inflammatory problems?

12. What negative feedback effect does exogenous cortisol administration have? (Fig. 7-19)

Cortisol Pathologies Result from Too Much or Too Little Hormone

Hypercortisolism
13. What are the effects of hypercortisolism? (Fig. 23-5a)

14. What are three causes of hypercortisolism?

Hypocortisolism
15. What are the cause(s) and effects of hypocortisolism?

CRH and ACTH Have Additional Physiological Functions
16. The association between stress and immune function appears to be mediated through the HPA pathway. What evidence supports this?

CRH Family
17. The CRH family includes:

18. What are some physiological effects of CRH?

ACTH and Melanocortins
19. ACTH is synthesized from a large glycoprotein called pro-opiomelanocortin (POMC). What, other than ACTH, is made from POMC? What are the physiological effects of these other substances? (Fig. 23-6; also see the Emerging Concepts box)

20. How many different melanocortin receptors have been identified?

THYROID HORMONES

21.   List the distinct cell types within the thyroid gland and give the hormones they secrete. (Fig. 23-7)

Thyroid Hormones Contain Iodine

22.   To which class of hormones do thyroid hormones belong? What makes them unusual? (Fig. 23-8)

23.   Draw the structure of thyroid follicles and describe the composition of colloid. (Fig. 23-7b)

24.   Diagram the process of thyroid hormone synthesis and secretion. (Fig. 23-9)

25.   Which is the more active thyroid hormone?

26.   How are thyroid hormones transported in the blood? Where in the cell are thyroid hormone receptors located?

Thyroid Hormones Affect Quality of Life

27.   List the actions of thyroid hormones in adults. (Table 23-2)

28.   What additional actions do they have in children?

Hyperthyroidism

29.   What are the effects of hyperthyroidism?

30. People with hyperthyroidism often have a rapid heartbeat. Explain how thyroid hormones cause this. (See Ch. 14.)

Hypothyroidism
31. What are the effects of hypothyroidism? (Fig. 23-10a)

TSH Controls the Thyroid Gland
32. Diagram the thyroid hormone control pathway. (Fig. 23-11)

33. Describe goiter formation by means of primary hypothyroidism and compare that to goiter formation by means of primary hypersecretion. (Figs. 23-10b, 23-12)

34. List the therapies for thyroid disorders.

GROWTH HORMONE

35. What factors are important for normal growth?

Growth Hormone Is Anabolic
36. Is growth hormone (GH) secreted throughout a lifetime?

37. When is peak GH secretion observed in humans?

38.   Diagram the GH control pathway. (Fig. 23-13)

39.   When is the daily GH peak observed?

40.   To which class of hormones does GH belong? Is it an exact fit to the model? Explain. (See Ch. 7.)

41.   What are the metabolic effects of GH?

42.   What are IGFs, and what role do they play? (Fig. 23-13; Table 23-3)

Growth Hormone Is Essential for Normal Growth in Children
43.   What are the effects of GH hypersecretion? (Fig. 23-14)

44.   What are the effects of GH hyposecretion?

Genetically Engineered hGH Raises Ethical Questions

▶   Take a moment to consider the ethical considerations involved with GH therapy.

TISSUE AND BONE GROWTH
45.   What are the two general areas of growth, and how are they measured?

Tissue Growth Requires Hormones and Paracrines
46.   What hormones/paracrines are required for soft tissue growth?

47.   Define hypertrophy and hyperplasia.

48. Describe the role of thyroid hormones in growth.

49. Describe the role of insulin in growth.

Bone Growth Requires Adequate Dietary Calcium

50. Describe the extracellular matrix of bone.

51. Why is bone considered a dynamic tissue? What do we mean by resorbed?

52. What are the two forms of bone? (Fig. 23-15)

53. Diagram the process of bone growth. (Fig. 23-16) Be sure to explain the differences between osteoblasts, osteoclasts, and osteocytes.

54. What factors control bone growth?

CALCIUM BALANCE

IP *Fluids & Electrolytes: Electrolyte Homeostasis*

55. Where is most of the body's calcium found?

56. List eight important functions of Ca^{2+} in the body (Table 23-4)

57. Describe effects of changes in plasma calcium levels on nerve cell function.

Plasma Calcium Is Closely Regulated

58. Describe how Ca^{2+} concentrations are regulated in terms of intake, output, and total body calcium.

59. Compare extracellular and intracellular Ca^{2+} concentrations.

60. Where does the body maintain a Ca^{2+} reservoir?

61. Describe osteoclast activity in bone resorption. (Fig. 23-18)

Three Hormones Control Calcium Balance

62. Which three hormones regulate Ca^{2+} movement between bone, kidney, and intestine? (Fig. 23-17)

63. Why are the parathyroid glands so important? Where are they found? (Fig. 23-19)

Parathyroid Hormone

64. What is the stimulus for parathyroid hormone (PTH) secretion? What is the negative feedback stimulus that shuts off PTH secretion?

65. What are the three actions of PTH? (Table 23-5)

66. If osteoclasts have no PTH receptors, how then are they directed in bone resorption?

Calcitriol

67. How does calcitriol enhance intestinal Ca^{2+} absorption? (Table 23-6)

68. How/where is calcitriol made? (Fig. 23-20)

69. How/where is calcitriol production regulated?

70. What role does prolactin play in calcitriol-mediated Ca^{2+} absorption?

Calcitonin

71. Where is calcitonin made and what is the nature of its structure? (Table 23-7)

72. What is calcitonin's role in the body?

73. What is a therapeutic benefit of calcitonin?

Calcium and Phosphate Homeostasis Are Linked

74. What functions, other than bone formation, involve phosphate?

75. Where is phosphate homeostasis carried out in the body?

Osteoporosis Is a Disease of Bone Loss

76. What is osteoporosis, and what are the effects of the disease? (Fig. 23-21)

77. Who in the population are more likely to develop osteoporosis?

78. List the risk factors for osteoporosis.

79. Which osteoporosis treatment is currently preferred? What is the mechanism of action for these drugs?

80. Why has estrogen/progesterone hormone replacement therapy been replaced as the leading osteoporosis therapy?

81. List preventative measures that young women should take to help stave off osteoporosis later in life.

TALK THE TALK

1,25-dihydroxycholecalciferol
acromegaly
Addison's disease
adenohypophyseal
adrenal cortex
adrenal medulla
adrenocorticotropic hormone
    (ACTH)
alpha-MSH
androgens
androstenedione
β-endorphin
bisphosphonates
C cells
Ca^{2+}-sensing receptor (CaSR)
calcitonin
calcitrol (1,25-dihydroxychole-
    calciferol)
calcium sensing receptors
cellular responses
chondrocytes
colloid
corticosteroid-binding globulin
    (CBG)
corticotropin-releasing
    hormone (CRH)
cortisol
coupling reactions
cretinism
crossover effect
Cushing's syndrome
dehydroepiandrosterone (DHEA)
deiodinase
diaphysis
diiodotyrosine
dwarfism
endocrine pathologies
epiphyseal plates
epiphysis
estrogens
exophthalmus
feedback patterns
follicles

follicular cells
giantism
glucocorticoids
goiter
Graves' disease
growth hormone (GH)
growth hormone binding
    protein
growth hormone-inhibiting
    hormone (GHIH)
growth hormone-releasing
    hormone (GHRH)
hormone receptors
hormone replacement therapy
hormone repressing element
hydroxyapatite
hyperplasia
hyperthyroidism
hypertrophy
hypocalcemia
hypoglycemia
hypothalamic-pituitary-
    adrenal (HPA) pathway
iatrogenic hypercortisolism
immunosuppressant effects
insulin-like growth factors (IGFs)
kyphosis
melanins
melanocortin receptors (MCRs)
melanocortins
melanocyte-stimulating
    hormone (MSH)
melanocytes
mineralocorticoid
mineralocorticoid receptors
    (MRs)
monoiodotyrosine
mucopolysaccharides
myxedema
nonsteroidal anti-inflammatory
    drugs (NSAIDs)
osteoid
osteoblasts

osteocalcin
osteoclasts
osteocytes
osteonectin
osteoporosis
osteoprotegerin
Paget's disease
pancreatitis
parathyroid gland
parathyroid hormone
pendrin (SLC 26A4)
permissive effect
primary hypercortisolism
primary hypothyroidism
progesterone
pro-opiomelanocortin (POMC)
protease
recombinant human growth
    hormone (rhGH)
secondary hypercortisolism
selective estrogen receptor
    modulator
sodium-iodine symporter (NIS)
teriparatide
tertraiodothyronine
tetany
thyroglobulin
thyroid peroxidase
thyroid-binding globulin
thyroid-stimulating hormone
    (TSH)
thyroid-stimulating
    immunoglobulins (TSI)
thyrotropin
thyrotropin-releasing
    hormone (TRH)
trabecular bone
triiodothyronine
urocortin
zona fasciculata
zona glomerulosa
zona reticularis

PRACTICE MAKES PERFECT

1. The active hormone of the thyroid gland is:

    a. thyroglobulin
    b. thyroxine
    c. diiodotyrosine

2. Classify each of the following hormones as anabolic (A) or catabolic (C):

    cortisol _____   growth hormone _____

3. TRUE/FALSE? Defend your answers.

    a. When we hear of athletes and body-builders taking "anabolic steroids," we know that they are taking glucocorticoids.

    b. A child who has a vitamin D deficiency will develop rickets (poor bone formation) because vitamin D plays a major role in the precipitation of calcium phosphate into bone.

4.   The diagram below represents calcium balance in the human body.

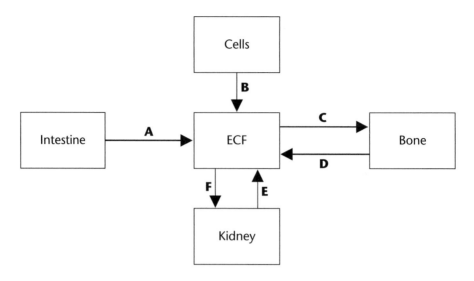

The labeled arrows represent calcium movement between the compartments. Answer the questions below that refer to the letters on the diagram.

a.   What hormone(s) directly or indirectly regulate movement at arrow A?

b.   Describe *as specifically as possible* the calcium movement at arrow F.

c.   What is the hormonal control over calcium movement at arrow E?

d.   Name two physiological functions of the calcium movement shown by arrow B.

e.   Name the hormone that controls calcium movement at arrow D.

5. On axis A below, plot the effect of plasma parathyroid hormone concentration on plasma Ca^{2+} concentration. On axis B below, plot the effect of plasma Ca^{2+} concentration on plasma parathyroid hormone concentration. Be sure to label the axes of each graph!

A

B

MAPS

Create pathway maps for each hormone introduced in this chapter. You should have details about trophic hormones and control pathways, as well as details at the cellular level where appropriate. The hormones presented were: cortisol, thyroid hormone, growth hormone, parathyroid hormone, calcitriol, and calcitonin.

BEYOND THE PAGES

TRY IT

Can you tie a bone in a knot? Place the leg bone of a chicken in commercial white vinegar and monitor what happens. The vinegar will dissolve the calcium phosphate of the bone, leaving behind the soft organic matrix. This is similar to the way osteoclasts secrete acid to dissolve bone in the body.

ETHICS IN SCIENCE

Should athletes be allowed to take prohormones like DHEA or androstenedione and synthetic growth hormone? Should children who are genetically short but have normal growth hormone secretion be given growth hormone? What are the potential risks of taking hormones if a person's own hormone production is normal?

24 The Immune System

LEARNING OBJECTIVES

When you complete this chapter, you should be able to:

- List the three major functions of the immune system.

- List the three categories of immune pathologies.

- Describe the differences between bacteria and viruses that require the body to have a variety of defense mechanisms.

- Identify the physical and chemical barriers that attempt to keep pathogens out of the body's internal environment.

- Describe and differentiate innate immunity and acquired immunity.

- Describe and differentiate cell-mediated immunity and humoral immunity.

- Describe and differentiate primary lymphoid tissues and secondary lymphoid tissues.

- Describe and differentiate encapsulated tissues and unencapsulated diffuse lymphoid tissues.

- Group leukocytes according to morphological and functional characteristics.

- Describe the relative prevalence in blood and the general immune functions of basophils, neutrophils, eosinophils, monocytes and macrophages, and lymphocytes.

- Diagram the nonspecific responses of innate immunity. Include in your diagram(s): phagocytes, NK cells, inflammatory response, histamine, interleukins, acute-phase proteins, bradykinins, and complement proteins.

- Diagram the humoral immune responses of B lymphocytes. Include in your diagram: antibodies, naive cells, clones, clonal expansion, memory cells, effector cells, plasma cells, primary immunity, secondary immunity, IgA, IgD, IgE, IgG, IgM.

- Draw and label an antibody.

- Describe how antibodies can make antigens more visible to the immune system, how they can enhance inflammation, and how they can activate other immune cells.

- Diagram the cell-mediated immune responses of T lymphocytes. Include in your diagram: T-cell receptor, MHC, MHC-antigen complex, MHC class I molecules, MHC class II molecules, cytotoxic T cells, and helper T cells.

- Map the immune responses to the following challenges: bacterial infection, viral infection, allergic reaction, and the transfusion of incompatible blood.

- Describe self-tolerance and how breakdown of this safeguard can lead to autoimmune diseases.

- Describe the relationships between the immune, nervous, and endocrine systems.

- Describe how stress can affect immunity.

SUMMARY

Immunity is the ability of the body to protect itself from pathogens. Anything that elicits an immune response is called an antigen, while a pathogen is anything that creates a pathophysiological condition. As you're probably beginning to see, the first step to understanding the immune system is conquering its vocabulary.

The immune system is spread throughout the body. The primary lymphoid tissues are the thymus and the bone marrow. These are places where immune cells are produced and mature. Secondary lymphoid tissues include encapsulated tissues (spleen and lymph nodes) and unencapsulated tissues (tonsils, GALT, skin, and respiratory clusters). There are several types of immune cells: eosinophils, basophils, neutrophils, monocytes, lymphocytes, and dendritic cells. Learn how each plays a different role in the immune response.

There are two types of immune responses: innate immunity and acquired immunity. Innate immunity involves a nonspecific immune response that reacts to all foreign particles. Inflammation is a hallmark reaction of innate immunity. If an antigen should get past the body's physical defenses, then it will initiate responses that lead to one of the following fates: the foreign particle could be phagocytized by tissue macrophages or neutrophils (thus being presented to other immune cells), or it could be destroyed via the complement system. At the site of antigen entry, chemicals are released by local cells that result in the attraction of immune cells and the formation of a warm, red, swollen area. You should learn the steps and chemicals involved in the innate immune response (Table 24-2).

Acquired immunity involves two responses: humoral immunity, brought about by B lymphocytes; and cell-mediated immunity, brought about by T lymphocytes. B lymphocytes (B cells) secrete antibodies (immunoglobulins) that recognize specific sites (epitopes) on specific antigens. B cells that have never been exposed to an antigen are called naive B cells. Once naive cells have encountered their antigen, they divide and produce memory cells and plasma cells in a process called clonal expansion. Memory cells stick around in the body and are responsible for the secondary immune response. Plasma cells secrete antibodies (Abs) into the humors or body fluids (hence the name humoral immunity). The primary function of Abs is to bind B cells and antigens and cause the production of more Abs, but Abs also have many secondary functions that target an antigen for destruction (Fig. 24-13). Abs are Y-shaped protein structures: the arms recognize the antigen, and the stem can attach to cell membranes. There are five classes of Abs: IgG, IgE, IgD, IgM, IgA.

T cells develop in the thymus. There, T cells that bind self proteins are destroyed. Mature T cells must have contact with their target cells to initiate cell-mediated immunity. T cell receptors (related, but not equivalent to Abs) must bind to a major histocompatibility complex (MHC) bearing a specific antigen fragment on a target cell surface. MHCs display protein fragments (either self or non-self) on the cell surface. If a T cell recognizes a foreign protein displayed in an MHC, it will initiate a response that either kills that cell or activates other immune cells. There are two types of MHC proteins: MHC-I is found on all nucleated cells; MHC-II is found only on antigen-presenting cells (APCs). There are two subtypes of T cells: cytotoxic T cells and helper T cells. Cytotoxic T cells kill a target cell by apoptosis when they recognize an MHC-antigen complex. When helper T cells recognize an MHC-antigen complex, they secrete cytokines that activate other T and B cells.

Natural killer cells are a distinct cell line. If they recognize an MHC-antigen complex on a cell, they will kill that cell. However, they can also kill an Ab-coated cell through a nonspecific process called antibody-dependent cell-mediated cytotoxicity. In addition to killing cells, they secrete a number of immune-activating and antiviral cytokines.

That's a mouthful! We suggest making maps or charts that show the differences between (1) innate and acquired immunity, and (2) humoral and cell-mediated immunity. Be sure to include information about the different types of cells and Abs involved.

Learn how the immune system responds differently to different types of antigens (bacterial, viral, allergenic).

The same processes we discussed earlier also apply to the body's recognition of self and non-self tissues. With RBC recognition, it involves surface antigens (because RBCs don't have MHC proteins). With other cases, like organ transplants, foreign MHCs and/or MHC-antigen complexes can cause an immune response that results in tissue rejection.

Finally, we are learning that the immune system is closely linked with the nervous and endocrine systems. It seems that some cytokines, neuropeptides, and hormones are secreted by some or all of these systems. Therefore, they all exert an integrated control over each other. Neuroimmunomodulation is the field that studies the brain-immune system interaction.

TEACH YOURSELF THE BASICS

OVERVIEW OF IMMUNE SYSTEM FUNCTION

1. Define immunity.

2. List three major functions of the immune system.

3. Immune pathologies typically fall into one of these categories:

PATHOGENS OF THE HUMAN BODY

4. List the types of infections common in the United States.

5. List other types of pathogens more common in other parts of the world.

Bacteria and Viruses Require Different Defense Mechanisms

6. Fill in the following table on the differences between bacteria and viruses: (Table 24-1)

	Bacteria	Viruses
Structure		
Living conditions		
Reproduction		
Susceptibility to drugs		

Viruses Can Replicate Only Inside Host Cells

7. Briefly describe ways that virus particles can enter a host cell. (Fig. 24-1b)

8. Next, describe how a virus takes over host cell resources. (Fig. 24-1b)

9. Describe two ways viruses are released from host cells.

10. What are potential behaviors of a virus once inside a cell?

THE IMMUNE RESPONSE

11. A distinguishing feature of the immune system is that it makes extensive use of

    _____ signaling.

12. List the four basic steps of all internal immune responses.

13. What are the two categories of the human immune response? Describe them.

14. Distinguish between cell-mediated immunity and humoral immunity.

ANATOMY OF THE IMMUNE SYSTEM

Lymphoid Tissues Are Everywhere

15. Name the two primary lymphoid tissues. (Fig. 24-2a; see Focus boxes, Figs. 16-4 and 24-14)

16. List the two types of secondary lymphoid tissues.

17. Describe the following encapsulated lymphoid tissues:

    spleen (Figs. 24-2a, 24-3)

    lymph nodes (Fig. 24-2b; see lymphatic circulation in Ch. 15)

18. Identify and describe diffuse lymphoid tissues. (Fig. 24-2)

19. What is the GALT? (See Ch. 21.)

Leukocytes Mediate Immunity

20. List ways that leukocytes (WBCs) differ from RBCs. (See descriptions of blood cells in Ch. 16.)

21. List and briefly describe the six basic types of leukocytes (and their derivative cells):

22. How are immune cells distinguished from one another in stained tissue samples?

Immune Cell Names

▶ Know the classification systems for immune cells. (Fig. 24-4)

23. What are granulocytes? Describe the different types of granulocytes and how they are distinguished.

24. What are phagocytes? (Review phagocytosis in Ch. 5.) What cells belong to this group?

25. Explain the following terms and give representative cell types, where possible:

   a.   cytotoxic cell

   b.   antigen-presenting cell (APC)

   c.   mononuclear phagocyte system

Basophils

26. Describe basophils. Include their relative abundance, location in the body, life span, and their role(s) in the immune system.

Neutrophils

27. Describe neutrophils. Include their relative abundance, location in the body, life span, and their role(s) in the immune system.

Eosinophils

28. Describe eosinophils. Include their relative abundance, location in the body, and their role(s) in the immune system.

Monocytes and Macrophages

29. Describe monocytes and macrophages.

30. How do macrophages participate as antigen-presenting cells? (Fig. 24-5)

Lymphocytes

31. Describe lymphocytes in terms of numbers, percentage in circulation, location, and role(s) in the immune system.

Dendritic Cells

32. Describe dendritic cells.

INNATE IMMUNITY: NONSPECIFIC RESPONSES

▶ Innate immunity either clears the infection or contains it until the acquired response is activated.

Physical and Chemical Barriers Are the Body's First Line of Defense

33. Describe three examples of physical barriers against foreign invaders. (See Ch. 3.)

34. Describe the specialized physical barriers of the:

    a.   respiratory system (See Ch. 17.)

    b.   stomach

    c.   tears

35. Lysozyme can only attack which type of bacterial cell wall?

Phagocytes Ingest Foreign Material

36. What are PAMPs and PRRs? What role(s) do they play in initiating the nonspecific immune response?

37. What are chemotaxins? Give some examples.

38. What are the primary phagocytic immune cells responsible for defense?

39. What is extravasation?

40. Diagram the process of phagocytosis. (Figs. 24-6, 24-7)

41. Some bacteria have evolved ways to hide from phagocytes. How have they accomplished this?

42. What are opsonins? What is their purpose?

43.   Why do sites of bacterial infection develop pus?

NK Cells Kill Infected and Tumor Cells

44.   How do NK cells work? (See information on apoptosis in Ch. 3.)

45.   Define interferon and distinguish between the alpha, beta, and gamma interferons.

Cytokines Create the Inflammatory Response

46.   What are three roles inflammation has in fighting infection?

47.   Which cells create the inflammatory response?

48.   What role do cytokines play in this response? (Table 24-2)

Acute-Phase Proteins

49.   When do we see an increased presence of acute-phase proteins?

50.   What are acute-phase proteins and where do they originate? Include examples.

51.   What happens to the level of acute-phase proteins as the immune response proceeds?

52.   What can happen to the level of acute-phase proteins in cases of chronic inflammatory conditions? (See information on atherosclerosis in Ch. 5.)

Histamine

53.   Where do we find histamine in the body?

54.   Describe the responses created by histamine action. What symptoms does it create near the infection site?

55. What is the purpose of the histamine reaction in the larger immune response?

56. How do antihistamines work? What chemicals, other than histamine, can be released in response to an antigen/allergen?

Interleukins

57. What are interleukins? What is their role in the immune response?

58. Interleukin-1 (IL-1) is secreted by _____. Its overall role is _____

_____

_____

Four specific actions of IL-1 include:

1. _____   3. _____

2. _____   4. _____

Bradykinin

59. What are kinins?

60. What are the physiological actions of bradykinin?

Complement Proteins

61. Complement is a collective term for:

62. Briefly describe the complement cascade. (Review the coagulation cascade in Ch. 16.)

63. What is membrane attack complex, and how does it cause cell lysis? (Fig. 24-8)

ACQUIRED IMMUNITY: ANTIGEN-SPECIFIC RESPONSES

64. How does acquired immunity differ from innate immunity?

65. Which immune cells primarily mediate the acquired response?

66. There are three main types of lymphocytes (and derivative cells). Identify and describe the role of each.

67. Describe the difference between active and passive immunity.

Lymphocytes Mediate the Acquired Immune Response

68. What gives different lymphocytes specificity for specific ligands? (Fig. 24-9)

69. What are lymphocyte clones? (Figs. 24-9, 24-10)

70. Diagram the lymphocyte life cycle, using the following terms: antigen, clonal expansion, effector cells, naive lymphocytes, memory cells, primary response, secondary response.

71. How do memory cells differ from effector cells?

B Lymphocytes Become Plasma Cells and Memory Cells

72. Where do B lymphocytes (B cells) develop?

73. What are antibodies (immunoglobulins)?

74. What purpose do surface antibodies on B cells serve? (Fig. 24-9)

75. What role do plasma cells play? What role do memory cells play?

76. What is humoral immunity?

77. Draw and label a graph that shows plasma antibody concentration during a primary response and during a secondary response. (Fig. 24-11)

78. Based on your graph and what you know so far, why do immunizations help fight against infection?

Antibodies Are Proteins Secreted by Plasma Cells

79. There are five general classes of Abs: IgG, IgA, IgE, IgM, IgD. Match the antibody group to its functions.

    a. Ig _____: allergic response

        Mast cell + Ig ___ + antigen = degranulation and histamine release

    b. Ig _____: B lymphocyte surface; an unclear physiological role

    c. Ig _____: found in external secretions (saliva, tears, mucus, breast milk); disables pathogens before body entry

    d. Ig _____: produced in secondary immune responses; 75% of adult serum Abs; can cross placenta and provide passive immunity for the infant

    e. Ig _____: B lymphocyte surface; primary immune responses; reacts to blood group antigens; activates complement

80. Define gamma globulins.

Antibody Proteins

81. Sketch a picture of the typical antibody molecule. Label all the parts and briefly describe the relevance of each part. (Fig. 24-12)

Antibody Functions

82. Where are most antibodies found? What percentage of plasma proteins do they account for in healthy individuals?

83. Humoral antibodies are most effective against:

84. How do antibodies make antigens more visible to the immune system? (Fig. 24-13)

85. How do antibodies enhance inflammation?

86. How do antigen-bound antibodies activate immune cells? (Fig. 24-13)

87. Specifically, describe antibody-dependent cell-mediated cytotoxicity.

B Cell Activation

88. Draw and label how antibodies are bound to the surface of B cell membranes. (Fig. 24-13)

89. How are B cells activated? (Fig. 24-13)

90. What happens after a B cell is activated?

T Lymphocytes Use Contact-Dependent Signaling

91. Where do T lymphocytes (T cells) develop? (Fig. 24-14)

92. What is the role of T lymphocytes?

93. Diagram how T cells carry out cell-mediated immunity. Identify T-cell receptors and MHC-antigen complexes. (Fig. 24-15)

MHC and Antigen

94. Define and describe the general function of major histocompatibility complexes (MHC).

95. Where do we find MHCs?

96. Why do MHC proteins vary from person to person? What about MHCs in identical twins?

97. Describe how MHC class I molecules work. (Fig. 24-16)

98. Describe how MHC class II molecules work.

99. Why are MHCs relevant to tissue transplants?

Cytotoxic T Cells

100. Describe how cytotoxic T cells recognize the infected cells and how they kill targeted cells.

Helper T (T$_H$) Cells

101. Describe the action of helper T cells.

102. List four cytokines secreted by helper T cells, including their functions.

IMMUNE RESPONSE PATHWAYS

▶ Once you've worked through these pathways the first time, go back and see if you can describe them completely without looking.

103. Map how the body responds to four challenges:

    a.   extracellular bacterial infection (Fig. 24-17)

    b.   viral infection (Fig. 24-18)

    c.   allergic response to pollen (Fig. 24-19)

    d.   transfusion of incompatible blood (Fig. 24-20; Table 24-3)

▶ These pathways show how the innate and acquired immune responses are interconnected processes.

Bacterial Invasion Causes Inflammation

104. Outline the integrated response to bacterial entry into the ECF. (Fig. 24-17) Feel free to do this as a map, but do not directly copy the one in the book (this is to give you more practice). Be sure to use all of the following terms:

acute phase proteins	chemotaxin	mast cell
antibodies	complement	membrane attack complex
antigen	extravasation	memory B cells
B lymphocytes	histamine	opsonins
bacteria, encapsulated	leukocyte	phagocytes
bacteria, not encapsulated	lyse	plasma cells
capillary permeability	lysozyme	plasma protein

Viral Infections Require Intracellular Defense

▶ Before viruses enter a host cell, innate and humoral defenses both control infection.

105. After viral entry into a host cell, what cell type is the main line of defense?

106. Map the steps of viral infection, assuming previous viral exposure/presence of Abs. (Fig. 24-18) Again, don't directly copy the figure in the book. Be sure to use all the following terms:

acute phase proteins	granzymes	MHC-II
antibodies	helper T cell	NK cell
antigen	host cell	opsonins
apoptosis	interferon-α	perforin
B lymphocytes	lyse	phagocytosis
cytokines	macrophage	T cell receptor
cytotoxic T cell	MHC-I	virus

Antibodies and Viruses

107. Why might Abs from one viral infection not be effective against subsequent viral infections? Give some examples of viruses for which this is TRUE.

Specific Antigens Trigger Allergic Responses

108. Define allergy and allergen.

109. How do immediate hypersensitivity and delayed hypersensitivity reactions differ?

110. What kinds of molecules can be allergens? Name some common allergens.

111. How are people exposed to allergens?

112. Name three effects of the massive release of histamine and other cytokines that cause the condition known as anaphylactic shock or anaphylaxis.

113. Map the steps in an immediate hypersensitivity reaction to an allergen. (Figs. 24-10, 24-11, 24-18, 24-19) Include the following terms:

allergen	degranulation	memory cell (T and B)
antibodies	first exposure	MHC-II
antigen-presenting cell	helper T cell	plasma cell
B lymphocyte	histamine	reexposure
complement	Ig __?__	T cell
complement proteins	inflammation	T cell receptor
cytokines	mast cell	

MHC Proteins Allow Recognition of Foreign Tissue

114. Which cell surface markers determine tissue compatibility? What's another name for these surface markers?

115. Use blood transfusion as an example of compatibility. Make a chart or table with the four blood groups, tell what antigens and antibodies a person with each group will have, and tell which groups will be compatible and incompatible for transfusions. (Fig. 24-20, Table 24-3)

The Immune System Must Recognize "Self"

116. Define self-tolerance.

117. How does self-tolerance arise?

118. What happens if self-tolerance fails? (Table 24-4)

Immune Surveillance Removes Abnormal Cells

119. Briefly describe the immune surveillance theory. What is the supporting evidence? What does it not explain?

NEURO-ENDOCRINE-IMMUNE INTERACTIONS

120. Define neuroimmunomodulation and describe some examples of current research into the area. (Fig. 24-21)

121. Describe the three known links between the immune, nervous, endocrine systems. (Fig. 24-21)

Stress Alters Immune System Function

122. Define stress and describe the body's responses to stress. (See information on glucocorticoids in Ch. 23.)

123. Where are most physical and emotional stressors integrated?

124. The nervous system responds to acute stress. Hormonal responses attend to chronic stresses. Based on what you learned in Chapter 6, why does this make sense?

125. Why is study of the stress response difficult? (Fig. 24-22)

Modern Medicine Includes Mind-Body Therapeutics

▶ Review some of the ways modern medicine is incorporating mind-body therapeutics.

TALK THE TALK

acquired immunity	antibodies	basophils
acquired immunodeficiency	antibody-mediated immunity	bone marrow
acquired immunodeficiency syndrome (AIDS)	antigen-presenting cells (APCs)	bradykinin
		cancer
active immunity	antigens	capsule
acute-phase proteins	antigen-specific responses	cell-mediated immunity
allergens	antihistamines	chemotaxins
allergy	antiviral drugs	chronic inflammatory disease
anaphylactic shock	apoptosis	clonal deletion
anaphylaxis	autoimmune disease	clonal expansion
antibiotics	B lymphocytes	clone

colony-stimulating factors (CSFs)
communication
complement
complement cascade
coordination
Corynebacterium diphtheriae
C-reactive protein
cytokines
cytotoxic cells
cytotoxic T cell
degranulation
delayed hypersensitivity reaction
dendritic cells
diffuse lymphoid tissue
effector cells
encapsulated
encapsulated lymphoid tissue
envelope
eosinophil-derived neurotoxin
eosinophils
extravasation
extravascularly
Fab regions
Fas
Fc regions
fight-or-flight
gamma globulin
general adaptation syndrome
granulocytes
granzyme
gut-associated lymphoid tissue (GALT)
heavy chain
helper T cells
heparin
hinge region
histiocytes
human immunodeficiency virus (HIV)
human leukocyte antigens (HLA)
humor therapy
humoral immunity
hypersensitivity

identification
IgA
IgD
IgE
IgG
IgM
immediate hypersensitivity reactions
immune response
immune surveillance
immunity
immunodeficiency disease
immunoglobulins
in vivo
incorrect response
inflammation
innate immunity
interferon-alpha
interferon-beta
interferon-gamma
interferons
interleukin-1 (IL-1)
islet cell antibodies
kinins
Kupffer cells
lack of response
Langerhans cells
leukocytes
light chain
lymph nodes
lymphocytes
macrophages
major histocompatibility complex (MHC)
mast cells
membrane attack complex
memory
memory cells
MHC class I molecules
MHC class II molecules
microbes
microglia
monocytes
mononuclear phagocyte system
naïve lymphocytes

natural killer cells (NK cells)
neuroimmunomodulation
neutrophils
nonspecific immune response
oncogenic viruses
opsonins
osteoclasts
overactive response
parasites
passive immunity
pathogens
pathogen-associated molecular patterns (PAMPs)
pattern recognition receptors (PRRs)
perforin
phagocytes
phagosomes
plasma cells
polymorphonuclear leukocytes
primary immune response
primary immunodeficiency
protease inhibitor
pus
recruitment
reticuloendothelial cells
reticuloendothelial system
retroviruses
reverse transcriptase
schistosoma
secondary immune response
self-tolerance
sensitivity
specific immune response
specificity
spleen
stress
stressors
suppression
T lymphocytes
T-cell receptors
thymus gland
thyroid-stimulating immunoglobulins
Toll-like receptors (TLRs)
tonsils

ETHICS IN SCIENCE

A mother brings her ill child to the doctor, who diagnoses a viral illness and sends the child home with instructions for his care. The mother is convinced that only antibiotics will cure her son, so she keeps calling and pestering the doctor's staff until the doctor finally prescribes antibiotics, despite knowing that they will not be effective. How is this scenario related to the development of drug-resistant strains of bacteria? What should doctors and the public be doing to prevent the development of more drug-resistant strains?

PRACTICE MAKES PERFECT

1. Match the appropriate letter to each description.

    _____produces antibodies; has large round nucleus with very little surrounding cytoplasm

    _____phagocytic; has nucleus with 3–5 segments, pale pink granules in cytoplasm

    _____releases histamine and heparin; dark blue-staining granules in cytoplasm

    a. neutrophil

    b. eosinophil

    c. basophil

    d. erythrocyte

    e. monocyte

    f. lymphocyte

2. Match the cell surface markers/receptors with the cells on which they are found.

    _____macrophage

    _____red blood cell

    _____B lymphocyte

    _____natural killer cell

    _____liver cell

    _____cytotoxic T cell

    _____plasma cell

    _____helper T cell

    a. MHC-I

    b. MHC-II

    c. T-cell receptors

    d. glycoprotein markers

    e. antibodies

    f. no receptors

3. List three functions of macrophages.

4. Can a mother with blood type A and a father with blood type B have a baby with blood type O? Explain. (Remember that each parent carries two alleles for the RBC surface antigens.)

5. A technician runs an ABO blood type test on Aparna's blood. Her blood agglutinated with anti-B serum but not with anti-A serum.

   a.   What is Aparna's blood type?

   b.   To what other ABO groups can she donate blood?

   c.   From what ABO groups can she receive blood?

6. You are walking barefoot through the cool spring grass, and you unconsciously step on a plant to which you are allergic. Your feet swell up in response. When you consider the physiology behind this reaction, you realize that there are seven kinds of immune cells involved in this hypersensitivity reaction. Name these cells.

7. Where do B lymphocytes develop? T lymphocytes?

8. Your friend Elizabeth knows that she is ABO blood type O. Why is the blood bank always calling her to ask for a blood donation?

9. You are the microscopic ace reporter for the news station Plasma 1, and you have just been alerted to a bacterial entry at the index finger. As you arrive on the scene, describe the events you observe.

10. What is the hallmark of the innate response?

11. Peng-Chai is feeling bad and thinks he is coming down with the flu virus that has been going around school, so he takes some old antibiotics that he has left over from a previous sinus infection. (Not something anyone should ever do!) Will the antibiotics help his viral infection? Explain.

12. What role do acute phase proteins play in an innate response?

13. What is the result of the terminal step in the complement system?

14. What is the difference between the Fc region and the Fab region of an antibody?

15. What is a major histocompatibility complex (MHC)? Describe the differences between the functions of type I MHC and type II MHC.

16. How do the body's white blood cells know to attack and destroy old red blood cells but not new ones?

MAPS

1. Create a map showing the process of clonal expansion.

2. Map the different groups of white blood cells; their structure and their function.

BEYOND THE PAGES

FOCUS ON:
PHYSIOLOGY

Biotechnology Focus: Stealth® Liposomes

Early in their history, liposomes seemed to be the ideal vehicle for drug delivery: magic bullets that could deliver drugs wherever they were needed. But the first animal trials were disappointing, since liposomes injected intravenously disappeared rapidly from the circulation without reaching their target cells. The body's immune system recognized them as foreign even though they were composed of biological phospholipids, and macrophages gobbled them up and digested them. The solution that occurred to researchers was to make the liposomes invisible to the immune system somehow so that they could slip by the macrophages and reach their intended targets. The first type of invisible liposome relied on carbohydrate groups attached to the liposome exterior that made the liposomes resemble red blood cells. The next generation of invisible liposomes was trademarked Stealth® liposomes. They are coated with a polyoxyethylene polymer that allows them to slip into the body undetected and remain in the circulation for as long as a week, delivering their encapsulated drugs to the targeted tissues.

See: Lasic, D., and Martin, F., eds. *Stealth® Liposomes*. Boca Raton: CRC Press, 1995.

25 Integrative Physiology III: Exercise

LEARNING OBJECTIVES

When you complete this chapter, you should be able to:

- Identify sources of ATP used by exercising muscle.

- Compare and contrast aerobic and anaerobic metabolism in exercising muscle.

- Describe an exercising muscle's use of energy substrates to create glucose for ATP production.

- Describe the changes in plasma glucose and plasma hormone levels during periods of exercise. What are the homeostatic challenges? What are the homeostatic goals?

- Explain how physiologists quantify the intensity of periods of exercise.

- Describe how oxygen consumption changes during and after exercise.

- Identify the factors that can limit exercise.

- Diagram the mechanisms and the outcomes of ventilatory responses to exercise. What are the homeostatic challenges? What are the homeostatic goals?

- Diagram the mechanisms and the outcomes of cardiovascular responses to exercise. What are the homeostatic challenges? What are the homeostatic goals?

- Describe the feedforward reflexes that anticipate homeostatic challenges of exercise.

- Diagram homeostatic thermoregulatory responses and mechanisms that take place during periods of exercise.

- Describe what has been proven and what has been suggested about the influence of exercise on health.

SUMMARY

Exercise is muscular activity that threatens homeostasis. ATP for muscle contraction comes from several sources: aerobic metabolism, phosphocreatine, and anaerobic (glycolytic) metabolism. Glucose and fats are the primary substrates for energy production, but fats can only be metabolized in aerobic conditions.

Remember from Chapter 22 that glucagon is a catabolic hormone. The catabolic hormones participating in exercise metabolism include glucagon, cortisol, catecholamines, and growth hormone. The action of these hormones raises the plasma glucose concentration. However, insulin concentrations do not rise during exercise. Remember that active skeletal muscle doesn't require insulin for glucose uptake. Therefore, low plasma insulin concentrations prevent other cells from taking glucose that could be used by the muscles.

The intensity of exercise is indicated by oxygen consumption. Oxygen consumption increases rapidly at the onset of exercise and this increase persists even after activity ceases. The ability of the muscle cells to consume oxygen is possibly a limiting factor in exercise capacity. This is reflected by the fact that mitochondria can increase in size and number with endurance training. Cardiovascular activity is the major factor limiting maximal exertion.

Respiratory and cardiovascular systems make adjustments in response to exercise. Feedforward signals and sensory feedback initiate exercise hyperventilation. Exercise hyperventilation maintains nearly normal P_{O_2} and P_{CO_2} by steadily increasing alveolar ventilation in proportion with exercise level. Cardiovascular responses include increased cardiac output, vasodilation and increased blood flow in skeletal muscles, and a slight increase in mean arterial blood pressure.

Exercise generates heat. In fact, most of the energy released during metabolism is not converted to ATP but is released as heat. There are two mechanisms by which the body regulates temperature during exercise: sweating and increased cutaneous blood flow.

Moderate exercise has been shown to affect health positively. It can alleviate or prevent the development of high blood pressure, strokes, and diabetes mellitus. However, a J-shaped curve (Fig. 25-10) relates immune function to exercise: only moderate exercise improves immune function. There has been little in the way of rigorously controlled evidence to support the theory that exercise improves the immunity of immunocompromised individuals.

TEACH YOURSELF THE BASICS

METABOLISM AND EXERCISE

1. Muscles require ATP for contraction. What are the sources of this ATP? (Figs. 25-1, 25-2; also see Ch. 12)

2. Without new ATP production, a muscle has enough ATP and phosphocreatine to supply energy for how many seconds of intense exercise?

3. What macromolecules are the primary substrates for energy production?

4. The most efficient ATP production is a result of the glycolysis-citric acid cycle pathway. Briefly describe or map these pathways in the presence and absence of oxygen. (Fig. 25-1; also see Ch. 4)

5. Discuss the advantages and disadvantages of anaerobic muscle metabolism versus aerobic metabolism. (Fig. 25-2)

6. Where does muscle obtain glucose for ATP production? (Fig. 25-1; also see Fig. 22-2)

7. TRUE/FALSE? Defend your answer.

   Aerobic exercise first uses glucose for ATP production, then turns to fatty acid metabolism. (Fig. 25-3)

8. Beta oxidation is (faster/slower?) than glycolysis.

9. What are the metabolic results of aerobic training?

Hormones Regulate Metabolism During Exercise

10. List hormones that affect glucose and fat metabolism during exercise and briefly describe their actions.

11. What happens to insulin secretion during exercise? Give the physiological mechanism and the adaptive significance for this pattern of insulin secretion.

Oxygen Consumption Is Related to Exercise Intensity

12. Exercise intensity is quantified by measuring oxygen consumption (V_{O_2}). Define oxygen consumption. In what units is oxygen consumption measured? (See oxidative phosphorylation in Ch. 4, and myoglobin/hemoglobin in Ch. 12.)

13. What is indicated by the maximal rate of oxygen consumption (V_{O_2max})?

14. Increased O_2 consumption persists even after activity ceases. (Fig. 25-4) Why?

Several Factors Limit Exercise

15. Describe some of the factors that can limit exercise capacity.

VENTILATORY RESPONSES TO EXERCISE

16. How do total pulmonary ventilation, alveolar ventilation, and rate and depth of breathing change in response to exercise? (See Ch. 18.)

17. Fill the gaps in the following pathways:

    a. Exercise begins → _____ (receptors) send signals to motor cortex.

    b. Motor cortex signals respiratory control center in the _____ to increase ventilation.

    c. As exercise continues, which sensory receptors send sensory feedback to the respiratory control center to ensure that O_2 use and ventilation are matched? What information is being reported by these receptors?

18. How does exercise hyperventilation nearly normal P_{O_2} and P_{CO_2}? (Figs. 25-5, 25-6)

CARDIOVASCULAR RESPONSES TO EXERCISE

19. The cardiovascular control center (CVCC) responds to exercise with (sympathetic/parasympathetic?) discharge. What effect does this have on cardiac output? peripheral arterioles?

Cardiac Output Increases During Exercise

20. Cardiac output increases dramatically with strenuous exercise. What factors influence cardiac output? (Fig. 14-31)

21. Looking at each of the factors that influence cardiac output, how is each factor affected by exercise (assuming a healthy heart)?

22. Describe how the autonomic nervous system influences heart function during exercise. (Fig. 14-31) Why are these changes necessary and important?

Peripheral Blood Flow Redistributes to Muscle During Exercise

23. During exercise, 88% of blood flow is diverted to exercising muscles. (Fig. 25-7) How is this different from resting muscle blood flow?

24. Describe the local and reflex processes that influence how the body redistributes blood flow during exercise. (Fig. 25-7)

Blood Pressure Rises Slightly During Exercise

25. What factors contribute to mean arterial blood pressure? (See Ch. 15.)

26. Total peripheral resistance decreases as exercise intensity increases. (Fig. 25-8a) This normally would be expected to (raise/lower?) arterial blood pressure. What factors contribute to changes in peripheral resistance during exercise?

27. What is the net result of the changes in cardiac output and peripheral resistance during exercise? (Fig. 25-8b) What factor offsets the decreased total peripheral resistance?

The Baroreceptor Reflex Adjusts to Exercise

28. Normally, increased blood pressure triggers a homeostatic effort to return blood pressure to normal. During exercise, though, there is no homeostatic decrease in BP. Why is this? Outline the possible mechanisms behind the exercise-related changes to the baroreceptor reflex. (See Ch. 8.)

FEEDFORWARD RESPONSES TO EXERCISE

29. Feedforward responses play a significant role in exercise physiology. For example, ventilation (increases/decreases?) upon beginning exercise, despite normal P_{CO_2} and P_{O_2}. (Figs. 25-5, 25-6; also see Ch. 6)

30. Diagram the feedforward response to exercise.

TEMPERATURE REGULATION DURING EXERCISE

31. What happens to most of the energy released during metabolism?

32. Endurance exercise events can create core body temperatures of _____ °C.

33. How does the body respond to this rise in temperature? (See Ch. 22.)

34. Both homeostatic responses to increased temperature can disrupt other homeostatic conditions. Describe the body's responses and the homeostatic challenges of:

    sweating (Fig. 20-17)

    increased cutaneous blood flow

35. Faced with maintaining either blood pressure or body temperature, which will the body select? What would cause the body to choose the other parameter? Why?

36. Describe the changes that take place with acclimatization to exercise in hot environments.

EXERCISE AND HEALTH

37. Exercise can improve several pathological conditions. List some of these conditions.

Exercise Lowers the Risk of Cardiovascular Disease

38   There is a relationship between exercise and cardiovascular disease. How does exercise affect:

BP

plasma triglycerides

HDL levels

▶ Even mild exercise can have significant health benefits.

Type 2 Diabetes Mellitus May Improve with Exercise

39. Regular exercise can improve type 2 diabetes mellitus. Briefly describe how this is so. (Fig. 25-9)

Stress and the Immune System May Be Influenced by Exercise

▶ Exercise is associated with a reduced incidence of disease and improved longevity.

40. Few rigidly controlled studies support that exercise boosts immunity, prevents cancer, or helps HIV-positive people fight AIDS. In fact, strenuous exercise can be detrimental. What is the physiology behind the potentially detrimental effects of strenuous exercise? (Fig. 25-10)

41. What does research indicate about the relationship of exercise and depression? Is this relationship proved by the evidence or overstated?

TALK THE TALK

acclimatization
adipose tissue
aerobic metabolism
alveolar ventilation
anaerobic pathways
ATP
baroreceptor reflex
beta cells of the pancreas
beta-oxidation
carbohydrates
cardiac output
cardiovascular control center
carotid body chemoreceptors
catecholamines
central chemoreceptors
cholinergic vasodilator system
citric acid cycle
contractility
convective heat loss
cortisol
dehydration
diabetes mellitus
endurance
epinephrine
evaporative cooling
excess postexercise oxygen
   consumption (EPOC)
exercise
exercise and depression

exercise and the immune
   system
exercise hyperventilation
fats
fatty acids
feedforward responses to
   exercise
force of contraction
glucagon
glucose
glucose tolerance test
glucose transporters
glycogen
glycolysis
glycolytic metabolism
growth hormone
heart rate
high blood pressure
homeostasis
hyperpnea
insulin
K^+
lactate
limbic system
liver
maximal rate of oxygen
   consumption ($V_{O_2\,max}$)
metabolic acidosis
mitochondria

motor cortex
muscle contraction
norepinephrine
oxidative phosphorylation
oxygen consumption (V_{O_2})
oxygen deficit
paracrines
parasympathetic output
P_{CO_2}
peripheral blood flow
peripheral resistance to blood
   flow
phosphocreatine
physiological integration
P_{O_2}
proprioceptors
pulmonary stretch receptors
pyruvate
respiratory control center of
   the medulla
Starling's law of the heart
sweating
sympathetic output
thermoregulation
vasoconstriction
venous return
ventilation

PRACTICE MAKES PERFECT

1. Which would you expect to change more during exercise: diastolic or systolic blood pressure?

2. TRUE/FALSE? Defend your answer.

    During exercise, cardiac output, stroke volume, and excess postexercise oxygen consumption all should be greater in a trained (physically fit) person than in an average person.

3. How is cardiorespiratory endurance (aerobic fitness) usually measured?

4. During exercise, inspiratory and expiratory reserve volumes decrease. Why do you think that is?

5. List two beneficial effects of exercise.

6. What physiological factors can limit exercise capacity?

7. Concisely describe the relationship between exercise and immunity.

MAPS

Design a map that integrates cardiovascular and respiratory responses to exercise. Include receptors, chemicals, calculations, etc.

26 Reproduction and Development

LEARNING OBJECTIVES

When you complete this chapter, you should be able to:

- Diagram the internal and external anatomy of both males and females.

- Diagram the processes of sexual differentiation that occur during male and female embryonic development.

- Contrast mitosis with meiosis, haploid with diploid, and autosomes with sex chromosomes.

- Diagram the common hormonal control pathway that governs reproductive function in both males and females. Include the feedback pathways on this diagram.

- Explain the pulsatile secretion of GnRH and its significance to reproductive physiology.

- Identify some of the environmental factors that can influence reproductive physiology.

- Diagram the process of spermatogenesis and explain the timeline on which this occurs. Identify anatomical structures and the roles of hormones involved.

- Diagram oogenesis and explain the timeline on which this occurs.

- Diagram the menstrual cycle and its complex hormonal control patterns.

- Diagram the erection reflex and describe the four phases of the human sexual response.

- Describe methods of contraception currently available.

- Diagram the process of fertilization.

- Diagram the process of embryo implantation in the endometrium.

- Describe the role of the hormones secreted by the placenta during pregnancy.

- Describe what we currently understand about the processes of labor and parturition.

- Diagram a mammary gland and describe milk and colostrum production.

- Diagram the let-down (milk ejection) reflex.

- Describe how the reproductive systems of males and females change with puberty and then how they change again with menopause and andropause.

SUMMARY

On a very general level, the human body contains two types of cells: diploid autosomal (somatic or body) cells and haploid sex (or germ) cells. The diploid number of chromosomes for humans is 46, and the haploid number is therefore 23 (exactly half the diploid number of chromosomes). In haploid cells, there are 22 autosomal chromosome pairs that control autosomal cell development and one pair of sex chromosomes that control germ cell development.

The male sex cells, spermatazoa, are produced by the testes; female sex cells, ova, are produced by the ovaries. Gametogenesis is the process by which sex cells are produced. The diploid primary sex cells (spermatogonia, oogonia) become secondary sex cells, and after the second meiotic division, these become haploid germ cells. While the process is similar for both male and female, the timing of gametogenesis is very different. Spermatogenesis begins in the embryo and stops just after birth. It resumes at puberty and continues throughout the male's lifetime. Oogenesis also begins in the embryo but stops before birth. At puberty, ovulation begins and occurs in a cyclical fashion until menopause. Both spermatogenesis and oogenesis are under hormonal control. Hormones involved include gonadotropins (FSH and LH from the anterior pituitary), sex hormones (androgens, estrogens, progesterone), inhibins, and activins.

Sexual differentiation takes place during the seventh week of embryonic development. Before differentiation, the gonadal tissue is considered bipotential. If the embryo has a functional Y chromosome, the SRY gene produces testis-determining factor that signals the testicular Sertoli cells to secrete Anti-Müllerian hormone. Anti-Müllerian hormone causes the degeneration of the Müllerian ducts. Leydig cells then begin to secrete testosterone and DHT, which cause the Wolffian ducts to develop into male accessory structures. Testosterone and DHT also create the external male genitalia. On the other hand, if the embryo does not have a functional Y chromosome, then none of the above happens, and the Müllerian ducts develop into female reproductive structures.

Male reproductive anatomy consists of the testes, accessory glands and ducts, and external genitalia. The urethra, which runs through the penis, is the common duct for sperm and urine movement (though not concurrently). Female reproductive anatomy consists of the ovaries, fallopian tubes, uterus, vagina, labia (majora, minora), and clitoris. The female urethra is completely separated from the reproductive structures.

The female reproductive cycle is perhaps one of the most complex patterns in physiology. It is under extensive hormonal control (Figs. 26-13, 26-14) and can be influenced by many emotional and physical factors. The menstrual cycle is composed of two concurrent cycles: the ovarian cycle and the uterine cycle. The ovarian cycle is further divided into phases: the follicular phase, ovulation, and the luteal phase. The uterine cycle is also divided into phases: menses, the proliferative phase, and the secretory phase.

During unprotected sex, the sperm and egg have the possibility of joining to form a zygote. Sperm must first be capacitated before they can make their way to the egg. If a sperm reaches an egg and successfully fertilizes it, only then will the egg complete its second meiotic division. Only one sperm is allowed to fertilize an egg, as the cortical reaction prevents polyspermy. Additionally, fertilization is species specific, meaning that only sperm and egg of compatible species can initiate procreation. If fertilization is successful, the two haploid germ cells create a diploid zygote that begins mitotic division to produce an embryo. The developing embryo attaches itself to the mother's uterine wall and parasitizes the mother's nutrient supply until birth. The placenta of the developing embryo secretes hormones that alter the mother's metabolism to ensure adequate nourishment for development.

After 38–40 weeks of pregnancy, the baby is born in a process called parturition. This process is under hormonal control, though the specific signaling is still unclear. The mother's mammary glands are activated under hormonal control and begin to secrete milk to feed the baby.

TEACH YOURSELF THE BASICS

1. What is a hermaphrodite? What is pseudohermaphroditism?

2. What is sexual dimorphism?

3. What is a zygote? An embryo? A fetus?

SEX DETERMINATION

4. Define the structures found in both male and female sex organs, then give the sex-specific names for each.

   a. genitalia

  b.   gamete

  c.   germ cell

5. Review of genetics: (See Appendix C.)

    Contrast the terms diploid and haploid.

    How many sets of chromosomes do nucleated body (somatic) cells have? (Fig. 26-1) _____

    How many autosomes in a somatic cell? _____ How many sex chromosomes in a somatic cell? _____

    How many autosomes do gametes have? _____ How many sex chromosomes? _____

    How is the X sex chromosome different from the Y?

Sex Chromosomes Determine Genetic Sex

6. XX individuals are usually _____ and XY individuals are usually _____. (Fig. 26-2)

7. There are many documented cases of abnormal sex chromosome distribution in humans. What are the roles of X and Y in determining the physical sex of an individual? What happens in the absence of the Y chromosome?

8. What are Barr bodies? How and why are they formed?

Sexual Differentiation Occurs Early in Development

9. Before the seventh week of development, it is morphologically difficult to determine an embryo's sex. Diagram the bipotential gonad, external genitalia, and internal genitalia, and identify the final male and female anatomical structures into which each develops. (Fig. 26-3; Table 26-1)

Male Embryonic Development

10. Explain the role of the following in sexual development of the genetically male embryo:

Y chromosome

SRY gene

testis determining factor (TDF)

Anti-Müllerian hormone (AMH)

Sertoli cells

Leydig cells

testosterone and DHT

11. Diagram the process of male sexual differentiation in the developing fetus.

12. What effect, if any, do sex hormones have on sexual behavior and gender identity?

Female Embryonic Development

13. Diagram the process of female sexual differentiation.

BASIC PATTERNS OF REPRODUCTION

14. Define gametogenesis.

15. The timing of gametogenesis is different for males and females. Briefly compare these timing differences.

Gametogenesis Begins in Utero

▶ For a summary of male and female patterns of gametogenesis, see Fig. 26-5.

16. Gametogenesis has some similar steps in both sexes. Briefly compare the steps of gametogenesis in males and females, describing the steps in terms of cell types, chronology, by whether DNA has replicated, whether the cell divides, and the number of chromosomes in the cell.

17. The first meiotic division creates two secondary gametes. Compare the fates of those gametes in males and females.

Male Gametogenesis

18. Diagram the process of male gametogenesis, beginning at puberty.

Female Gametogenesis

19. Diagram the process of female gametogenesis, beginning at puberty.

20. What are polar bodies?

21. Define ovulation.

22. What happens if an egg is not fertilized? What happens when an egg is fertilized?

The Brain Directs Reproduction

23. What roles do the hypothalamus and anterior pituitary play in sex hormone production?

24. List the steroid sex hormones. (Fig. 26-6)

25. Which hormone is dominant in males? Where is it secreted? What is its more potent derivative hormone?

26. Which hormones are dominant in females? Where are they secreted?

27. What does the enzyme aromatase catalyze?

Control Pathways

28. Diagram the basic pattern of hormonal control of reproduction in both sexes. Identify all the hormones involved. (Fig. 26-7)

29. What are the generalized actions of FSH? LH?

30. Describe the roles that gonads, inhibins, and activins play in regulation of reproductive function.

Feedback Pathways

▶ Review feedback pathways and up-regulation in Ch. 6.

31. Sex hormones follow general short-loop and long-loop feedback patterns. Describe the feedback effect of each of the following. (Fig. 26-7; Table 26-2)

    a.   gonadal steroids

    b.   low estrogen levels

    c.   sustained elevated estrogen

Pulsatile GnRH Release

32. Describe the tonic, pulsed GnRH release in both sexes. What is the physiological mechanism for this pulsing? What are some therapeutic applications of this pulsatile nature of GnRH secretion?

Environmental Factors Influence Reproduction

33. What are some of the environmental factors that influence reproductive hormones and gametogenesis? (Fig. 7-22)

MALE REPRODUCTION

34. Male anatomy: (Fig. 26-8)

    The external genitalia are the _____ and _____.

    What is the common passageway for urine and sperm?

    What two tissues form the erectile tissue?

    The tip of the penis is called the _____ penis. It is covered by tissue called the _____.

    unless this tissue has been surgically removed in a procedure known as _____.

35. What is the function of the scrotum?

36. What is cryptorchidism and why is it usually corrected?

37. List the three male accessory glands.

Testes Produce Sperm and Hormones
38. Human testes: (Fig. 26-9)

The tough outer fibrous capsule encloses the _____ tubules.

What structures/cells are found between these tubules?

What is the epididymis? (Fig. 26-9b)

What is the vas deferens?

Seminiferous Tubules
39. Describe the two cell types that compose the seminiferous tubules. (Fig. 26-9c, d)

40. What is the function of the basal lamina that surrounds the outside of the tubule? (Fig. 26-9d)

41. List the three functional compartments in the testes.

42. Describe the composition of the fluid in the lumen of the seminiferous tubules.

Sperm Production
43. Diagram the anatomical arrangement of the Sertoli cells and the spermatogonia. (Fig. 26-9c, d)

44. Diagram the process of sperm maturation. (Fig. 26-10)

45.   Diagram the parts of a mature sperm and indicate their functions. (Fig. 26-10)

Sertoli Cells

46.   What role do the Sertoli cells play in sperm maturation?

47.   List the substances made or secreted by the Sertoli cells.

48.   What is the function of androgen-binding protein? (Fig. 26-11) Why is it necessary?

Leydig Cells

49.   What is the function of the Leydig cells? (Fig. 26-9c, d)

50.   Leydig cells are active in the fetus but inactive after birth until puberty. What role do they play in the fetus?

51.   The bulk of the body's testosterone is produced in the Leydig cells. Into what other hormones can testosterone be converted?

Spermatogenesis Requires Gonadotropins and Testosterone

52.   What is the target tissue of FSH and what effect does the hormone have on this tissue? (Fig. 26-11)

53.   What is the target tissue of LH and what effect does the hormone have on this tissue?

54.   Diagram the hormonal control patterns for FSH and LH secretion. (Fig. 26-11)

Male Accessory Glands Contribute Secretions to Semen

55. What is semen?

56. What substances, other than sperm, are found in semen? Give their source and their function. (Table 26-3)

57. What percentage of semen volume is from the accessory glands?

Androgens Influence Secondary Sex Characteristics

58. List the primary sex characteristics of males.

59. List the secondary sex characteristics of males.

60. Androgens are (anabolic/catabolic?) steroids. Why?

61. What are some of the side effects and behavioral changes brought about by steroid hormone abuse?

FEMALE REPRODUCTION

Female Have Ovaries and a Uterus

▷ See Pap smear in Ch. 3 Running Problem Conclusion.

62. List the parts of external female genitalia that make up the vulva (or pudendum). (Fig. 26-12a)

63. Diagram the internal female anatomy.

64. How do the fallopian tubes move an egg to the uterus? (Fig. 26-12c)

The Ovary Produces Eggs and Hormones

65. Draw the anatomy of an ovary and its follicles. (Fig. 26-12d)

66. Distinguish between the following structures: primary oocyte, primary follicle, and theca. (Fig. 26-12e)

A Menstrual Cycle Lasts About One Month

67. How many days is the average menstrual cycle?

68. What is the purpose of the menstrual cycle?

69. The menstrual cycle is described according to changes in what two structures?

70. Name and describe the three phases of the ovarian cycle. (Fig. 26-13)

71. Name and describe the phases of the uterine cycle. (Fig. 26-13)

Hormonal Control of the Menstrual Cycle Is Complex

72. List the hormones involved in control of the ovarian and uterine cycles. (Fig. 26-13)

73. Which hormone dominates the follicular phase?

74. Which hormones dominate the luteal phase?

Early Follicular Phase

75. What event marks day 1 of a cycle?

76. Hormones:

    What happens to FSH and LH secretion just before the cycle begins?

    What changes does FSH cause? (Fig. 26-13; Table 26-4)

    Which hormones do the granulosa cells and theca secrete? (Fig. 26-14a)

    Why do FSH levels decline as the follicular phase progresses?

77. Now, in more detail, explain the role of each of the following in follicle development. (Table 26-4)

    granulosa cells

    thecal cells

78. What is the antrum?

79. What is atresia?

80. What happens to the endometrium during the follicular phase?

Late Follicular Phase

81. What happens to ovarian estrogen levels as the follicular phase nears its end? (Fig. 26-13)

82. What hormones are the granulosa secreting at this point? (Fig. 26-14b)

83. What effects do the granulosa hormones exert on the gonadotropin/sex hormone control pathway?

84. Describe the LH surge.

85. Meiosis resumes just before ovulation. When did meiosis pause for this oocyte?

86. What is the result of this meiotic division? (Fig. 26-5)

87. Antral volume now reaches its maximum or minimum?

88. How do the high estrogen levels prepare the uterus for pregnancy?

Ovulation

89. At which point of the cycle does ovulation occur?

90. What enzyme does a mature follicle secrete that promotes ovulation?

91. The breakdown products of collagen create an inflammatory reaction. What role does this reaction play in ovulation?

92. Describe the process by which an egg is ejected from its follicle.

93. Describe luteinization.

94. What happens to estrogen synthesis at ovulation?

Early to Mid-Luteal Phase

95. Describe corpus luteum secretion of progesterone and estrogen secretion following ovulation.

96. How do these hormones affect FSH and LH production? (Fig. 26-14c)

97. What are the effects of progesterone on the uterus?

98. What additional effects does progesterone have on a woman's body?

Late Luteal Phase and Menstruation

99. What happens to the corpus luteum in the absence of pregnancy? (Fig. 26-14d; also see Ch. 3) How does this affect circulating hormone levels? How does this affect the FSH/LH control pathway?

100. When progesterone secretion decreases, what change does this initiate in the endometrium?

101. Describe menstruation. (See Ch. 16.)

Hormones Influence Female Secondary Sex Characteristics

102. What are some female secondary sex characteristics?

103. Which part of the adrenal gland secretes androgens?

104. What effect do these androgens have in women?

PROCREATION

105. What types of structures and behaviors have evolved to ensure reproductive success in humans and many other terrestrial vertebrates?

106. List the two stages of the male sex act.

The Human Sexual Response Has Four Phases

107. Describe the four stages of the human sex act (coitus).

The Male Sex Act Includes Erection and Ejaculation

108. What happens physiologically to allow erection?

109. Diagram the erection reflex. (Fig. 26-15; also see urination reflex in Fig. 19-18 and defecation reflex in Ch. 21)

110. What is the climax of the male sexual act? What does this accomplish?

111. The average semen volume is 3 mL. What percentage of that volume is sperm?

112. Define emission and ejaculation.

113. What structure prevents the mixing of sperm and urine?

Sexual Dysfunction Affects Males and Females

114. What are some factors involved in erectile dysfunction (ED)?

115. What treatment options are available for sexual dysfunction? Briefly describe their mechanisms of action.

Contraceptives Are Designed to Prevent Pregnancy

Barrier Methods

116. List and describe barrier method contraception options available today.

Implantation Prevention

117. List and describe implantation prevention options available today.

Hormonal Treatments

118. List and describe hormone-based contraceptive options available today.

119. Why has the development of male hormonal contraceptives been slow?

120. How do contraceptive vaccines work?

Infertility Is the Inability to Conceive

121. Define infertility.

122. List some potential causes of infertility.

123. What percentage of all pregnancies terminate spontaneously?

124. Describe the process of *in vitro* fertilization.

PREGNANCY AND PARTURITION

Fertilization Requires Capacitation

125. What is capacitation?

126. For how long following ovulation can the egg be fertilized?

127. How long is sperm viable in the female reproductive tract?

128. Where does fertilization take place? (Fig. 26-16)

129. Diagram the process of fertilization. (Figs. 26-16, 26-17) Include the following terms: acrosome, acrosomal reaction, capacitation, cortical granules, cortical reaction, enzymes, granulosa cells, meiosis, polyspermy, second polar body, sperm-binding receptor, sperm nucleus, zona pellucida, zygote nucleus. (Use a separate sheet of paper.)

130. Fertilization creates a zygote with a (diploid/haploid?) set of chromosomes.

The Developing Zygote Implants in the Endometrium

131. When does the dividing zygote move into the uterine cavity? How is this movement accomplished? (Fig. 26-18)

132. In which developmental stage is the embryo at this point?

133. What structure does the outer blastocyst become? (Fig. 26-19a)

134. Into what structures does the inner blastocyst cell mass develop?

135. How long after fertilization does implantation into the uterine wall usually take place?

136. Describe the process of implantation.

137. What are chorionic villi and what is their function?

138. Explain the relationship between the mother's blood supply and that of the fetus. (Fig. 26-19b)

139. Why is the abnormal separation of the placenta from the endometrium a medical emergency?

The Placenta Secretes Hormones During Pregnancy

140. When an embryo has implanted, what prevents menstruation from occurring?

Human Chorionic Gonadotropin (hCG)

141. To what other hormone is hCG structurally related? Can hCG bind to this other hormone's receptors?

142. Describe the role of hCG on progesterone secretion early in pregnancy. Why is progesterone secretion needed?

143. What is the second function of hCG?

Human Placental Lactogen (hPL)

144. What is the older name for hPL?

145. To what two other hormones is hPL structurally related?

146. What are the postulated roles for hPL during pregnancy? What is gestational diabetes mellitus?

Estrogen and Progesterone

147. What effect does continuous secretion of estrogen and progesterone during pregnancy have on FSH and LH secretion? What is the goal of suppressing FSH and LH secretion?

148. What are the functions of estrogen and progesterone during pregnancy?

149. What other hormones are secreted?

Pregnancy Ends with Labor and Delivery

150. How long is gestation in humans?

151. Define parturition and labor.

152. What are some of the potential triggers responsible for initiating parturition?

153. What position is the fetus normally in at the time of labor? (Fig. 26-20a)

154. Diagram the positive feedback loop of parturition. (Fig. 26-21)

155. What two hormones promote uterine contractions? (Fig. 26-21)

156. What happens to the placenta when the baby is born? (Fig. 26-20d)

The Mammary Glands Secrete Milk During Lactation

157. Describe mammary gland structure. (Fig. 26-22)

158. Which hormone influences breast development at puberty?

159. How do the breasts change during pregnancy? What hormones induce these changes?

160. Estrogen and progesterone (inhibit/stimulate?) milk secretion by the mammary gland epithelium.

161. Prolactin controls milk production. (Fig. 7-13) Draw the prolactin control pathway.

162. Compare the composition of colostrum and breast milk.

163. What causes milk production to increase after pregnancy?

164. How does suckling act as a stimulus for milk production and release? Draw the reflex. (Fig. 26-23)

165. How is milk ejection accomplished?

Prolactin Has Other Physiological Roles

166. What are the roles of prolactin in men and non-nursing women? (See neuroimmunomodulation in Ch. 24.)

GROWTH AND AGING

Puberty Marks the Beginning of the Reproductive Years

167. When does puberty begin for girls? For boys? (See leptin in Ch. 22.)

168. Puberty requires the maturation of which control pathway? What is one explanation for how this happens? What are some of the signals responsible for the onset of puberty?

Menopause and Andropause Are a Consequence of Aging

169. What changes happen in the female reproductive system after about 40 years of menstrual cycles? Why?

170. What changes can accompany the postmenopausal lack of estrogen? (See Chs. 22–23.)

171. What is the current drug therapy for treating menopause-related conditions? How do these drugs work?

172. What physiological changes are seen in reproductive function as a man ages?

TALK THE TALK

5α-reductase
abstinence
acrosomal reaction
acrosome
activins
allantois
amnion
amniotic fluids
anabolic steroids
androgen receptor
androgen-binding protein (ABP)
androgens
andropause
anovulatory
Anti-Müllerian hormone (AMH)
antrum
apoptosis
aromatase
assisted reproductive technology (ART)
atresia
autosomes
Barr body
basal compartment
bipotential
blastocyst
blood-testis barrier
bulbourethral (Cowper's) glands
capacitation
centromere
cervical cap
cervix
chorion
chorionic villi

clitoris
coitus
collagenase
colostrum
condom
contraception
contraceptive sponge
corpora cavernosa
corpus albicans
corpus luteum
corpus spongiosum
cortex
cortical granules
cortical reaction
corticotropin-releasing hormone (CRH)
cryptorchidism
diaphragm
dihydrotestosterone (DHT)
ductus deferens
egg (secondary oocyte)
ejaculation
emission
endometrium
environmental estrogens
epididymis
erectile dysfunction (ED or impotence)
erection
erection reflex
erogenous zone
estradiol
estrogens
external genitalia
extraembryonic membrane
fallopian tubes
fimbriae

finasteride
first meiotic division
first polar body
flaccid
follicle-stimulating hormone (FSH)
follicular phase
foreskin
gametes
gametogenesis
genital tubercle
germ cells
gestation
gestational diabetes mellitus
glans
gonadotropin-releasing hormone (GnRH)
gonadotropins
gonads
granulosa cells
human chorioic somatomammotropin (hCS)
human chorionic gonadotropin (hCG)
human placental lactogen (hPL)
hymen (maidenhead)
hypoprolactinemia
in vitro
in vivo
infertility
inhibins
internal genitalia
interventional methods
intrauterine devices (IUDs)
kisspeptin
labia majora

labia minora
labioscrotal swellings
labor
lactation
leptin
let-down reflex
Leydig cells
LH surge
libido
luteal cells
luteal phase
luteinization
luteinizing hormone (LH)
medulla
meiosis
meiotic
melatonin
menarche
menopause
menses
menstrual cycles
menstruation
Müllerian ducts
myoepithelial
myometrium
oocytes
oogonia
oral contraceptives
orgasm
ova
ovarian cycle
ovaries
oviducts
ovulation
parturition
penis
perimenopause
phosphodiesterase-5 (PDE-5)

phytoestrogens
placenta
plasmin
polycystic ovary syndrome
 (PCOS)
polyspermy
postovulation
postpartum
prepuce
primary follicle
primary gametes
primary oocytes
primary sex characteristics
primary spermatocytes
progesterone
progestins
prolactin-inhibiting hormone
 (PIH)
proliferative phase
prostate gland
prostatic hypertrophy
pseudohermaphroditism
puberty
pudendum
pulse generator
relaxin
scrotum
second meiotic division
second polar body
secondary gametes
secondary sex characteristics
secondary spermatocytes
secretory phase
selective estrogen receptor
 modulators (SERMs)
semen
seminal vesicles
seminiferous tubules

Sertoli cells
sex chromosomes
sex-determining region of the
 y chromosome (SRY gene)
sexual dimorphism
sister chromatids
sperm
spermatids
spermatogonia
spermicides
SRY protein
sterilization
stimulated
stroma
suckling
sustentacular cells
testes
testis-determining factor
testosterone
theca
transforming growth factor-β
tubal ligation
urethra
urethral folds
urethral groove
uterine cycle
uterus
vagina
vas deferens
vasectomy
vulva
Wolffian ducts
womb
yolk sac
zona pellucida
zygote

ETHICS IN SCIENCE

In recent years, the use of donor eggs and hormonal therapy has made it possible for postmenopausal women to bear children through *in vitro* fertilization. For example, in California a 53-year-old woman gave birth to quadruplets. When these children are 20, their mother will be 73, if she is still alive. Older men, such as the former South Carolina Senator Strom Thurmond, had children with younger women for years. Should an age limit be placed on women who desire *in vitro* fertilization?

PRACTICE MAKES PERFECT

1. Anatomical structures are considered *homologous* if they have the same origin and *analogous* if they are similar in function but do not have the same origin. Using the information in Fig. 26-3, match the following parts of the male and female reproductive tract and mark them as homologous or analogous. Not every part will have a corresponding part in the opposite sex. In that case, mark them "unique."

Male	Corresponding female part?	Analogous or homologous?	Female
bulbourethral gland			a. clitoris
ductus deferens			b. fallopian tube
penis			c. ovary
prostate gland			d. labia majora
scrotum			e. labia minora
seminal vesicle			f. uterus
testis			g. vagina

2. Which set of terms below corresponds to the blanks in the sentence?

    The _____ later develops into the _____ which secretes _____ until the _____ takes over the role of maintenance of pregnancy.

    a. ovary, corpus luteum, estrogen and progesterone, endometrium
    b. follicle, corpus luteum, luteinizing hormone, placenta
    c. ovary, placenta, follicle-stimulating hormone, corpus luteum
    d. ovary, corpus luteum, estrogen and progesterone, placenta
    e. follicle, corpus luteum, estrogen and progesterone, placenta

3. You are a researcher who has just discovered the hypothalamic-pituitary control axis for the ovary. Now you have conducted some experiments to see if menopause is due to a failure of the ovarian cells themselves, or a failure of one of the trophic hormones. Your working hypothesis is that menopause is a result of pituitary failure. If this is true, what results do you expect to obtain in your tests?

    a. levels of pituitary gonadotropins will be: *elevated   normal   decreased*

    b. estrogen levels will be: *elevated   normal   decreased*

    c. administering estrogen: *will restore normal menstrual cycles   will have no effect on cycles*

    d. administering GnRH: *will restore normal menstrual cycles   will have no effect on cycles*

    e. administering FSH and LH: *will restore normal menstrual cycles   will have no effect on cycles*

    d. administering progesterone: *will restore normal menstrual cycles   will have no effect on cycles*

4. Explain the function of the corpus luteum during early pregnancy.

5. What is the adaptive value of androgen-binding protein?

6. Fill in the following chart on reproduction.

	Male	Female
What is the gonad?		
What is the gamete?		
Cell(s) that produce(s) gametes		
What structure has the sensory tissue involved in the sexual response?		
What hormone(s) control(s) development of the secondary sex characteristics?		
Gonadal cells that produce hormones and the hormones they produce		
Target cell/tissue for LH		
Target cell/tissue for FSH		
Hormone(s) with negative feedback on anterior pituitary		
Timing of gamete production in adults		

7. Name the hormone(s) in women that is (are) the primary control for the following events:

proliferation of the endometrium

initiates development of follicle(s)

ovulation

development of endometrium into a secretory structure

keeps corpus luteum alive in early pregnancy

keeps endometrium from sloughing (i.e., menstruating) in early pregnancy

MAPS

1. Create a map showing the events determining the sex of an embryo, starting with the sex chromosomes.

2. Outline the uterine and ovarian cycles and include the major hormones active during each phase.

3. Draw a reflex map that shows how the female birth control pill that contains estrogen and progesterone works to stop the production of ova.

Answers to Workbook Questions

CHAPTER 1

PRACTICE MAKES PERFECT

1. Physiology is the study of body function; anatomy is the study of body structure.
2. a. The hypothesis might be "Pravistatin lowers cholesterol in rats."
   b. An appropriate control would be to administer an inert substance in the same fashion that the Pravistatin is administered.
3. As the concentration of nerve growth factor (NGF) increases, the number of migrating cells increases. NGF has its minimum effect at a concentration of 10^{-5} M and its maximum effect at a concentration of 10^{-3} M.
4. The independent variable, which is controlled by the experimenter, is the extracellular concentration of glucose; it goes on the x-axis. The dependent variable, which is measured by the experimenter, is the intracellular concentration of glucose; it goes on the y-axis. The data do not begin at low values, so the origin is not given the value of 0,0. On the x-axis, the first heavy line is given the lowest value (80), with each heavy line after that representing an increase of 10 mM. The x-axis data are evenly spaced. The y-axis values are not evenly spaced, so the person constructing the graph must select a range of values that includes all points. The lowest data point is 52, so the first heavy line on the y-axis is given a value of 50, with each heavy line after that representing an increase of 10 mM (60, 70, etc.). The y-axis data points can then be plotted according to their actual values. The data points do not fall exactly on a line, so a best-fit line should be drawn (see Fig. 1-7d). The x-axis is a continuous function, so individual points should be connected. The graph flattens out at x = 130 and the line changes slope from diagonal to horizontal. Summary of results: The intracellular concentration of glucose increases linearly as extracellular glucose concentration increases, up to an extracellular concentration of 100 mM. At that concentration, the intracellular concentration of glucose reaches a maximum value of about 100 mM.
5. a. Bar graphs: x-axis has the three temperatures, y-axis is oxygen consumption. You would use different colors/patterns of bars for summer- and winter-collected animals.
   b. Growth is a continuous function, therefore this data set is appropriate for line graphs. Height and weight should be plotted on separate graphs, and each graph will have two lines; one for girls and one for boys.
   c. Data points from a population of people selected at random are graphed on a scatter plot. In this example, the x-axis is food intake and the y-axis is glucose concentration.

CHAPTER 2

TEACH YOURSELF THE BASICS

67. To make a 1 molar (1 mole/L) solution of NaCl, weigh out 58.5 g NaCl and add water until the total volume is 1 liter.
69. One mole of magnesium ions (Mg^{2+}) contains 2 equivalents.
70. A 10% (wt/vol) solution contains 10 g of solute per 100 mL solution. To make 250 mL of 10% NaCl, weigh out 25 g of NaCl and add water until the final volume is 250 mL.

71. A solution with 200 mg NaCl/dL = 200 mg/100 mL. This is equal to 2000 mg/1000 mL, or 2 g/L.

QUANTITATIVE THINKING

Task 2:   6 moles NaCl + 6 moles glucose = 12 moles total solute in 3 L volume = 4 mol/L = 4 M
Task 3:   300 mM glucose = 0.3 mole glucose/liter. Mol. weight glucose = 180.

   180 g/mole = ? g/0.3 mole = 54 g

   54 g glucose/1 liter = ? g/0.6 L = 32.4 g glucose into 600 mL solution.

PRACTICE MAKES PERFECT

1. The carbon atom has 6 protons and 6 neutrons in the nucleus. There are 2 electrons in the first electron shell and 4 electrons in the outer shell. See Fig. 2-2 for oxygen. The oxygen atom has 8 protons and 8 neutrons in the nucleus. There are 2 electrons in the first electron shell and 6 electrons in the outer shell.

2. :Ö::C::Ö:

3. b, d, a, c, f, g, h

4. (answers in bold)

ELEMENT	SYMBOL	ATOMIC NUMBER	PROTONS	ELECTRONS	NEUTRONS*	ATOMIC WEIGHT
Calcium	**Ca**	20	**20**	**20**	20	40.1
Carbon	C	**6**	6	6	**6**	12
Chlorine	**Cl**	17	17	**17**	18	35.5
Cobalt	Co	27	27	27	**32**	58.9
Hydrogen	**H**	1	1	1	0	1
Iodine	I	53	**53**	**53**	74	**127**
Magnesium	**Mg**	12	12	12	12	**24**
Nitrogen	N	7	**7**	7	**7**	14
Oxygen	O	**8**	8	**8**	**8**	16.0
Sodium	**Na**	11	11	**11**	**12**	23
Zinc	**Zn**	30	30	**30**	35	65
Copper	**Cu**	29	**29**	29	35	**64**
Iron	Fe	26	**26**	**26**	30	55.8
Potassium	**K**	19	19	**19**	20	**39**

*Number of neutrons is the most common isotope.

5. Water = H_2O. (2 H × at. wt. 1 = 2) + (1 O × at. wt. 16 = 16) = 18 daltons

6. f, e, b, d, a, c

7. Both a and c are correct. They have the correct number of carbon, hydrogen, oxygen and nitrogen atoms. The COO⁻ group on the right can rotate around the C–C bond, so both drawings are correct. Figures b and c both have 5 bonds on the second-to-last carbon instead of 4.
8. a. Covalent
   b. Hydrogen
   c. Water is the only solvent in biological systems.
   d. The charged ions will be attracted to the partial charges in the polar regions of the water molecule. This allows the ions to dissolve in the water.
9. If an oxygen atom gains a proton, it becomes a fluorine atom.
10. Chlorine has 7 electrons in its 8-place outer shell; potassium, like sodium, has only one electron in its outer shell. Thus, chlorine takes an electron from potassium, creating Cl⁻ (chloride) and K⁺. Both ions are stable as their outer shells are now filled. See Fig. 2-4.
11. Polar molecules are hydrophilic because their regions of partial charge will interact with the polar regions of water, allowing the polar solute to dissolve. Nonpolar molecules are hydrophobic and will not dissolve because they have no regions of charge to disrupt the hydrogen bonds between adjacent water molecules.
12. The molecular weight of sodium chloride (NaCl) is 58.5. One mole weighs 58.5 grams, so 0.5 mole will weigh 29.25 g.
13. A 0.5 M NaCl solution has 0.5 mole/L, or 29.25 g/L. To make 0.5 L, put 14.625 g into 500 mL solution.
14. A 0.1 M solution has 100 mmoles/L.
15. A 50 mM solution contains 50 mmoles glucose/liter, or 5 mmoles/100 mL. One mole contains 180 g, so 1 mmole contains 0.180 g. Therefore 5 mmoles = 0.9 g.
16. To make 100 mL of a 3% glucose solution, take 3 g glucose and add water to give 100 mL final volume. For molarity: 1 mole/180 g = ? mole/3 g = 0.017 mole. 0.017 mole/100 mL = 0.17 mole/L = 170 mM.
17. Mixed solution contains 2 L with 400 mmoles glucose and 800 mmoles NaCl.
   a. Total concentration is 1200 mmol/2 L or 600 mmol/L = 0.6 M.
   b. The NaCl concentration is 400 mmol/2L or 200 mmol/L = 0.2 M.
   c. The glucose concentration is 800 mmol/2L or 400 mmol/L = 0.4 M.
18. Na⁺ contains 1 mEq/mmol because the ion has a single charge. Therefore, 142 mEq/L = 142 mmol/L.
19. Ca²⁺ contains 2 mEq/mmol because the ion has a charge of 2+. Therefore, 5 mEq/L = 2.5 mmol/L.
20. c; d and e; a; f; b and d
21. On the left side of the reaction, water (H_2O) is the acid because it donates an H⁺, and the amine is the base. On the right side of the reaction, the amine donates the H⁺ and acts as the acid, while the hydroxide ion (OH⁻) is the base.
22. H_2CO_3 is carbonic acid, which dissociates into H⁺ and bicarbonate ion, HCO_3^-, which acts as a base.
23. A 0.5 M solution equals a 500 mM solution, which is more acidic than a 50 mM solution.
24. c, b, e, a, f
25. d; a and e; a, b, c; a and c; f
26. See Fig. 2-13 for possible organization. Consult with classmates or your instructor to see if your map includes all the pieces it should include.

CHAPTER 3

PRACTICE MAKES PERFECT

1. Mitochondria would have the highest probability of existing independently and evolving because they have their own DNA with which they can reproduce and make new proteins. Mitochondria also contain the enzymes and proteins needed to make ATP.
2. Compartmentation of the nucleus allows the cell's control center to operate without being greatly affected by conditions in the cytoplasm. For example, cytoplasmic enzymes cannot enter the nucleus.

3. In the absence of a cytoskeleton, intestinal cells could not link to each other at cell junctions. They would also not have microvilli that increase the surface area for absorption of nutrients.

4. Many epithelial cells are exposed to chemical and mechanical stress, so they are constantly undergoing mitosis to make new cells. Many environmental chemicals can damage chromosomes, leading to abnormal (cancerous) daughter cells. In addition, the constant reproduction of these cells increases the probability of genetic mutations during cell division.

5. You would expect the solutions to be different because tight junctions are used to create a barrier between the compartments on each side of the cell layer.

6. The surface layer of the epidermis is composed of mats of keratin fibers and extracellular matrix that are left behind when keratinocytes die. This layer acts as a waterproof layer to prevent loss of water and heat.

7. c, d, a, e, g, f

8. Under skin: a, b, c, d; sheaths: a; cartilage: a; adipose: e; tendons and ligaments: a; blood: d; lungs and blood vessels: a, b, c, d; bones: a.

9. Packages proteins: c; modifies proteins: c; series of tubes: c; protein synthesis: a.

10. a. See pp. 72–75; b. see pp. 65–66; c. see pp. 66–67; d. see p. 76; e. see p. 55; f. see pp. 67–68; g. see pp. 69–70; h. see p. 69; i. see pp. 76–77; j. see p. 77; k. see p. 76; l. see pp. 78–79; m. see p. 79; n. see pp. 80–81; o. see p. 80; p. see Fig. 3-30; p. 81.

CHAPTER 4

FOCUS ON PHYSIOLOGY

f. The scale is balanced again with 6 blocks on the left and 12 blocks on the right.

i. The scale is balanced again with 2 blocks on the left and 4 blocks on the right. If more A is added to the reaction, A, C, and D all increase. If C and D are converted into E, the amount of A will decrease.

PRACTICE MAKES PERFECT

1. Entropy: d; potential energy: a, c, e; kinetic energy: b; exergonic: f; endergonic: g.

2. Vitamins and ions act as cofactors or coenzymes. Cofactors must bind to the enzyme before substrates will bind to the active sites. Coenzymes act as receptors or carriers for atoms or functional groups.

3.

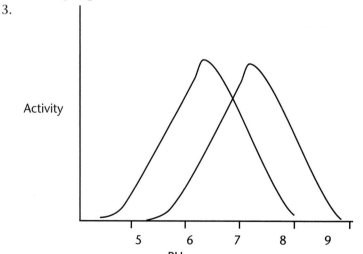

4. a. Graph A is exergonic and graph B is endergonic.
    b. Graph A is more likely to go in the forward direction because it has a lower activation energy.
    c. In graph A the products have a lower free energy (stored energy) than the substrates, therefore more energy was released.
    d. Graph A belongs with reaction 2 and 4. Graph B belongs with reactions 1 and 3.

5. c, b, a, d, g, h, f
6. a and c

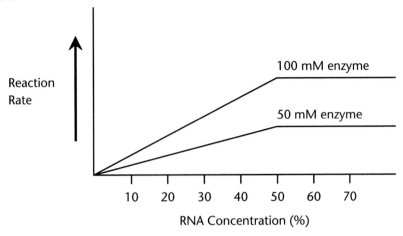

b. When the reaction rate is maximal, all active sites on the enzymes are occupied with substrate, i.e., the enzyme is saturated. See pp. 39–45 for review of the underlying explanation.

7. NADH is oxidized (loses an electron) to become NAD^+. H^+ is reduced because it gains an electron to become an H that combines with carbon.

8. This is a dehydration reaction because water is removed.

9. d, a, g, f, k, c, b, h, e, i

10. a. NADH donates high-energy electrons.
    b. $FADH_2$ donates high-energy electrons.
    c. Oxygen combines with H^+ and electrons to form water.
    d. ATP synthase transfers the kinetic energy of electrons moving down their concentration gradient to the high-energy bond of ATP.
    e. H^+ is concentrated in the intermembrane space, storing energy in its concentration gradient.
    f. The inner membrane proteins convert energy from high-energy electrons into either the work of moving H^+ ions against their concentration gradient or into heat.

CHAPTER 5

QUANTITATIVE PHYSIOLOGY

Volumes of distribution: interstitial volume = ECF – plasma;
ICF volume = total body water – ECF volume.

PRACTICE MAKES PERFECT

1. Time values are distance^2, or 1, 4, 9, and 16.

   The graph has distance on the x-axis and evenly spaced values from 0 at the origin to 4 mm. The y-axis is time and the values range from 0–16, with evenly spaced tic-marks.

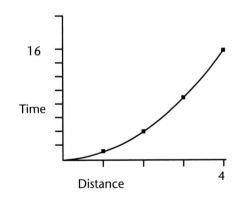

2. According to Fick's law, the rate of diffusion $\propto \dfrac{\text{surface area} \times \text{concentration gradient}}{\text{membrane resistance} \times \text{membrane thickness}}$

   So as membrane thickness increases by a factor of 2, the rate of diffusion will decrease by half.

3.

Method	Movement Relative to Concentration Gradient	Energy Source	What Affects Rate?	Through Membrane Bilayer or Through Protein Transporter?	Exhibits Specificity?	Exhibits Competition?	Exhibits Saturation?	Examples
Simple Diffusion	down	[ ] gradient	[ ] dependent	diffuse across lipid bilayer	no	no	no	gases, small nonpolar molecules, urea
Facilitated Diffusion	down	[ ] gradient	[ ] dependent. has maximum rate	on protein carrier	yes	yes	yes	glucose into cells from ECF
Primary Active Transport	against	ATP	[ ] dependent. has maximum rate	via protein carrier	yes	yes	yes	Na^+-K^+-ATPase
Secondary Active Transport	one or more down, one or more against	[ ] gradient, usually of Na^+	[ ] dependent	via protein carrier	yes	yes	yes	Na^+-glucose symport

[ ] = concentration. ECF = extracellular fluid

4. Na^+/K^+-ATPase: b, c, d; Na^+-glucose: a, d, e; Ca^{2+}-ATPase: c; $Na^+/K^+/2\ Cl^-$: a, d, e; Na^+/H^+: b, d, e.
5. The membrane permeability to Ca^{2+} appears to change with light. Light is a form of energy, so it could be doing one of several things: (1) opening a Ca^{2+} channel, (2) providing energy for the conformation change of a Ca^{2+} carrier protein, or (3) changing the structure of the membrane in some other way to make it permeable to Ca^{2+}. The first two explanations are more specific and relevant to the material in this chapter.
6. See pp. 159–171 and Figs. 5-31–5-33.
7. 1 OsM/1.86° = ? OsM/0.55° = 0.296 OsM or 296 mOsM plasma osmolarity. Intracellular osmolarity is always the same as plasma at equilibrium.
8. 1 mL sample/1 mg glucose = ? mL total/5000 mg glucose. The beaker has 5000 mL or 5 liters.
9. Absorption rate decreased as apical Na^+ concentration decreased, so transport appears to be Na^+-dependent. At the Curesall concentrations tested, transport rate was always proportional to concentration. The Na^+-dependence of the transport suggests that the transporter may be a secondary active transporter similar to the Na^+-glucose transporter even though transport did not show saturation.
10. Molecular weight of NaCl = 58.5. A 0.9% solution contains 0.9 g/100 mL solution or 9 g/liter. 1 mole NaCl/58.5 g NaCl = ? mole/9 g NaCl = 154 mmoles NaCl × 2 osmoles/mole = 308 mosmoles/L
11. Molecular weight of glucose = 180. 5% solution = 5 g/100 mL or 50 g/liter. 1 mole glucose/180 g = ? moles/50 g = 0.278 mole × 1 osmole/mole = 0.278 osmole or 278 mosmoles/liter
12. a. Rate is measured as mg/min, mg/sec, mmole/min, or mmole/sec.
    b. In graph #1, the concentration inside the cell reaches a maximum value over time, presumably showing that the system has come to equilibrium. If concentration inside equals concentration outside, the movement of glucose must be by diffusion. You cannot tell from this graph if the diffusion is simple diffusion across the phospholipid bilayer or

if it is facilitated diffusion, a passive process. In graph #2, the rate of movement does not reach a maximum, suggesting that there are no transporters to become saturated. This graph would support a hypothesis of simple diffusion. However, it is possible that the extracellular glucose concentration did not reach a high enough value to cause saturation of the carriers in the artificial membrane.

   c.   The line in graph #1 leveled off because there was no more glucose entering the artificial cell.

13.

Solution A	Membrane	Solution B	Osmolarity of A Relative to B
100 mM glucose	no net movement	100 mM urea	isosmotic
200 mM glucose	no net movement	100 mM NaCl*	isosmotic
300 mOsM NaCl	no net movement	300 mOsM glucose	isosmotic
300 mM glucose	$\Rightarrow$	200 mM CaCl$_2$**	hyposmotic

*100 mM NaCl = 200 mOsM NaCl. **200 mM CaCl$_2$ = 600 mOsM CaCl$_2$

14. The IV solution is isosmotic so there will be no change in the osmolarity of either the ECF or ICF. The infusion goes into the plasma (ECF), so ECF volume will increase initially. Because the solution is all NaCl, a nonpenetrating solute, the solution will all remain in the ECF. There is no concentration gradient to cause water to move into or out of the ICF, so ICF volume will remain unchanged.

15. The solution is isosmotic and hypotonic to the cell. The graph should be labeled with "time" on the x-axis and "cell volume" on the y-axis. Cell volume increases at the arrow then levels off at a new, larger volume.

16. a.   340 mosmoles/L × 13 L = 4420 mosmoles in ECF

     b.   160 mosmoles/L × 1 mmole NaCl/2mosmoles = 80 mmoles

     c.   New ECF volume = 13 L + 1 L = 14 L. Solute = 4420 + 160 mosmoles = 4580 mosmoles. Osmolarity = 4580 mosmoles/14 L = 0.327 OsM or 327 mOsM

17. Cl$^-$ will move from side two to side one because of a concentration gradient for Cl$^-$. As the ions move, side one will develop a net negative charge while side two will develop a net positive charge, creating a membrane potential. Net Cl$^-$ movement will stop when the positive charge on side two that holds Cl$^-$ in that compartment is equal in magnitude to the concentration gradient driving Cl$^-$ into side one.

18. When ECF K$^+$ increases, an equal amount of some anion has also been added to the ECF, so there is no change in the charge on the ECF. When ECF K$^+$ increases, the membrane potential of a liver cell depolarizes because less K$^+$ leaks out of the cell. Liver cells are not excitable and do not fire action potentials but can experience changes in membrane potential.

19. The membrane potential will slowly depolarize to zero as K$^+$ leaks out and Na$^+$ leaks into the cell. Normally the Na$^+$/K$^+$-ATPase removes ions that leak across the membrane.

CHAPTER 6

PRACTICE MAKES PERFECT

1. Local communication with neighboring cells is carried out by chemical communication (autocrines, paracrines, cytokines). Long-distance communication is carried out by the nervous system, hormones, and cytokines. Local communication takes place by diffusion, which limits its speed. Neural signals are the fastest means of communication. Long-distance chemical communication relies on the circulatory system.

2. Different types of receptors for one chemical signal allow different responses to a single signal. In addition, there are many different chemical signals, each with its own receptor or receptors. Receptor number is not constant. Cells can alter their receptor number by adding or withdrawing receptors (up- and down-regulation).

3. Cascades allow amplification of signals. A single event could not elicit as large a response without a cascade.

4. You would not expect the effects of growth hormone to exert negative feedback because then growth would not take place. On the other hand, without some form of control, growth would continue unchecked. By having growth hormone concentration act as the negative feedback signal, the body can keep the secretion of growth hormone in a desirable range.

CHAPTER 7

PRACTICE MAKES PERFECT

1.

	Peptide	Steroid
Transport in plasma	dissolved	bound to carrier proteins
Synthesis site in endocrine cell	rough endoplasmic reticulum	smooth endoplasmic reticulum
Method of release from endocrine cell	exocytosis from secretory vesicles	simple diffusion across phospholipid bilayer
General response of endocrine cell	modification of existing proteins	transcription, translation, and synthesis of new proteins

2. Connected by nerve fibers: P; connected by blood vessels: A; hormones made in hypothalamus: P; under influence of hypothalamic hormones: A; secretes peptide hormones: A, P.
3. e
4. cortisol: a; aldosterone: a and d; growth hormone: b, c; vasopressin: d; prolactin: b; thyroxine: c.
5. a. False. Steroid hormones cannot be stored in vesicles because they are lipid-soluble and would diffuse out across the vesicle membrane.
   b. It is true that steroid-secreting cells have lots of smooth endoplasmic reticulum, but false that they have lots of Golgi. The Golgi complex is used to package peptides into vesicles (see answer for part a).
6. a. 2; b. 1—mRNA carries the code for peptide synthesis.
7. The tissue has lots of membrane-bound secretory vesicles, indicating that the cells synthesize and store peptides. Insulin is a peptide hormone, so the tissue is endocrine pancreas.
8. For the pathway, see Fig. 7-15. Secondary hypocortisolism originating at the pituitary means that both ACTH and cortisol secretion are down. If you administer ACTH, her cortisol secretion should increase because there is nothing wrong with the adrenal cortex; it has simply lacked stimulation by ACTH.
9. The antibodies stimulate the thyroid gland, so you expect thyroid hormone (thyroxine) levels to be elevated. This eliminates patients A and C. The elevated thyroxine will have a negative feedback effect on the pituitary and shut off endogenous TSH production; therefore TSH should be below normal. Only patient B has elevated thyroxine and decreased TSH.

Graph

The data for the patients in problem #9 are best shown with a bar graph. There are two parameters being measured for each patient, so you would need two sets of values on the y-axis. This can be done by placing the scale for thyroxine on the left side of the graph and the scale for TSH on the right side of the graph. On the x-axis, each patient is represented by a pair of bars, side by side, one bar for each hormone. Distinguish the hormones by using different colors or fill patterns in the bars.
See the graph that follows:

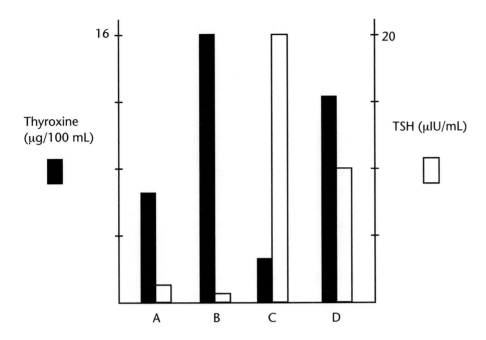

CHAPTER 8

PRACTICE MAKES PERFECT

1. speed
2. ECF (left to right): Na, Cl, K; ICF: K, Na, Cl
3. Dendrite $\Rightarrow$ cell body $\Rightarrow$ trigger zone (axon hillock) $\Rightarrow$ axon $\Rightarrow$ axon terminal $\Rightarrow$ releases neurotransmitter by exocytosis $\Rightarrow$ neurotransmitter diffuses across synaptic cleft to combine with receptor on postsynaptic cell
4. a. K^+ will leave the cell due to the electrical gradient (positive inside) repelling it. Na^+ will leave because of both the electrical gradient and the concentration gradient favoring its movement out.

   b. $$E_{Na} = \frac{61}{z} \log \frac{[Na]out}{[Na]in} \quad z = +1, [Na]out = 15, [Na]in = 175$$

   $$= -65.1 \text{ mV}$$

5. a
6. b, d
7. b
8.

Location of channel	Ion(s) that move(s)	Chemical or volt gating?	Physiological process in which ion participates
dendrite	Na^+	chemical	graded potentials
axon	Na^+	electrical	action potentials
axon	K^+	electrical	action potentials
axon terminal	Ca^{2+}	electrical	exocytosis of neurotransmitter

9. The ground electrode is set to zero millivolts and the potential difference is measured between this electrode and the inside of the cell.
10. Stretching could destroy the integrity of the cell membrane or could pop open channels that would allow ions to move between the cell and the extracellular fluid.
11. See Fig. 8-16.

CHAPTER 9
PRACTICE MAKES PERFECT

1. a. The cerebrum contains the motor cortices for voluntary movement (walking to class) and the cerebral cortex for higher brain functions (pondering physiology). It also contains the visual cortex for integrating visual information from the eyes, and association areas that integrate sensory information into the perception of tripping and falling. The basal ganglia are associated with control of movement. The amygdala of the limbic system is linked to the emotions of anger and embarrassment.
    b. The cerebellum coordinates the movement of walking.
2. An electroencephalogram measures the summed electrical activity of brain neurons. The wave patterns of the EEG vary with different states of consciousness and can be used to assess if brain activity is following normal patterns.
3. a. higher thought processes
    b. centers for homeostasis, neurohormone production
    c. centers for involuntary functions (breathing, heart rate) and eye movement
    d. coordination and control of body movement
4. b
5. c
6. c

CHAPTER 10
PRACTICE MAKES PERFECT

1. Rapidly adapting phasic receptors adapt to constant pressure from clothing and stop sending information to the brain.
2. A salty solution cannot be tasted at the back of the tongue because there are no salt receptors there.
3. True. The layers of the retina are arranged so that light must pass through layers containing nerves and blood vessels before it strikes the photoreceptors.
4. e
5. In nearsightedness (myopia), the light focuses in front of the retina. In astigmatism, the cornea is not a perfect dome, so the light does not focus evenly on the retina.
6. b
7. Loss of air conduction but not bone conduction suggests a problem with the middle ear or a plugged ear canal.
8. The spot of bright light appears black on the paper. Light causes retinal to release from the opsin portion of rhodopsin and be transported out of the photoreceptor. The slow recovery results from time needed to transport the retinal back into the rod and also the time needed for the rod membrane potential to return to its former state.
9. The horizontal canal is involved in movement that changes the left-right position of the head.
10. Rods are responsible for low-light monochromatic vision, so a nocturnal animal would have more rods and fewer cones.
11. d

CHAPTER 11
PRACTICE MAKES PERFECT

1. a
2. a. antagonist; b. agonist

3. Autonomic neurotransmitters (NTs) are released from varicosities along the axon as well as at the axon terminal, so their release is less specifically associated with receptors on the target cell. Autonomic NT release can be modulated by many chemical factors such as hormones and paracrines. In addition to being broken down by synaptic enzymes, autonomic NTs may diffuse away from the receptors or be transported intact back into the neuron. A major difference between autonomic and somatic motor NTs is that the amount of NT will vary the response of the target cell. Each action potential in a somatic motor neuron creates one muscle twitch.

4. Monoamine oxidase (MAO) is the enzyme that breaks down catecholamines at autonomic synapses, therefore inhibition of MAO will prolong or enhance target responses to autonomic signals.

5. False. The onset of steroid action takes at least 30 minutes, which is too slow for most fight-or-flight situations.

CHAPTER 12

PRACTICE MAKES PERFECT

1. Muscle contraction requires Ca^{2+} and uses ATP.
2. Somatic motor neurons secrete acetylcholine.
3. True.
4. Na^+ entry is greater because Na^+ has both concentration and electrical gradients favoring its movement. K^+ has concentration favoring its efflux but an electrical gradient opposing efflux.
5. d
6. Contraction is more rapid because Ca^{2+} floods the cytoplasm as it moves down its concentration gradient. Removal of Ca^{2+} for relaxation requires that the cell use ATP to pump Ca^{2+} against its gradient.
7. Temporal summation in muscle results when multiple action potentials repeatedly stimulate a muscle fiber, causing an increase in its force of contraction, up to some maximum value. Temporal summation in a neuron occurs when multiple subthreshold stimuli arrive at the trigger zone and create a suprathreshold signal. In the neuron, the result is an action potential of constant amplitude, in contrast to the increasing force observed in the muscle.
8. Fatigued muscles have less ATP with which to pump Ca^{2+} back into the sarcoplasmic reticulum.
9. If skeletal muscles had gap junctions, all fibers would contract simultaneously. This would prevent us from regulating which muscles are contracting and would prevent us from regulating the force of contraction.
10. True. Muscles shorten during isotonic contractions.
11. a
12. a
13. Motor proteins in the nervous system include the microtubules and foot proteins used for axonal transport and the cytoskeleton that helps growing neurons find their targets. Motor proteins in Chapter 3 include microtubules that move cilia and flagella, microtubules associated with centrioles for chromosome movement during cell division, fibers that help mobile white blood cells (phagocytes) move, and fibers that move vesicles and other organelles around the cytoplasm.
14. Fewer ACh receptors would mean that a muscle would not respond as strongly to neurotransmitter release. This would probably manifest itself as muscle weakness or paralysis because the muscle fiber would not fire action potentials and excitation-contraction coupling would not occur.
15. You would expect the latency period to be longer when the nerve is stimulated because the additional steps of neurotransmitter release, diffusion across the synapse, opening of ACh-gated channels, and ion movement must take place.

CHAPTER 13

PRACTICE MAKES PERFECT

1. The steps are stimulus, receptor, afferent pathway, integrating center, efferent pathway, tissue response, systemic response. Examples of reflexes in Ch. 13 include the muscle spindle and Golgi tendon reflexes, the flexion reflex, and the crossed extensor reflex.
2. a
3. In Fig. 13-9, sensory input diverges to excite multiple CNS locations. In Fig. 13-11, sensory neurons diverge, efferent pathways from the motor cortex diverge, neural pathways from the motor cortex and motor association areas converge, and multiple neural pathways converge on the prefrontal cortex and motor association areas. There could be additional examples within these figures. See how many you can list.

CHAPTER 14

TEACH YOURSELF THE BASICS

24. Tube B has the larger pressure gradient, 65 mm Hg (75 – 10), so it will have the greatest flow. The pressure gradient in Tube A is only 50 mm Hg.

QUANTITATIVE THINKING

1. Resistance should decrease and radius increase.
2. Flow $\propto \Delta P/R$, where $R \propto L/r^4$. So Flow $\propto \Delta P \times r^4/L$
   Flow A $\propto (50 \times 16)/16 \propto 50$.   Flow B $\propto (72 \times 1)/2 \propto 36$
3. $CO = HR \times SV$ so $SV = CO/HR$. Before Romeo, SV = 69.4 mL/beat. After Romeo, SV = 125 mL/beat.

PRACTICE MAKES PERFECT

1. acetylcholine, atrioventricular, cardiac output, electrocardiogram, end-diastolic volume, end-systolic volume, sinoatrial, stroke volume

2.

	Electrical Event	**Mechanical Event**
P wave	atrial depolarization	atrial contraction follows
QRS complex	ventricular depolarization	ventricular contraction and atrial relaxation
T wave	ventricular repolarization	ventricular relaxation
PQ segment	AV node delay	atrial contraction
ST segment	some parts of ventricle depolarizing, some repolarizing	end of ventricular contraction
TP segment	none	atrial and ventricular diastole

3. If heart rate speeds up, there is less time for passive (gravity-assisted) ventricular filling, so atrial contraction takes on increasing importance.
4. See Table 14-3.
5. Kinetic energy is unchanged but the hydrostatic pressure decreases. See Fig. 14-3.
6. Several mechanisms to alter would include Ca^{2+} entry from the extracellular fluid, Ca^{2+} release from the sarcoplasmic reticulum, the regulatory protein phospholamban, the activity of the Ca^{2+}-ATPase, and anything that would alter binding of Ca^{2+} to troponin or the interaction of actin and myosin.

7. As venous return increases, the muscle fibers stretch more and therefore contract more forcefully (the Frank-Starling law of the heart). This increases stroke volume and therefore increases cardiac output. Sympathetic input increases ventricular contractility and therefore stroke volume. Sympathetic input onto veins will cause venous constriction and increase venous return, also increasing stroke volume.

8. See Fig. 14-18.

9. Cardiac output by the right heart must equal cardiac output by the left heart (4.5 L) or blood will begin to collect on one side of the circulation.

10. At point A, the atrial pressure exceeds ventricular pressure. Ventricular pressure equals aortic pressure at point C. The highest pressure in the aorta is about the same as the highest ventricular pressure, 120 mm Hg.

11.

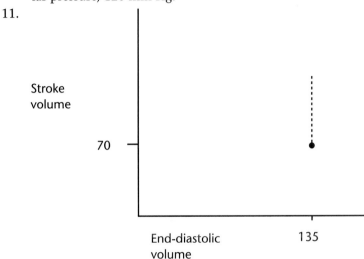

1. a. mm Hg

CHAPTER 15

QUANTITATIVE THINKING

1. a. mm Hg
    b. At age 20, her pulse pressure is 40 mm Hg and her MAP is 83 mm Hg. At age 60, her pulse pressure is 43 mm Hg and her MAP is 96.
    c. Postmenopausal women are more likely to develop atherosclerosis, which increases blood pressure when arteries stiffen and lose elastance.

2. If the radius goes from 2 to 3, radius^4 goes from 16 to 81, about a 5-fold increase. Resistance therefore falls by the same factor. As resistance decreases, blood flow increases.

3. Mean arterial pressure will increase.

4. CO = HR × SV. SV = EDV – ESV, or 110 mL. CO = 140 bpm × 110 mL/beat = 15,400 mL/min or 15.4 L/min.

PRACTICE MAKES PERFECT

1. arteries: a, b; arterioles: a; capillaries: e; veins: a, d.

2. For physical characteristics, see Fig. 15-2. Functions: arteries and arterioles carry blood to tissues. Arteries store and release energy created by the heart. Arterioles are the site of variable resistance to regulate blood flow to individual tissues and to help maintain blood pressure. Capillaries are the site of exchange between blood and cells. Veins carry blood from tissues to heart and serve as a volume reservoir.

3. The cuff must be inflated to at least 130 mm Hg in order to stop flow through the artery. By the time blood reaches the cuff, the pressure will be less than 130 due to friction loss.

4. Healthy arteries expand during ventricular systole and store energy that they slowly release as they recoil during diastole. Hardened atherosclerotic arteries are unable to stretch, forcing the heart to work harder to inject the same amount of blood into them. Narrowing of the arteries also slows outflow so that diastolic pressures will be higher than normal. This will decrease pulse pressure.

5. Resistance to blood flow depends on the total cross-sectional area at any level of the circulatory system. Although individual capillaries are very narrow, their total cross-sectional area is very large, so their resistance is low.

6. Flow to different regions is altered by constricting or dilating the arterioles leading into the different tissues. Resistance of skeletal muscle arterioles decreases to increase flow, while flow to the digestive system is decreased by increasing arteriolar resistance.

7. SA node: d, e; ventricle: e; skeletal muscle capillary: f; cardiac vasculature: a, c; renal arterioles: a; brain arterioles: f.

8. Blood flow through kidneys decreases. Mean arterial pressure increases. Blood flow through skeletal muscle increases. Cardiac output does not change. Total resistance increases. Blood flow through the venae cavae and lungs does not change.

9. False. Epinephrine comes from the adrenal medulla and combines with β_2 receptors to cause vasodilation. Epinephrine on β_1 receptors will cause increased heart rate and force of contraction.

10. In smooth and cardiac muscle, contraction is due to calcium entry. By blocking the Ca^{2+} channels through which Ca^{2+} enters, we can keep your blood vessels (arterioles) relaxed and decrease the volume of blood pumped in a single contraction (stroke volume), both of which should decrease your blood pressure.

11. Figure should include carotid and aortic baroreceptors, sensory neurons to the CVCC, parasympathetic neurons to the SA node (ACh on muscarinic receptors), sympathetic neurons to SA node and ventricles (norepi on β_1), sympathetic to α receptors on arterioles, epinephrine from the adrenal medulla to all adrenergic receptors.

12. See Figs. 15-22–15-24.

13. a. Blood flow in his legs is decreased because of increased resistance in the leg arteries. The pain comes from hypoxia when muscles are unable to get sufficient blood flow and oxygen.

    b. A peripheral vasodilator would have no effect in the legs because the arterioles that dilate would be beyond the obstructed arteries. If peripheral vasodilation decreased blood pressure, flow into the legs would decrease even more.

    c. Sympathetic nerves primarily affect arterioles, so there would be no effect for the same reason as in b above.

CHAPTER 16

PRACTICE MAKES PERFECT

1. Two main functions of blood are to transport substances throughout the body and to help protect the body from foreign invaders.

2. Leukocytes can be identified by the shape of the nucleus, the presence and staining color of granules in the cytoplasm, and by whether they can carry out phagocytosis.

3. Reticulocytes are immature red blood cells. Their presence in the circulation suggests that the bone marrow is responding to loss of circulating red blood cells by stepping up marrow production of new cells. This would mean that the marrow is responding normally but that something in the circulation is destroying the circulating red blood cells.

4. If hematocrit is 40%, then 60% of the sample is plasma. Sixty percent of a total blood volume of 4.8 L = 2.88 L plasma.

5. b

6. EPO is more accurately described as a cytokine because it is synthesized on demand rather than being stored in secretory vesicles. Because EPO is not made in advance and stored, it was only recently that the EPO-secreting cells in the kidney were identified.

CHAPTER 17

QUANTITATIVE THINKING

1. (approximate values) TLC = 4 L; VC = 2.75 L; ERV = 1 L; RV = 1.25 L
   Total pulmonary ventilation = 3 breaths/15 sec × 60 sec/min × 500 mL/breath = 6000 mL/min = 6L/min

2. 720 mm Hg × 0.78 = 562 mm Hg P_{N_2}

3. Pulmonary ventilation = tidal volume × rate = 300 mL/breath × 20 breaths/min = 6000 mL/min
   Alveolar ventilation = (V_T – dead space) × rate = (300 mL – 150 mL/breath) × 20 breaths/min = 3000 mL/min

4. If you assume that both patients have dead space = 150 mL, patient A has alveolar ventilation of 350 mL/breath × 12 br/min = 4.2 L/min, while patient B has alveolar ventilation of 150 mL/br × 20 br/min = 3 L/min.

PRACTICE MAKES PERFECT

1. tidal volume; partial pressure of oxygen; residual volume; inspiratory reserve volume

2. Diaphragm and external intercostals are skeletal muscles, so answer is b. Bronchioles: c and d.

3. a. The P_{O_2} increases and P_{CO_2} decreases as fresh air comes in to the alveoli, but no gases exchange with the blood.
    b. Tissue P_{O_2} decreases and P_{CO_2} increases as there is no gas exchange.
    c. The bronchioles constrict in response to decreased P_{CO_2} in an attempt to send ventilation to alveoli with better perfusion. Likewise, the pulmonary arterioles constrict in response to decreased P_{O_2} in an attempt to send blood to better ventilated alveoli. The effectiveness of these responses will depend on the size of the affected area.

4. Expired air from the alveoli is mixing with atmospheric air in the dead space, increasing the P_{O_2} of the expired air and decreasing its P_{CO_2}.

5. <, See Fig. 17-11.
   >, Peripheral arterioles dilate when P_{O_2} decreases but pulmonary arterioles constrict.
   <, Bronchioles have a much greater total cross-sectional area.
   >, Surfactant decreases the surface tension of fluid lining the alveoli and therefore increases compliance.

CHAPTER 18

QUANTITATIVE THINKING

1. The oxygen consumed by the tissues is extracted from the blood. Therefore the difference in oxygen content between arterial and venous blood represents oxygen that went into the tissues. If you know how much oxygen is extracted per liter of blood, then you simply need to find how much blood must flow past the tissues to supply 1.8 L of oxygen per minute. Blood flow past the tissues is the cardiac output.

   1.8 L O_2/min = (190 – 134 mL O_2/L blood) × blood flow past tissues (CO)
   CO = 32.1 L blood/min

2. Total oxygen = amount dissolved + amount bound to hemoglobin. (See Fig. 18-6.)
    a. Dissolved in plasma = plasma volume × plasma concentration. If hematocrit is 38%, 62% of total blood volume is plasma, or 4.2 L × 0.62 = 2.6 L plasma × 0.3 mL O_2/100 mL plasma = 7.8 mL O_2 dissolved.
    b. Bound to hemoglobin (Hb) depends on amount of hemoglobin and the percent saturation of that amount. Maximum O_2-carrying capacity = O_2 carried by Hb at 100% saturation.

13 g Hb/dL whole blood × 4.2 L blood = 546 g Hb in blood of this person. At 100% saturation, 1.34 mL O_2/g Hb × 546 g Hb = 731.64 mL O_2 carried on Hb. But this person's Hb is only 97% saturated: 731.64 mL O_2 × 0.97 = 709.7 mL O_2 carried bound to Hb.

709.7 mL O_2 carried bound to Hb + 7.8 mL O_2 dissolved = 717.5 mL O_2 in this person's blood.

(This problem does not take into account the fact that venous blood will have a lower oxygen content than arterial blood.)

PRACTICE MAKES PERFECT

1. a. T; b. F ;c. F
2. a. F; b. T; c. T
3. a. T; b. T; c. T; d. F; e. F
4. a, e
5. <, See Fig. 18-10.
   <, The percent saturation of hemoglobin is essentially identical in these two people, so the amount of hemoglobin becomes the important factor for how much oxygen is being transported.
   =, Arterial P_{O_2} is determined by alveolar P_{O_2} and is not affected by the amount of hemoglobin in the blood.
6. a. and b. P_{O_2} decreased because barometric pressure decreased due to increased altitude.
   c. Arterial P_{O_2} decreased because P_{O_2} of inspired air is down.
   d. Arterial P_{CO_2} is down because you begin to hyperventilate.
   e. Arterial pH is up because of decreased P_{CO_2}. By law of mass action, the equilibrium between CO_2 and H^+/HCO_3^- is disturbed, converting more H^+/HCO_3^- to CO_2.
   f. They will have more hemoglobin because their state of chronic hypoxia triggered erythropoietin synthesis and new RBC synthesis.
   g. The best course of action is to take the person back to lower altitudes to remove the source of the stress (hypoxia).
   h. Initially the hypoxia triggers a hyperventilation response. But hyperventilation decreases arterial P_{CO_2} and increases arterial pH, both of which will tend to decrease ventilation and offset the hypoxic response. However, the central chemoreceptors are able to adapt to chronic changes in P_{CO_2}, so over the course of several days, the ventilation rate increases as the hypoxia effect is less opposed by the low P_{CO_2}. Review Ch. 18's Running Problem and its conclusion for additional information.
7. Hyperventilation will increase plasma P_{O_2} but have minimal effect on the total oxygen content because so little oxygen is carried dissolved in plasma. Hemoglobin is already carrying nearly maximal amounts of oxygen, and the saturation curve is nearly flat at these values of P_{O_2}, so no more oxygen will be transported by hemoglobin.

CHAPTER 19

QUANTITATIVE THINKING

1. Clearance of χ = excretion rate of χ / plasma concentration of χ
   creatinine clearance = (276 mg/dL urine × 1100 mL urine/day)/1.8 mg/dL plasma = 168.7 L plasma/day
   GFR = creatinine clearance, so GFR = 168.7 L plasma/day
2. Filtration rate of χ = GFR × plasma concentration of χ
   1 mg X/mL plasma × 125 mL plasma/min = 125 mg X filtered /min. Same values for inulin. Excretion rate of inulin = filtration rate, or 125 mg inulin excreted/min. Cannot say what the excretion rate of X is because there is insufficient information. We must know whether X is reabsorbed or secreted by the tubule in order to estimate its excretion rate.

3. a. When reabsorption of glucose reaches the transport maximum, 100% of what is filtered is being reabsorbed. Therefore, the transport rate of glucose at the T_m is identical to the glucose filtration rate. The renal threshold is the plasma concentration at which the T_m is reached. By substitution:

   Filtration rate of χ = GFR × plasma concentration of χ (or) T_m = GFR × renal threshold
   90 mg glucose/min = GFR × 500 mg glucose/100 mL plasma, or GFR = 18 mL plasma/min

   b. Creatinine clearance = GFR = 25 mL/min (You cannot calculate this because you are not given plasma creatinine.)

   phenol red clearance = (5 mg/mL urine × 2 mL urine/min)/2 mg/mL plasma = 5 mL plasma/min

4. Graphing question. To calculate the points for the filtration graph of Z, multiply various plasma concentrations in the range of 0–140 mg Z/mL plasma times the GFR. The line will be a straight line beginning at the origin and extending upward to the right. For secretion, you know that at a plasma concentration of 80 mg Z/mL plasma, secretion reaches its maximum rate of 40 mg/min. Plot that point. Draw the secretion line from the origin to that point. At plasma concentrations above the renal threshold, secretion rate does not change, so the line becomes horizontal. To draw the excretion line, add the filtration rate and secretion rate at a number of plasma concentrations of Z. The excretion line will extend upward with a steeper slope than that of the filtration line from the origin to the renal threshold. At that point, the slope of the line changes and the line runs parallel to the line of filtration rate.

PRACTICE MAKES PERFECT

1. Proteinuria suggests that the filtration barrier in the glomerulus has been disrupted or that proximal tubule cells are no longer able to reabsorb the small filtered proteins.
2. Glucose is usually absent because 100% of what is filtered is reabsorbed.
3. See Fig. 19-2.
4.

	Hydrostatic pressure	Fluid pressure	Osmotic pressure	Net direction of fluid flow
Glomerular capillaries	55	15	30	into Bowman's capsule
Peritubular capillaries	10	negligible	30	into capillaries
Systemic capillaries	32–15 (see p. 507)	negligible	25	out of capillaries into interstitial fluid
Pulmonary capillaries	< 14 (see p. 552)	negligible	25	into capillaries

5. a. Renal threshold: plasma concentration at which transport maximum for a substance is reached (mg/mL plasma).
   b. Clearance: volume of plasma cleared of a substance per unit time (mL plasma/min).
   c. GFR = glomerular filtration rate. Volume of plasma filtered into Bowman's capsule per unit time (mL plasma/min).
6. a. Both terms refer to a volume of plasma per unit time. Clearance is specific for a single substance that is removed from the plasma, while GFR is the bulk filtration of plasma and almost all its solutes.
   b. These terms are related (see definition in 5.a above), but one deals with a transport rate while the other is the plasma concentration associated with that transport rate.
7. K^+ movement across the apical membrane is against the gradient, so it must be by some form of active transport. The $Na^+/K^+/2\,Cl^-$ symporter is one way to bring K^+ into the cell. On the basolateral side, K^+ can leave by moving passively down its concentration gradient. One way to do this would be to have K^+ leak channels on the basolateral membrane.

8. Capillary in hand ⇒ venule, vein ⇒ inferior vena cava ⇒ right atrium ⇒ tricuspid valve ⇒ right ventricle ⇒ pulmonary valve ⇒ pulmonary artery, arteriole, capillary, venule, vein ⇒ pulmonary vein ⇒ left atrium ⇒ mitral valve ⇒ left ventricle ⇒ aortic valve ⇒ aorta ⇒ renal artery ⇒ afferent arteriole ⇒ glomerulus ⇒ Bowman's capsule ⇒ proximal tubule ⇒ loop of Henle ⇒ distal tubule ⇒ collecting duct ⇒ renal pelvis ⇒ ureter ⇒ urinary bladder ⇒ urethra ⇒ leaves body in the urine

9. Inulin clearance (= GFR) at MAP = 100 will be less than GFR at MAP = 200 because the kidney autoregulation of GFR is only effective to a MAP of 180.

10. a. According to the graph, phenol red filters, but then additional is secreted. The secretion line can be calculated by subtracting the filtration rate from the excretion rate at a series of plasma concentrations. The line goes from the origin to the point (0.05, 0.2), then changes slope and runs horizontally because the rate has reached its maximum.

    b. The slope changes because secretion has a transport maximum.

    c. If phenol red secretion is inhibited, less would be excreted and more would stay in the plasma. Thus, plasma clearance would decrease.

CHAPTER 20

QUANTITATIVE THINKING

Osmotic diuresis: To calculate the volume of fluid passing any point, use the equation: solute/volume = osmolarity. You know that solute = 150 milliosmoles and you know the osmolarity at points A, B, and C. Solving for volume, you get volume going past A = 1.5 L, past B = 0.5 L, and past C = 0.125 L.

When the solute amount doubles, the volume doubles: A = 3 L, B = 1 L, and C = 0.25 L. From this, you can see that glucose remaining in the tubule lumen will cause additional water to be excreted.

PRACTICE MAKES PERFECT

1. =, Urine osmolarity cannot be greater than the medullary interstitial osmolarity. With excreted glucose, the *volume* will increase.

    <, Aldosterone secretion is directly inhibited by high osmolarity, even if ANGII is present.

    >, Respiratory acidosis is characterized by elevated P_{CO_2}.

    >, Renal reabsorption of buffer will be greater in acidosis.

    <, Ventilation will be reduced in alkalosis and elevated in acidosis.

    <, Renin is secreted in response to low blood pressure.

    =, Filtration is not regulated and depends only on the plasma concentration of a substance and the GFR.

2. a. no change; b. decrease; c. decrease or no change; d. no direct effect; e. no change; f. will increase eventually as a result of homeostatic compensation for decreased MAP

3. An increase in plasma osmolarity will cause an increase in vasopressin secretion, so the x-axis is osmolarity and the y-axis is vasopressin.

4. $CO_2 + H_2O \Leftrightarrow H^+ + HCO_3^-$. If the green fruit you ate represents bicarbonate or H^+, you have sent the equation out of balance relative to its equilibrium, making the CO_2 side appear to be too large. Convert some of the red fruit (CO_2) into green fruit to balance out the equation again.

5. Elevated P_{CO_2} and bicarbonate suggests a respiratory acidosis. Therefore, answers a, b, and c are all correct.

6. Normal pH is 7.4 (Mr. Osgoode is in acidosis) and normal arterial P_{CO_2} is 40. It appears that Mr. Osgoode is hyperventilating in an attempt to compensate for a metabolic acidosis; the hyperventilation is elevating his pH (decreasing his H^+) but also decreasing his P_{CO_2} and plasma HCO_3^-.

7. Ari is breathing through an extended dead space (the tube), so he is experiencing alveolar hypoventilation, resulting in respiratory acidosis. You would expect his pH to be lower than normal, and both HCO_3^- and P_{CO_2} to be greater than normal, due to the retention of CO_2.

TRY IT

Kitchen Buffers

The baking soda and vinegar combination foams up, producing CO_2. This simple reaction has been used for centuries by cooks to make batters rise. If you taste the solution after adding baking soda, it will not be nearly as sour because the H^+ have been buffered and converted to water.

CHAPTER 21

PRACTICE MAKES PERFECT

1. stomach: 2; ileum: 5; esophagus: 1; ascending colon: 6; pyloric sphincter: 3; duodenum: 4.
2. Salivary amylase is denatured by the acidic conditions in the stomach.
3. Pepsin is produced by the stomach and is active at low pH; trypsin is produced by the pancreas and is active at high pH.
4. The stomach prevents autodigestion by (1) secreting inactive enzyme, (2) secreting mucus, and 3) secreting a layer of bicarbonate under the mucus to neutralize acid.
5. a. False. Short-chain fatty acids are not incorporated into chylomicrons. All other parts are true.
   b. False. Most bile salts are reabsorbed and used again.
   c. False. Much more water is reabsorbed than ingested because large volumes of water are present in secreted fluids.
   d. True. Without oxygen, transporting epithelia cannot carry about aerobic metabolism and make enough ATP to support the secondary active transport of glucose into the cell.
6. a. Refer back to Figs. 20-21, 20-22.
   b. Excess vomiting would cause metabolic alkalosis because of the bicarbonate that is reabsorbed, which acts as a buffer.
7. The enteric nervous system can act as its own integrating center, without communicating with the CNS.
8. Glutens are protein that can cause allergies if absorbed intact, and the intestines of infants have the ability to absorb intact molecules via transcytosis. Delaying exposure to glutens can reduce chances of developing allergies.
9. receptors: eyes (sight of test), ears (sound of people talking about the test)
   afferent path: sensory neurons
   integrating center: cerebral cortex, with descending pathways through limbic system
   efferent path: parasympathetic neurons to enteric nervous system
   effector: parietal cells
   cellular response: second-messenger initiated modification of proteins
   tissue response: acid secretion
10. Both systems secrete material from ECF into the lumen. Most of the GI secretion is reabsorbed while most of the renal secretion is excreted. Reabsorption in the kidney is equivalent to absorption in the GI tract. New material in the GI tract is absorbed, but many components of secretion are reabsorbed. Both systems excrete material to the external environment; GI excretion is primarily solid while urine is liquid. Movement through the systems is quite different. Filtration, created by hydraulic pressure, creates bulk flow of fluid through the tubule, unaided by muscle contraction until the urine has left the renal pelvis. Motility in the GI tract is completely dependent on coordinated smooth muscle contraction.
11. Test the pH optimum for activity of the two enzymes. Gastric lipase will be most active in low pH, while lingual lipase is more active at higher pH.

CHAPTER 22
PRACTICE MAKES PERFECT

1. c is incorrect.
2. True. See the section in Ch. 22 titled "Energy Use Is Reflected by Oxygen Consumption."
3. Glucagon is catabolic; insulin is anabolic.
4. The body is able to make glucose from nonglucose precursors through the pathways of gluconeogenesis.
5. Insulin secretion will be greater when glucose is given orally because GIP will be secreted due to insulin in the intestine and because all absorbed glucose goes into the hepatic portal system. Glucose given by IV will be taken up by cells before it gets to the pancreas, so the pancreas will not sense as large an increase in blood glucose.
6. a
7. $(9 \times 7) = 63$ calories from fat; $63/190 = 33\%$, about the maximum recommended fat content for food
8. Humans with obesity may have defective leptin receptors or a defective pathway for leptin action in target cells, so that they are actually leptin-deficient even though plasma levels of the protein are high.
9. a. True. These are catabolic effects and can elevate blood glucose concentrations.
    b. False. The glucose transporters of liver are not insulin-dependent. However, insulin does increase liver metabolism of glucose so that glucose utilization is increased.
    c. True. All cells use a GLUT-family transporter to take up glucose, and all mediated transport systems show saturation.
10. If people ingest more amino acids than they need for protein synthesis, the excess is turned into glucose or fat.
11. Alpha cells secrete glucagon.

Plasma concentration	Change: ↑, ↓, or N/C	Rationale for your answer
Glucose	↑	Glucagon increases plasma glucose.
Insulin	↓	The increase in plasma glucose will decrease insulin secretion.
Amino acids	↓	Glucagon stimulates conversion of amino acids into glucose.
Ketones	N/C	Although fats are broken down, the cells can use them and there is no excess ketone production.
K^+	N/C	These patients do not have an acid-base disorder and there is no other association of K^+ with glucagon.

CHAPTER 23
PRACTICE MAKES PERFECT

1. b
2. Cortisol is catabolic; growth is anabolic.
3. a. False. Cortisol is net catabolic and causes muscle breakdown.
    b. False. Vitamin D deficiency causes rickets because the children are unable to absorb dietary Ca^{2+}.
4. a. Calcitriol
    b. Ca^{2+} is filtered into the tubule at the glomerulus if it is not bound to plasma proteins.
    c. Renal reabsorption is enhanced by PTH and calcitriol.
    d. Ca^{2+} entry into cells initiates exocytosis of vesicles and muscle contraction in smooth and cardiac muscle.

e.   PTH; calcitriol to a lesser extent.
5. a.   x-axis is PTH, y-axis is plasma Ca^{2+}. Line goes upward to the right.
    b.   x-axis is Ca^{2+}, y-axis is PTH. The line goes downward from left to right: as plasma Ca^{2+} increases, PTH secretion decreases.
6. Alpha cells secrete glucagon.

CHAPTER 24

PRACTICE MAKES PERFECT

1. f, a, c
2. macrophage: a, b; red blood cell: d; B lymphocyte: a, b, e; natural killer cell: a, c; liver cell: a; cytotoxic T cell: a, c; plasma cell: a; helper T cell: a, c
3. Macrophages ingest and destroy material; present antigens on MHC-II; secrete cytokines to initiate the inflammatory response and activate T cells.
4. If the mother has alleles AO and father has alleles BO, they could have a child with blood group O, alleles OO.
5. a.   If Aparna's blood reacts to anti-B serum but not the anti-A, she has B antibodies on her cells and type B blood.
    b.   She can donate to type B or type AB.
    c.   She can receive blood from type B or type O.
6. antigen-presenting cells (macrophages, B cells), helper T cells, B cells, plasma cells, memory B cells, memory T cells, basophils (mast cells)
7. B cells develop in bone marrow; T cells develop in the thymus gland.
8. ABO blood type O has no antigens on the RBC membrane and will therefore not react with any antibodies in the recipient's plasma, so it is the universal donor. However, anti-A and anti-B antibodies in blood type O mean that type O patients can only receive type O blood, hence the high demand.
9. Your description should include the events shown in Fig. 24-17.
10. Inflammation is the hallmark of the innate immune response.
11. Antibiotics will not help his viral infection because those drugs only act on bacteria. In addition, antibiotics stay in the extracellular fluid; viruses hide inside host cells.
12. Acute phase proteins act as opsonins and promote inflammation.
13. The complement cascade ends with production of membrane attack complex that causes pathogens to lyse.
14. The Fc region (stem) binds to immune cells; the Fab region recognizes and binds to antigen.
15. MHC is a family of membrane proteins that the body uses for presenting foreign antigens and for recognition of self. All nucleated cells have MHC-I but only antigen-presenting immune cells have MHC-II.
16. Old red blood cells lose the membrane markers that identify them as "self," which then leads the immune system to attack them.

CHAPTER 25

PRACTICE MAKES PERFECT

1. Systolic blood pressure, because the force of cardiac contraction is increased but peripheral resistance is decreased.
2. True
3. by the maximum rate of oxygen consumption
4. Inspiratory and expiratory reserve volumes decrease because tidal volume increases.
5. improved glucose tolerance, higher HDL and lower triglycerides, improved cardiovascular function, weight loss and/or improved muscle tone

6. The factors that can limit exercise capacity include the ability of the muscle to provide ATP, the ability of the respiratory system to provide oxygen, and the ability of the cardiovascular system to supply the muscles with oxygen and nutrients.
7. Too much or too little exercise decreases immunity.

CHAPTER 26

PRACTICE MAKES PERFECT

1.

Male	Corresponding female part?	Analogous or homologous?
bulbourethral gland	unique	
ductus deferens	b	analogous
penis	a	homologous
prostate gland	unique	
scrotum	d	homologous
seminal vesicle	unique	
testis	c	homologous
Female		
E. labia minora		homologous to parts of the penile shaft
F. uterus	unique	
G. vagina	unique	

2. e
3. a. decreased; b. decreased; c. will have no effect on cycles; d. will have no effect on cycles; e. will restore normal menstrual cycles; f. will have no effect on cycles
4. The corpus luteum during early pregnancy secretes estrogen and progesterone to prevent the endometrium from sloughing.
5. Androgen-binding proteins keep androgens inside the tubule so that their concentration there is much higher than outside the tubules.

6. Fill in the following chart on reproduction.

	Male	Female
What is the gonad?	testis	ovary
What is the gamete?	sperm	ovum
Cell(s) that produce gametes	spermatogonia	follicle
What structure has the sensory tissue involved in the sexual response?	glans penis	clitoris
What hormone(s) controls development of the secondary sex characteristics?	androgens	estrogens and androgens
Gonadal cells that produce hormones and the hormones they produce	Sertoli: inhibin, activin Leydig: testosterone, DHT, estradiol	Theca: androgens Granulosa: estrogens, progesterone, inhibin luteal: estrogen, progesterone, inhibin
Target cell/tissue for LH	Leydig cells	thecal cells; oocyte just before ovulation
Target cell/tissue for FSH	Sertoli cells	granulosa cells
Hormone(s) with negative feedback on anterior pituitary	inhibin, testosterone	inhibin, progesterone, sustained high estrogen
Timing of gamete production in adults	constant	cyclic until menopause

7. Proliferative endometrium: estrogen
initiates development of follicle(s): FSH
ovulation: LH, estrogen
secretory endometrium: progesterone
keeps corpus luteum alive: human chorionic gonadotropin
keeps endometrium from sloughing (i.e., menstruating) in early pregnancy: estrogen and progesterone